Table of contents

Introduction 5
Paul Griffin
Allen & Overy LLP

Structuring LNG projects – evolution or revolution in the LNG supply value chain? 13
Joanna Kay
Tullow Oil plc
Peter Roberts
Ashurst LLP

LNG sale and purchase agreements 29
Susan H Farmer
Holman Fenwick Willan LLP
Harry W Sullivan, Jr
ConocoPhillips

LNG trading 53
Anthony Patten
Allens Arthur Robinson
Philip Thomson
Ashurst

LNG shipping 65
David Gardner
Energy Transact LLP

Coal bed methane for LNG 85
Daniel Gosewich
Queensland Treasury Corporation
Toby Hewitt
Dart Energy Limited

Floating LNG 107
Matthew Griffiths
Royal Dutch Shell plc

Natural gas price reopeners and English law 121
Paul Griffin
Allen & Overy LLP

US LNG import terminals: the perfect storm 163
Donna J Bailey
Chevron Gas and Midstream, Chevron USA Inc

Shale gas for LNG 175
Vivek Bakshi
Jeff Scobie
Ron Stuber
Fraser Milner Casgrain LLP

LNG master sale and purchase agreements 187
Steven Paul Barra
Eni SpA

Financing LNG projects 203
James Douglass
Linklaters LLP

LNG – a minefield for disputes? 237
James Baily
Paula Hodges
Herbert Smith LLP

LNG regulation 251
Garry Pegg
Philip Weems
King & Spalding

About the authors 289

Liquefied Natural Gas

The Law and Business of LNG, Second Edition

Consulting Editor **Paul Griffin**

Consulting editor
Paul Griffin

Publisher
Sian O'Neill

Editor
Carolyn Boyle

Marketing manager
Alan Mowat

Production
Russell Anderson

Publishing directors
Guy Davis, Tony Harriss, Mark Lamb

Liquefied Natural Gas: The Law and Business of LNG, Second Edition
is published by
Globe Law and Business
Globe Business Publishing Ltd
New Hibernia House
Winchester Walk
London SE1 9AG
United Kingdom
Tel +44 20 7234 0606
Fax +44 20 7234 0808
Web www.globelawandbusiness.com

Printed and bound by CPI Group (UK) Ltd., Croydon, CR0 4YY

ISBN 978-1-905783-64-9

Introduction

Paul Griffin
Allen & Overy LLP

Only five years ago, the first edition of this book was published. It was the first publication to look broadly at the law and business of LNG and it covered the major areas and issues at that time. The first sentence of the introduction asserted that the LNG business was then at a time of unprecedented change. In the short intervening period, that change has accelerated and broadened, necessitating this second edition which includes some new sections on topics which were scarcely in contemplation just that short time ago: shale gas for LNG, coal bed methane/coal seam gas for LNG and floating LNG. Today also sees a time of discord in many LNG markets with an unprecedented number of disputes, particularly in relation to the nature and operation of price reopener provisions. In this second edition, the sections on LNG sale and purchase agreements and LNG disputes have been revised to reflect these circumstances, and there is a new section which seeks to put in context the making and disputing of these provisions.

Before looking at some of the reasons for these changes and some of the themes which are current across all elements of the LNG business, it is helpful to put today's circumstances in context. Put simply, it is difficult to see where the law and business of LNG are now (and will be in the future) without first seeing where they have come from.

Regardless of the pace of this evolutionary change, the LNG business still sees the linking (one way or another) of upstream oil and gas projects; liquefaction plants for the production of liquefied natural gas for loading onto LNG tankers; and regasification plants for the receipt of LNG and its regasification so as to make natural gas available for delivery to consumers. One of the characteristics of the LNG business is the remote nature of the sources of production and the markets of consumption. Another is the high capital cost of these developments. The requirement of project participants to recover these expenditures and generate profits resulted originally in rigid and dedicated contractual structures over long timescales.

Until recently, there was also a comparative scarcity of participants. The producers of natural gas for the purposes of liquefaction tended to be major oil and gas companies, often together with an entity of the host state. The developers of regasification facilities and the purchasers of LNG tended to be large utility businesses (whether of gas or power) in the ownership of the host state, or operating under preferential rights granted by the host state.

The traditional structures of the LNG business were documented in comparatively inflexible contracts of lengthy duration. The scale of the capital commitments and the nature of the transactions resulted in long-term contractual arrangements which enshrined the essentially exclusive nature of the transactions and the intended fidelity between the producers as sellers of LNG and the buyers of LNG. A consequence of the LNG chain being a composition of several separate infrastructure projects is that there may have been a greater or lesser extent of integration of participants in each link of the overall chain.

Early LNG sale and purchase agreements provided for prices in the money of the day. Prices became more usually linked to crude oil following the oil crisis of the 1970s, with many adjustment formulae having effect by reference to a number of crude oils or other petroleum products. The periodic adjustment of pricing also often referred to inflation. In some cases the contracts for sale and purchase of LNG presumed that such prices would be capable of calculation throughout the contract period by reference to the identified adjustment mechanism. In other cases it was contemplated that changes would occur over time which might necessitate the revisiting or reopening of the pricing provisions so as to realign them with the original intention of the parties. These provisions might include periodic price reopeners, floor and cap prices and certain provisions generally seen in contracts of international trade, such as most favoured nation clauses. With the deregulation and general liberalisation of several markets, it was also necessary for buyers and sellers to contemplate alternative pricing mechanisms which had reference not only to the prices of competing fuels over time, but also to competing gas over time in the consumers' markets. Perhaps the first project to address these matters extensively was Atlantic LNG, under which part of the output was intended to be subject to delivery into the North American market in Boston (on terms of 'net-back' pricing, under which the price under the LNG sale and purchase agreement was calculated by reference to the prevailing price of gas in the purchaser's market) and part was for delivery into Spanish markets on traditional European terms. The fate of this approach is now well known and is studied in more than one section of this book.

In the traditional structure of an LNG project, the financing of the upstream development, the liquefaction project and the regasification project was carried out primarily with equity funds. To the extent that debt was raised for the purposes of the development, it tended to be raised at a corporate level by individual participants rather than by the projects specifically. The same was the case for most shipping financing and the development of regasification and downstream facilities.

Those structures tended to apply until the 1990s and a number of changes can be identified in more recent times. Perhaps the primary development has been a loosening of the contractual structures of the LNG chain and the introduction of flexibility and more diverse participation in a number of the elements of LNG projects. These changes have continued to gather pace, to the effect that the traditionally inflexible and long-term arrangements have in some markets given way to liquid arrangements that more closely resemble those applicable to the trading of oil or other commodities. But some markets – particularly in Asia – remain broadly tied to the traditional model and, even in these changed markets, many of the older contracts and structures remain in place and will do so for many years to come. A number of factors have contributed to today's circumstances.

There have also been developments in the manner of financing LNG projects. As an early example, the sponsors of the Ras Laffan liquefaction project in Qatar made use of a then-buoyant bond market in the United States for the raising of finance, as well as obtaining debt on a project finance basis. But while project financing may provide a diversity of financing options for the liquefaction project, this diversity may come at the expense of lesser flexibility for the liquefaction project in relation to its overall structuring and the terms on which LNG is sold. For example, in relation to the Ras Laffan project, the bonds were based on an underlying sales contract, with a specified minimum price. When this minimum price was renegotiated in relation to increased volumes, additional financial support was required from some of the project sponsors.

For many years, shipping arrangements in relation to LNG were fixed, with specific ships being dedicated to journeys between a specific liquefaction project and a specific regasification project. The vessels were more like a pipeline than a means of moving a product from a point of production to diverse markets. Until the mid-1990s, most deliveries were made ex-ship at the port adjacent to the regasification terminal. Since then, most deliveries have been made free-on-board adjacent to the liquefaction terminal. This change was led by the determination of the then-emerging Korean buyers of LNG to develop their shipping industry and there are now various models of ownership and use, with today's aggregators of LNG requiring flexibility of shipping arrangements as an enhancement of their optimisation planning.

A common element which has contributed to many of these developments is the changing nature of gas markets, particularly by reference to their deregulation or liberalisation over more recent times. While the upstream gas production businesses have remained largely monopolised, many established purchasers of LNG have been subject to changes of structure, constitution and operation. Among the factors which have contributed to changed methods of buying and selling (and the creation of opportunities for flexible supplies and diversity of markets) have been disputes under existing LNG sale and purchase agreements. The contracts of the LNG business are typically long term and tend to be written under governing laws which seek to enforce the bargain made between the parties regardless of changing circumstances over time. The contract concluded between Nigeria LNG and ENEL of Italy in May 1992 provided for the delivery of cargoes of LNG to a port to be developed in Italy. For a number of reasons (including environmental concerns), the port was not developed as expected and the purchaser sought to claim relief from the delivery and take-or-pay obligations under the principle of *force majeure*. This dispute was referred to arbitration in 1996 pursuant to the terms of the LNG sale and purchase agreement. However, before the arbitration panel had issued a decision the dispute was resolved by reason of an ancillary arrangement agreed between Gaz de France and the purchaser. Under this ancillary arrangement, the absence of the port and regasification terminal in Italy was overcome by means of a swap transaction under which Gaz de France agreed to take delivery of the contracted cargoes at one of its existing regasification terminals in France and to make available to the purchaser in Italy corresponding quantities of natural gas by means of delivery by pipeline.

Another example concerned the Dabhol project in India. The consortium developing a power project in Maharashtra State in India adopted a diversified strategy of LNG supply, purchasing quantities of LNG from producers in Oman and Abu Dhabi, and signing a 'confirmation of intent' to take from Malaysia. The power project was beset by difficulties and was not in a position to take the quantities of LNG to which it had committed. Whatever the provisions of the gas supply agreement concerning the project's take-or-pay commitments, the practical result was the availability of LNG from the supplying liquefaction plants which was available for disposal to other markets.

The traditional pricing structure of LNG sale and purchase agreements provided for a linkage of the LNG price over time to the prevailing price of oil products. To a large extent, this linkage reflected the price movement of the competing fuel from the purchaser's point of view. Also, in the largely monopolised markets into which LNG may have been supplied, the purchaser was likely to enjoy pass-through pricing for its gas supply to consumers and little competition in gas supply. Moves towards competitive and increasingly short-term markets created tensions with existing, inflexible LNG sale and purchase agreements.

Among the steps towards liberalisation, governments or regulators often require the removal of exclusive or protective rights to markets and the reduction of the buyer's ability to pass through its costs of acquisition of LNG. Also, the measures used to develop and hasten competition will include the making available of existing facilities for the use of others. Opportunities for short-term LNG sales and purchases resulted in market movements. For example, in the mid-1990s BOTAS of Turkey issued a tender for the supply of a specific number of LNG cargoes over a fairly short period of time. This approach was more akin to a procurement exercise than the typical long-term relationship building of an LNG supply arrangement and was followed by a similar approach (in the buyer's market of the day) by China National Offshore Oil Corporation (CNOOC) in relation to its Guangdong development. This was perhaps an early indication of a growing trend of national oil company participation in the LNG business. CNOOC was also a pioneer of integrating its interests all along the LNG chain, including acquiring upstream interests.

One result of these many changes was a move away from the traditional fidelity of the relationship between the participants in the several stages of the LNG chain. With new friends came the emergence of market prices and market terms, necessarily different from those applicable under the long-term contractual arrangements that originally supported the development of the facilities and markets which have now become available to others.

For many, these developments offered new opportunities. But these same circumstances represented a threat to those who remained contractually bound to the rigid and fixed structures put in place to facilitate the initial project developments. These opportunities were unavailable to them.

These processes of change occur at different rates in different markets. Although circumstances will always differ, there has tended to be a 'time line' to the development of gas industries which reflects an early concentration on the development of infrastructure and markets against the background of exclusive rights and protected markets to a later concentration on the separation of the ownership and use of facilities, and the creation of competition in contestable areas and the application of regulation in those areas which are not. While the legal and regulatory regime in many producing states will have remained reasonably stable (with the result that markets and long-term contracts will have remained consistent with the legal and regulatory regime), the same cannot readily be said of downstream markets. Laws and markets will necessarily influence the form of contracts which are used to document commercial arrangements, and it is inevitable that laws, markets and contracts will be changing and inconsistent over time. In some circumstances these changes militate against the original intentions of the parties, or are at least inconsistent with them. The differing interests of the many participants in the separate stages of the LNG chain will be aligned at the outset, but are likely to move in different ways.

In short, the inflexible project structures and the long-term contracts have not gone away as new markets and opportunities have emerged. One result of this is that the commercial benefits to be gained from new trading opportunities and commercial arrangements may diminish when compared with the commercial value to be obtained from a close monitoring of existing contractual commitments and their diligent management over time.

The development of gas industries has seen the creation of infrastructure and businesses under legal or contractual exclusivity, with the result of monopoly or dominant positions for those initial incumbents. With the maturity of industries have come the separation of ownership and use of facilities and the development of regulation towards common access to facilities. Central to this maturity is the matter of access for new participants to the existing facilities of others. The consequences of this have been the need to make contractual provision for the shared use of facilities and the growing prevalence of allocation and access arrangements for pipelines and for vessels.

The emergence of the LNG markets of the early 2000s blurred the focus on access to existing facilities in favour of encouraging new investment as the consumer markets in North America and Europe became deficient in indigenous gas supply. The European Union has had to recognise that in order to encourage capital expenditures on new regasification facilities, the rules of open access had to be relaxed. As a market emerged for LNG in the Atlantic basin, it was possible to see liquid markets and transparent prices in the United States and the United Kingdom, but the respective costs of capacity in reception facilities and pipelines distorted the commercial analysis of prices of natural gas as a commodity alone. The development of trade in the Atlantic basin saw the emergence of market prices for LNG cargoes, and also for related rights such as terminal and system capacity. But the respective producer markets were very different and the main consumer markets were subject to differing pressures. Concerns over security and diversity of energy supply have become tempered by broader trade interests and political considerations. The recent mood of resource nationalism among some producer states has not left existing LNG contracts unaffected and is likely also to affect the development of new and enduring contractual arrangements. Correspondingly, the moves of the European Union towards creating a single market in natural gas (a commodity on which it is import-dependent) exacerbated issues of politics and economics relating to all gas, but particularly LNG.

But since the publication of the first edition of this book, many of these considerations have been turned on their heads as the impact of shale gas and coal bed methane/coal seam gas has come to be recognised. The fundamentally short position of the United States in relation to gas has become one of surplus, and seemingly for many years to come. The many intended regasification terminals (and the contracts on which they were structured) seem now to have no purpose in these changed circumstances and many of the intending developers of these facilities are now reversing their business models to contemplate LNG liquefaction and export, or mixed use. This change has had a radical effect on the markets and prices of the Atlantic basin and more widely, and has provoked physical responses (eg, the reconfiguration of terminals and pipelines), as well as contractual ones. It has also had considerable geopolitical impact in relation, of course, to US self-sufficiency and also in relation to Russia's reorientation away from the planned LNG export facilities of only a few years ago.

While the markets of Asia-Pacific have not (yet) seen such an impact of shale gas, they have nonetheless witnessed radical change. Queensland's projects concerning coal seam gas for LNG have prompted massive foreign investment (whether into the upstream, downstream, marketing or shipping arrangements), and the making of many new, long-term contracts for the sale of LNG, particularly to Asian markets. At the same time, conventional LNG developments in Western Australia have created similar effects. As coal bed methane developments get underway in Indonesia, it is perhaps a portent of a wider trend that each of Malaysia and Indonesia (both early and substantial participants in the international LNG export business) are now moving towards LNG imports into certain areas, while continuing exports from others. With many high-cost LNG developments targeting North Asian markets, who is to say how much pressure will be applied to those contractual arrangements if abundant unconventional sources are discovered locally, and how these contracts and their pricing and dispute resolution provisions will respond? These developments have arisen largely from the application of new technologies in the context of populated, onshore areas. This has often resulted in a clash of the prevailing legal and regulatory regimes for land access, mineral recovery and hydrocarbon production, as well as a tension between hydrocarbon recovery and environmental sustainability. These tensions have, in a number of jurisdictions, necessarily moved on to reflect the rights or positions of native titleholders and original landholders.

More generally, LNG trading has become increasingly diverse, with supplies between a multitude of different points at a time when LNG prices in the Americas and Europe have diverged from those in Japan and much of Asia. But one pricing effect that now seems to be an enduring one is the separation of gas prices from oil prices in many European markets, as well those of North America.

These are exciting times for many in the LNG business, with new trading opportunities in the developing shorter-term markets and the potential for enhancements of existing liquefaction and regasification facilities to the economic benefit of existing owners and new investors. For many, though, these new opportunities are tempered by the presence of existing contractual arrangements, whether in relation to the development or operation of facilities or the buying and selling of LNG under long-term contracts on terms which are broadly inflexible to new market opportunities. Many contracting parties can readily identify the extent to which their embedded and long-term contractual arrangements are now at odds with the markets and their prices. For some producers, this may be the difference between prices of LNG under their sales contracts and the costs of producing LNG at a time of creeping measures by host states, or the increased costs and time delays in an overstretched construction and service sector. For others (both buyers and sellers), the transparent market prices of natural gas in a number of jurisdictions provide a ready measure of the extent to which long-term commitments are now 'out of the money'. Projections of forward prices may exacerbate that unease.

More substantially, these changes not only cause price imbalance over time, but may also call into question the continuing enforceability of these contractual commitments. Whatever the basis of the change, it is likely that the commercial effect will be to the advantage of one party and the disadvantage of the other. The gas sector has seen perhaps two main periods of contractual discord under long-term contracts. The first was the so called 'take-or-pay wars' in the United States in the mid-1980s, when the regulatory bargain was undone to the detriment of the pipeline companies. The consequential disputes were resolved against the background of potential litigation in the local courts. The second example was the restructuring of the gas market in the United Kingdom in the mid-1990s. Again, the resulting disputes were resolved against the background of litigation in the English courts. But similar circumstances in the international LNG sector require dispute resolution against the background of arbitration in accordance with the terms of the contractual arrangements rather than in local courts. The necessarily international nature of the LNG business has led to the parties specifying a chosen compromise law and arrangements for arbitration, including the venue and process. Almost every chapter of this book contains the themes of contract review and restructuring.

While the renegotiation of long-term gas contracts is never straightforward, the LNG sector has some characteristics which suggest that renegotiation of these contractual arrangements may be more than usually difficult. Things have moved a long way from the early structures on which the LNG business was built. Matters of politics and economics are invariably relevant to the LNG business, and the extent of foreign investment and international trade attracts the application of state-to-state treaties and bespoke private law arrangements. The implementation of projects along the LNG chain attracts considerable scrutiny from local interests and non-governmental organisations, particularly during the implementation stages. The granting of necessary licences or other grants entails the close participation of governments or government entities over time. Also, the requirements of broad international relations and the presence of trading restrictions or sanctions may impinge on contractual performance over time. The success of project financing in recent times has led to a greater presence of export credit agencies, multilateral agencies and commercial banks within the arrangements of the LNG sector. These participations not only increase the complexity of the arrangements at the outset, but may also make more difficult the revision or change of these contracts over time. With growing demands for funds to service sovereign debts and the declining availability of predictable, long-term LNG developments on take-or-pay terms, it may be wondered how many commercial lenders see this sector as one for further growth, or even retention of existing business. As some utility purchasers of LNG have made their way upstream towards liquefaction or production interests and some producers have made their way downstream towards regasification projects and consumer markets, the alignment of interests required for periodic contract restructuring looks likely to be more and more difficult to achieve. As well as contractual responses to changing markets, recent times have seen physical responses including the revision of existing pipelines and terminals and the proving of floating liquefaction as well as regasification facilities.

These themes and others are among those examined by the contributors to this book, which I hope will serve as a practical and informative guide to some of the diverse considerations of politics, economics and law which are part of the LNG business.

Paul Griffin is recognised as one of the world's leading energy lawyers, with over 25 years' experience in the international oil, gas and LNG sectors. He focuses on M&A transactions and large-scale commercial agreements for energy clients. He has also been involved in disputes and matters of public law in relation to the oil and gas sector.

Mr Griffin was named the Global Oil and Gas Lawyer of the Year 2010 (second consecutive year) by the Who's Who Legal Awards and the World's leading Energy Lawyer 2010 by Expert Guides: Best of the Best.

Structuring LNG projects – evolution or revolution in the LNG supply value chain?

Joanna Kay
Tullow Oil plc
Peter Roberts
Ashurst LLP

1. Introduction

Much has been written about the rapid growth of the LNG market. From the building of the first commercial liquefaction plant in Cleveland, Ohio in 1941, it has grown into a global industry which is now expected to meet at least half of the anticipated 520 billion cubic metres growth in demand for gas by 2035.

Alongside the rapid growth of the industry has come much evolution. The first-generation projects of the 1970s (eg, the sales of LNG from Indonesia to Japan) look very different from more recent projects such as the Equatorial Guinea LNG project, which delivered its first cargo in May 2007.

Those first-generation Indonesian projects were characterised by a high degree of government involvement – the LNG seller was Pertamina, the former Indonesian government oil and gas agency, with private sector participation in the upstream functions of natural gas production and liquefaction limited to contractor roles. By contrast, under the Equatorial Guinea LNG project, the LNG seller is a joint venture comprising not only traditional upstream participants such as an international oil company (Marathon) and the national gas company (Sociedad Nacional de Gas de Guinea Ecuatorial), but also two trading companies (Mitsui and Marubeni). In the Indonesian project, Pertamina supplied LNG under a long-term, take-or-pay-based sale and purchase agreement to an identified buyer or group of buyers (all Japanese gas or electricity utilities) for delivery to nominated delivery points on an ex-ship basis. LNG produced by the Equatorial Guinea LNG project is sold under a long-term sales agreement to a dedicated marketing entity, BG Gas Marketing (BGGM), on a free on board (FOB) basis. This arrangement provides BGGM with a baseline LNG supply in its portfolio of short, mid and long-term sales agreements to support its committed regasification slots; but with destination flexibility under the buyer's control, BGGM has the ability to trade cargoes in the market, taking advantage of arbitrage opportunities in the spot market.

The past 40 years have seen much change in the LNG industry, leading some commentators to question whether the projects of today bear any resemblance to the traditional project structures of the past. Nonetheless, the fundamental characteristics of the LNG industry remain basically the same today as they did before.

2. Understanding the supply chain

Any attempt to describe the structural options for the development of an LNG project would be difficult to undertake without consideration of what is commonly referred to as the 'LNG supply value chain'.

The supply chain comprises five key elements:

- the production of natural gas (and transportation for liquefaction);
- its liquefaction;
- the transportation of the resultant LNG from the liquefaction facility to where it is needed;
- regasification of the LNG and storage; and
- the supply of the regasified LNG to the end user for consumption or further processing.

The following schematic represents one way of ordering the supply chain structure (although a number of variations on this theme are possible):

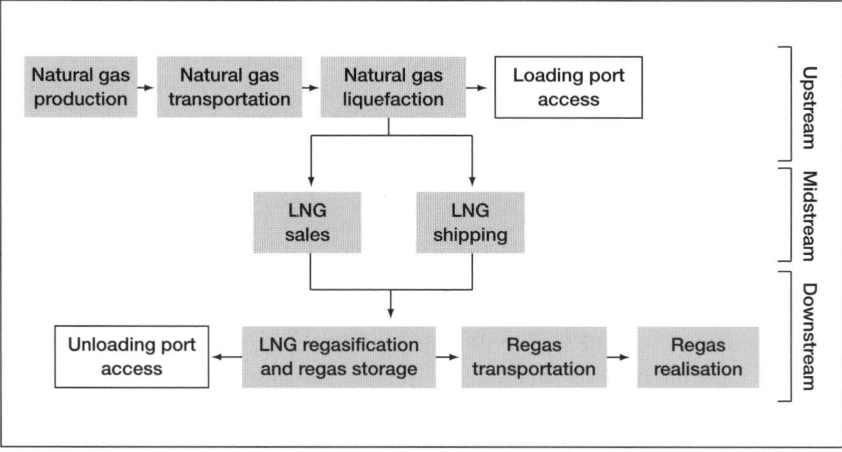

Because of the essential interdependency of the components illustrated above, the structure can properly be described as a chain. At its head are the upstream functions of natural gas production, pipeline transportation of the natural gas to the gas liquefaction facilities and liquefaction to give LNG. In the centre of the supply chain are the midstream functions of selling and shipping LNG. At the end of the supply chain are the downstream functions of LNG regasification and storage, pipeline transportation, supply and consumption of the resultant regasified LNG.

In terms of physical infrastructure, the supply chain can be represented as follows on the next page.

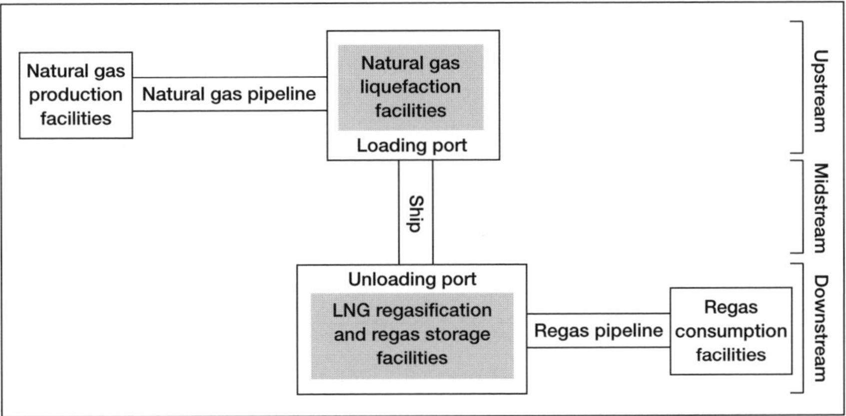

The supply chain could also be described as a series of free-standing projects which are carried out under the umbrella of an overall project. Each of the major project components – natural gas production, liquefaction, LNG shipping, LNG regasification and regasified LNG consumption – would constitute a sizeable infrastructure project in its own right, and each has its own unique characteristics and complexities. Unifying the parallel development, financing and completion of each of these projects through integration into an overall LNG project is the essential additional step which brings significant further complexity.

Therefore, the concept of the supply chain manifests itself in several ways:

- contractually, through the interlinking of the contractual commitments between the various participants;
- economically, through the addition of value at each step of the chain as natural gas is produced, processed into LNG, transported and sold as regasified LNG; and
- physically, through the interconnection of the various infrastructure items needed to deliver regasified LNG ready for consumption.

There is no single definition of the supply chain: it is whatever it needs to be in order to give the requisite definition to any particular LNG project.

3. Supply chain components

The function and application of the individual components of the supply chain are further described below.

3.1 Natural gas production

This is the function of exploring for and producing natural gas, which will then be liquefied to give LNG. This function may be undertaken directly by the host government of the state in which the natural gas deposits reside or by a private sector gas producer under a form of concession (whether a production sharing arrangement, a licence or a service contract) which has been granted by the host government.

If the gas producer is also the entity responsible for natural gas liquefaction (as described below), there might not be a separate arm's-length agreement for the sale and supply of natural gas for liquefaction. However, such an agreement will be required where the gas producer and the gas liquefaction entity are separate entities (whether affiliated or entirely unconnected) and the gas producer sells and supplies natural gas to the gas liquefaction entity.

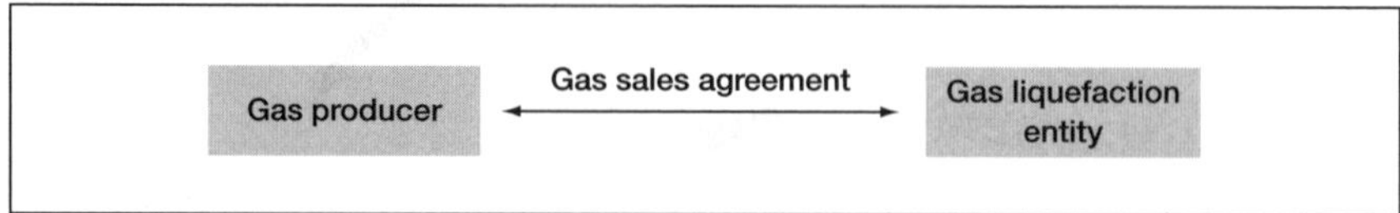

Non-associated natural gas will generally be preferred as feedstock for gas liquefaction, as its production will be decoupled from reliance on the vagaries of associated crude oil production. Rich gas (ie, natural gas high in the heavier hydrocarbon fractions) could present a challenge as a feedstock. Stripping and selling these liquid fractions to deliver the necessary methane for liquefaction could provide a significant additional revenue stream, which could be used to underpin the economics of the LNG project. However, this approach also entails incurring additional capital and operating costs for the necessary liquids production, storage and sales infrastructure, and requires the existence of a proximate demand market for these liquid fractions. An ancillary domestic gas consumption project (where natural gas is supplied locally to a proximate market by pipeline) could also provide a useful economic underpinning to the LNG export project.

3.2 Natural gas transportation

This is the function of transporting natural gas from the point of production to the gas liquefaction facilities, typically by pipeline. This function may be undertaken by:

- the gas producer (as part of the natural gas production function);
- the gas liquefaction entity (as part of the gas liquefaction function); or
- an independent gas pipeline owner (transporting natural gas for a tariff on behalf of either the gas producer or the gas liquefaction entity as the shipper).

In the case of transportation by an independent pipeline owner, there will be a separate arm's-length gas transportation agreement (entered into between the pipeline owner as transporter and whichever entity has elected to be the shipper), for which a tariff is payable.

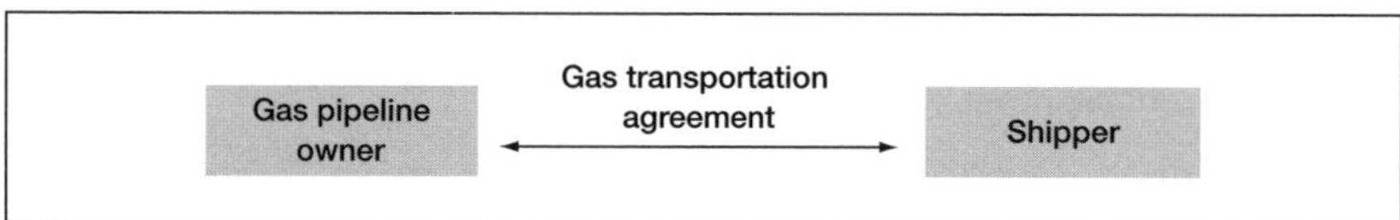

Where the point of production and the gas liquefaction facilities are adjacent – for example, in a floating LNG project (see below) – this element of the supply chain will not be required.

3.3 Natural gas liquefaction

This is the function of liquefying natural gas to give LNG at the gas liquefaction facilities. This function may be undertaken by the gas producer as part of a complete natural gas-to-LNG production cycle or by an independent gas liquefaction entity.

The independent gas liquefaction entity could buy natural gas from the gas producer under a separate arm's-length natural gas sales agreement and then produce and sell LNG for its own account.

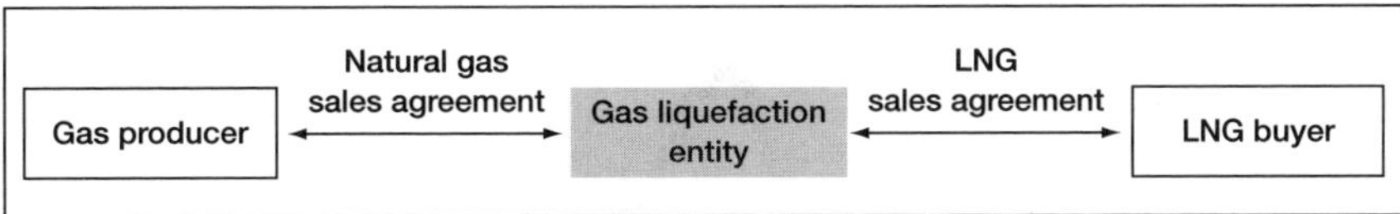

Alternatively, the gas liquefaction entity might not be involved in the business of buying natural gas or selling LNG, and instead will aim only to provide a gas liquefaction service on behalf of a third party (eg, the gas producer) under a separate arm's-length gas liquefaction (tolling) agreement, for which a tariff would be payable. Natural gas will be supplied by the third party, liquefied by the gas liquefaction entity and then returned to the third party as LNG (in exchange for a fee).

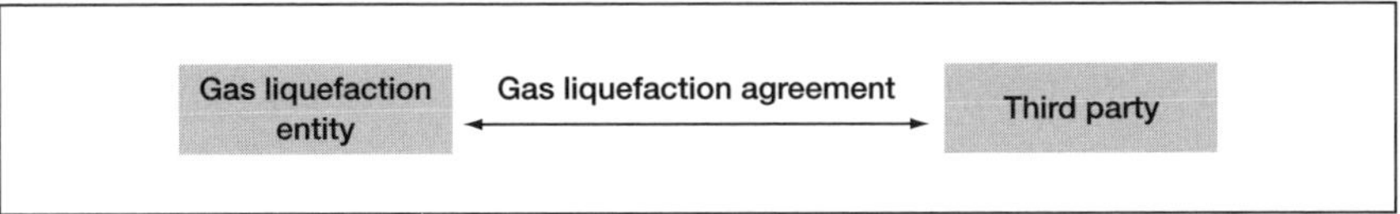

The resultant LNG is then loaded onto a LNG ship at the loading port (in respect of which consideration of the arrangements for access to the port will be essential, although this does not constitute a supply chain component in its own right).

3.4 LNG sales and shipping

Whether LNG is sold on a delivered at place (DAP) basis (ie, on an ex-ship basis whereby the LNG seller assumes responsibility for shipping LNG from the loading port to the unloading port – formerly 'delivered ex-ship' (DES) under the Incoterms 2000 formulation) or on an FOB basis (ie, the LNG buyer assumes that responsibility), there will be a contractual commitment regulating how the sale and purchase of LNG is structured and who has responsibility for the performance of the LNG shipping function. The functions of LNG sales and LNG shipping are so inextricably linked that they should properly be considered as a single component.

The primary responsibility for shipping LNG is allocated between the LNG seller and the LNG buyer. The LNG shipping function could be undertaken directly by the party which assumed this responsibility if that party owns the necessary LNG ships. Alternatively, the responsible party may appoint an independent LNG shipowner under a separate arm's-length charterparty arrangement for the discharge of this function as represented overleaf.

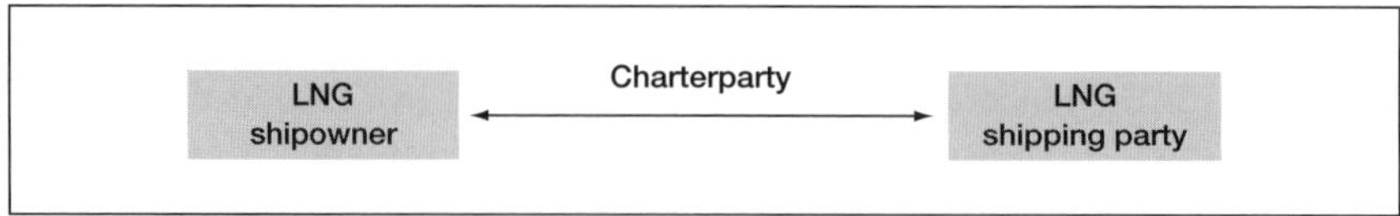

The LNG shipowner entity could be independent of the LNG shipping party or it could be an affiliated entity (the latter case is common even where a party owns its LNG ships within a wider corporate group).

3.5 LNG regasification storage

This is the function of regasifying LNG upon its arrival and discharge from the LNG ship at the unloading port (although this does not constitute a supply chain component in its own right, consideration of the arrangements for access to the unloading port will be essential), whereby the resultant regasified LNG can be stored and distributed for consumption. LNG could also be stored cryogenically as LNG prior to being regasified (and stored as a gas).

The LNG regasification/regasified LNG storage function may be undertaken by the LNG buyer or the LNG seller; in each case, the relevant party would be the owner of the LNG regasification facility.

Alternatively, the LNG regasification/regasified LNG storage entity might not be involved in the business of buying or selling LNG or regasified LNG. Instead, it could offer to regasify LNG and store regasified LNG on behalf of a third party under a separate arm's-length LNG regasification/regasified LNG storage (tolling) agreement, for which a tariff is payable.

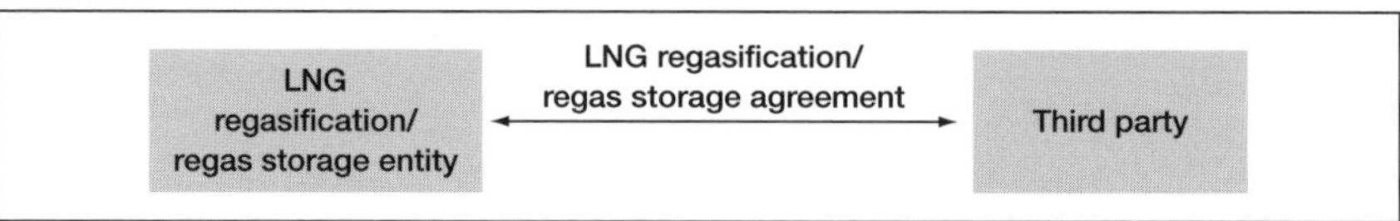

3.6 Regasified LNG transportation

This is the function of transporting regasified LNG from the LNG regasification facility to the regasified LNG consumption facilities (where those facilities are not adjacent), typically by pipeline. This function may be undertaken by:

- the LNG seller (where it has undertaken to deliver regasified LNG to the final regasified LNG buyer);
- the regasified LNG buyer (where it has undertaken to take delivery of regasified LNG at the tailgate of the LNG regasification facility); or
- an independent gas pipeline operator (transporting regasified LNG for a tariff on behalf of either the LNG seller or the regasified LNG buyer as shipper).

In the case of transportation by an independent pipeline owner, there will be a separate arm's-length regasified LNG transportation agreement (entered into between the pipeline owner as transporter and whichever entity has elected to be the shipper), for which a tariff is payable as represented overleaf.

3.7 Regasified LNG consumption

This is the end of the supply chain. Consumption may be realised through the use of the regasified LNG as feedstock for power generation or petrochemical production. Alternatively, the regasified LNG could be consumed directly by industrial or domestic customers.

The regasified LNG consumption function is arguably the most important component of the supply chain: without sustainable demand for the volume of regasified LNG at a price which underpins the production of natural gas and its subsequent liquefaction, shipping and regasification, there would be no LNG project. The revenues which result from regasified LNG consumption will need to flow back up the supply chain in order to cover all of the upstream LNG project costs.

4. Assembling the supply chain

At the outset, the analysis of the supply chain could assume a separation of functions such that each supply chain component is self-standing and undertaken by discrete entities. However, in reality these components may be assembled as a whole in many ways; moreover, several components (eg, natural gas production, transportation and liquefaction together, or LNG regasification, regasified LNG transportation and consumption together) may be bundled and undertaken effectively as a single function by a single entity.

There is no best or preferred manner for structuring an LNG project. Rather, the structure of each project will be determined on an individual basis reflective of the particular characteristics of the project. Each project participant will have a unique risk profile and will require a project structure which offers the best protection to its interests.

In the simplest case, there would be an identified group of project participants; each participant would have the same equity share in each component of the supply chain, such that there would always be perfect alignment of their interests. In reality, project participants are unlikely to be invested in all of the components and may have varying equity interests where they are invested. In turn, this could lead to conflicting commercial interests between some of the project participants. A participant in the natural gas production function would ideally prefer to charge the highest possible sale price for that natural gas where it is sold as feedstock for liquefaction, without regard for downstream economics; on the other hand, a participant invested moderately in the natural gas production function, but heavily in the regasified LNG consumption function, would prefer to see a low natural gas sale price, since this would translate into lower regasified LNG costs. These interests must be accommodated in the contractual matrix that the supply chain creates.

When putting together an LNG project, parties are often tempted to focus most interest on the development and financing of the infrastructure, notably the hardware associated with natural gas production and liquefaction, LNG shipping, LNG regasification and regasified LNG consumption. However, the primary focus should be on the metrics of the regasified LNG consumption market, since it is the essential revenue generator without which the LNG project would be unsustainable. Any LNG project is essentially demand driven; without a meaningful commitment of a buyer (or buyers) to purchase the requisite volumes of LNG at an acceptable price and for the intended project duration, there would be no project. This reality is clearly illustrated by the emergence of competing energy sources in the end-user market; a reduced demand for gas from LNG because of the availability of energy alternatives will have a ripple effect all the way back upstream.

5. Evolution of the supply chain

Some commentators have suggested that the traditional supply chain structure is in danger of fragmenting (or disappearing) as a consequence of the rapid emergence of an increasingly global LNG industry. It will come as no great surprise that there have been considerable technical advances in the gas industry over the past 40 years. While significant evolution has also taken place in the LNG industry over that period, none of these changes threatens the traditional components of the supply chain – but the changes are still noteworthy because they will change the customary way of things.

5.1 Unconventional gas

One development which could have a particular impact on the LNG industry has been the emergence of technologies which allow for the commercialisation of unconventional gas reserves. These unconventional gas reserves include tight gas, coal bed methane and shale gas.

Access to new natural gas reserves has had both a positive and negative effect on the LNG industry. By 2014, Australia will have almost 56 million tonnes per annum of LNG export capacity, much of which will be based on coal bed methane projects as feedstock such as BG's Queensland Curtis project and Santos's Gladstone project. With 20 million tonnes per annum of existing export capacity and an additional 36 million tonnes per annum currently under construction, Australia will soon rival Qatar's export capabilities. By contrast, in 2008 shale gas production swamped the domestic market in the United States, depressing gas prices by 60% in the period 2008 to 2010 and dealing LNG imports to the United States a major blow. Unconventional resources are now projected to account for 73% of US natural gas production by 2030.

Thus, the success of these new technologies has propelled new countries centre stage in the LNG industry, while simultaneously significantly reducing the presence of other participants in the industry.

5.2 Floating LNG

Another technical innovation has been the development of floating LNG (FLNG) projects. As the name suggests, these are liquefaction or regasification facilities constructed on new or converted vessels rather than fixed facilities built onshore.

FLNG allows access to natural gas reserves and markets previously considered too remote or low value to warrant the development of traditional liquefaction or regasification facilities. The relative complexity of floating liquefaction technology means that it is still very much in its infancy. Nonetheless, many of the majors are investing in these nascent technologies in anticipation of the rewards they will offer in the future. The most advanced of these projects is Shell's Prelude project in offshore Western Australia, which is expected to commence production in 2016. Located in the remote Browse Basin, Shell's 2007 and 2009 discoveries of the Prelude and Concerto fields were considered too small (they are expected to yield 3 trillion cubic feet of gas) and too remote to warrant the significant capital investment required for a traditional liquefaction facility. However, the use of floating liquefaction technologies will unlock these discoveries, and other stranded reserves, in the future.

One reason why FLNG is so attractive is that it introduces mobility to two of the key infrastructure components of the LNG supply chain. These facilities can be deployed to new reserves and markets as and when supply or demand (as the case may be) wanes. It is too early to say what will happen in the floating liquefaction market, but floating storage and regasification units are already offering developers this flexibility. Excelerate's Excelsior EBRV is used by Repsol at Bahia Blanca GasPort during the peak demand period of May to August, giving Repsol additional regasification capacity to meet domestic supply spikes during the southern hemisphere's winter. FLNG means that the LNG supply chain can now be more reactive to supply/demand changes than ever before.

The development of FLNG facilities brings with it many advantages, but also necessarily new challenges. By their nature, floating facilities carry with them an increased risk of supply disruption. FLNG facilities tend to be smaller than onshore liquefaction or regasification facilities as they are seafaring vessels; they are more affected by adverse weather conditions; and they offer less opportunity to incorporate alternative LNG or natural gas supplies (eg, through pipelines) to mitigate delay or interruption events when they occur. This will need to be factored into the overall project economics and the legal framework of the project as a whole adapted accordingly.

5.3 Integration and specialisation in the LNG chain

One of the most striking changes in the LNG industry has been the change in the traditional allocation of responsibility for the exercise of the supply chain components. Increasingly, prospective LNG project participants are now willing to undertake what for them might be regarded as new areas of project participation.

The supply chain components have traditionally been allocated between the principal LNG project participants, LNG sellers and LNG buyers, as represented overleaf.

As part of the process of evolution, LNG sellers are increasingly participating in the downstream functions of LNG regasification and regasified LNG transportation and consumption; on the other hand, LNG buyers are increasingly participating in the upstream functions of natural gas production, transportation and liquefaction. For example, the Korean state gas agency KOGAS, a major buyer of LNG for delivery

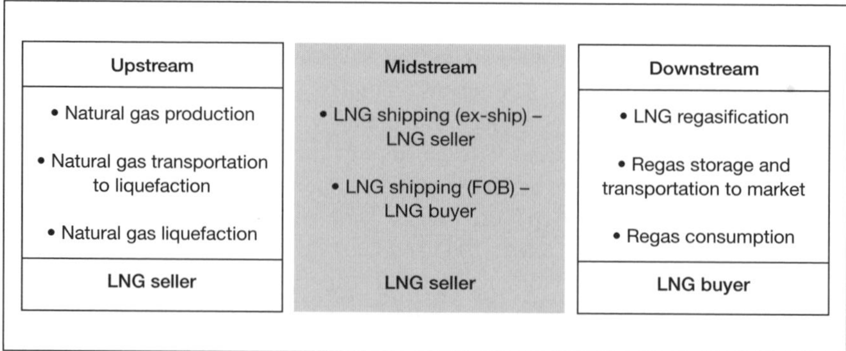

into Korea, has structured upstream investments into LNG export projects in Australia, Oman, Qatar and the Yemen; moreover, the Malaysian state oil and gas agency Petronas, a major LNG exporter, has taken downstream positions in regasified LNG facilities (eg, the Dragon LNG project in the United Kingdom) and gas supply businesses in the United Kingdom. There is increasing lateral integration across the supply chain, coming from both ends and eroding the traditional demarcation of the LNG seller's and the LNG buyer's provinces.

That said, certain functions within the supply chain are becoming the preserve of niche operators. LNG shipping is increasingly being undertaken by specialist shipping companies (eg, Golar LNG and Bergesen Worldwide) which are not directly connected to the LNG seller or buyer and undertake this function in the expectation of securing an economic return in their own right from the carriage of LNG under charter. It is increasingly less typical that the LNG seller or buyer automatically undertakes the LNG shipping function as a direct extension of its respective sale or purchase commitments.

There is also increasing reliance on the model whereby independent third parties undertake the LNG regasification function on a tolling basis (typically in the United States, but also in the United Kingdom), rather than the LNG buyer performing this function directly as part of the LNG import function. By the same token, there may be increased interest in seeing the natural gas liquefaction function undertaken on a tolling basis. First-generation Indonesian LNG export projects did in some respects establish the principle of a service fee-based gas liquefaction function, but this was not a true tolling activity in the way that LNG regasification is emerging to be. Nevertheless, an early form of precedent has been around for some time.

While the supply chain components have not changed, the allocation of these components could therefore be redrawn as represented on the following page.

5.4 Participants

The character of the parties participating in the supply chain is also changing.

The LNG seller is less typically a state or parastatal entity (eg, a national oil company). The process of liberalisation of the energy sector means that the LNG seller is more often a private entity, albeit one in which the state might hold a certain percentage shareholding interest. For example, the Nigerian National Petroleum

Upstream	Natural gas liquefaction	Midstream	LNG regasification	Downstream
• Natural gas production • Natural gas transportation to liquefaction	• Natural gas liquefaction	• LNG shipping	• LNG regasification	• Regas storage and transportation to market • Regas consumption
LNG seller LNG buyer	LNG seller LNG buyer Third party	LNG seller LNG buyer Third party	LNG buyer LNG seller Third party	LNG buyer LNG seller

Corporation, which is owned by the government of Nigeria, holds a significant minority interest in several projects for the export of LNG from Nigeria (together with various private sector participants).

The LNG buyer is also less typically a state, parastatal entity or national utility interest holding a strong local franchise. The process of liberalisation of the energy sector (and, particularly, the advent of the private sector independent power producers and merchant gas and power markets) increasingly means that the LNG buyer may be a private entity, selling its resultant products into a competitive merchant market without the comfort of a monopoly franchise and/or long-term sales agreements. This has been a feature of LNG sales into China and could become a feature of LNG sales into Japan, as Japanese downstream markets continue to deregulate. This evolution results in greater uncertainty in respect of the continuing covenant of the buyer and will necessitate a reassessment of the bankability of the LNG project.

The changing character of the participating parties also poses particular challenges to third-party debt financiers of any aspect of the supply chain, where question marks over the sustainable creditworthiness of these parties can require innovative techniques for project structuring and protection.

5.5 The role of host governments

The role of host governments has also evolved. Host governments – on both the LNG export and LNG import sides – historically played an essential role in the development of any LNG project. From the LNG export perspective, the host government had a keen interest in developing and selling the state's natural gas reserves in a manner which best promoted and protected the returns to the state; from the LNG import perspective, the host government wanted to ensure a secure supply of LNG, such that the best possible foundation was given to a downstream market which relied greatly on regasified LNG.

As a consequence, under the traditional model, prospective participants in an LNG project often did not have the luxury of selecting their preferred options for project participation. Rather, the relevant host government chose to engineer the sector for which it had responsibility according to the national interest, shaping the interface of the supply chain components. The host government could:

- be a direct or indirect project investor or operator (at the expense of private sector investment or operational participation, which might be prohibited in whole or in part);
- reserve certain supply chain components for its preferred agencies;
- aggregate several supply chain components as a single unit; or
- elect to apply (or withdraw) various economic incentives intended to promote the greater use of LNG.

The traditional interventionist role taken by host governments in the development of LNG projects has by no means disappeared. The Qatari government's role in directing the strategic development of its estimated 800,000 billion cubic feet of reserves is still very evident. However, as the LNG market has grown and diversified so the role of host governments has changed. This is particularly noticeable in some of the newer Western entrants to the LNG industry. The Australian government has taken no direct participation in the numerous LNG projects currently in development in the country. It has stepped back from a strategic role driving the development of these projects and has instead assumed much more of an administrative or regulatory role, granting the environmental and other permits required under state and national law. This is not to say that host governments which assume this regulatory role have no control over the LNG supply chain. The granting or withholding of vital permits and consents can, in itself, make or break a project. However, with the arrival of new free market economies into the LNG industry, the day-to-day strategic direction of interventionist states under the traditional model is being eroded.

5.6 Market liberalisation

LNG projects have traditionally been engineered on the foundation of (at least) a 20-year project cycle; this approach necessitates an attempt to predict the evolution (in terms of volume demand, price certainty and intervention from competing fuels) of the downstream demand profile for regasified LNG. In the traditional LNG project construct, such predictions have been relatively comfortable because of the existence of heavily regulated markets, monopoly franchise areas and the comparatively limited number of participants in the LNG sector. If commercial difficulties arose during the currency of these long-term arrangements, the parties would work together to overcome these difficulties; this goodwill would often be expected to remedy any gaps in the underlying project contracts.

Worldwide energy sector liberalisation (and, in particular, the advent of merchant markets for gas and electricity) has introduced significant levels of uncertainty, reducing the ability of LNG sellers to predict downstream market price and volume demand movements accurately. LNG project agreements have become longer, more detailed and more complicated, as they try to cater for every commercial contingency. Yet even the most complicated agreement cannot provide for every development in the industry. The collapse of the market for LNG imports in the United States on the back of the shale gas revolution in 2008 has resulted in a market swing whereby the United States is increasingly looking to become an LNG exporter rather than an importer. Thus in 2010 the Cheniere Sabine Pass regasification terminal received its first approvals for a liquefaction expansion project which will transform it into the world's first bi-directional LNG processing facility, giving it both import and export capabilities. Market liberalisation and the influx of new, more adaptable market-orientated participants is therefore transforming the LNG 'tramlines' of the past into a highly flexible and dynamic industry which can react to changes in supply and demand.

5.7 LNG sales arrangements

The evolution of LNG sale and purchase arrangements over the years also reflects the growing flexibility and dynamism of the LNG industry. The increasing reliance in LNG sale and purchase arrangements on market-out rights (ie, rights to redirect cargoes, to suspend the obligation to deliver or take delivery of LNG or to terminate the arrangement, which arise in the event of demand market price movements that make the sale and purchase arrangement economically unpalatable for the LNG seller or buyer) illustrates the uncertainty caused by the unpredictability of market evolution and the desire of the parties to afford themselves the greatest possible protection against the risks represented by this uncertainty.

The impact of the growing gap in global gas prices is a prime example of this. In 2009 the Henry Hub and National Balancing Point gas prices averaged $4/million British thermal units (mmBtu) and $5/mmBtu, compared to approximately $9/mmBtu in Japan and Continental Europe. The price divergence between the spot market in the United States and United Kingdom and oil-linked gas prices elsewhere – combined with shock factors in other markets, such as the nuclear disaster at Fukushima in March 2011 which saw governments globally pull back from the nuclear agenda – has had a profound impact on both LNG sellers and LNG buyers. LNG buyers are facing pressure from their consumers to reduce prices to a more competitive level, but are often constrained by the high prices negotiated in their long-term sales agreements when the gas supply was lower. This tension has fed back up the chain and increasingly LNG sellers are looking to renegotiate price and volume under their long-term sales agreements. Yemen LNG, which had developed projects on the assumption of high demand and prices in the United States, has recently been rerouting cargoes to more lucrative markets such as Japan.

The exposure of the LNG industry to market fluctuations means that there are an increasing number of LNG price and volume renegotiations, and wholesale restructuring of existing long-term LNG projects may become a frequent occurrence.

6. The fundamental nature of the supply chain structure

The structural representation of an LNG project as a supply chain presents several noteworthy characteristics which remain as relevant today for LNG project developers as they did for first-generation project participants.

6.1 The need to view the supply chain holistically

Like any chain, the supply chain is only as strong as its weakest link. While developments in the LNG industry have made the supply chain more flexible and consequently more robust, the risk remains that the failure of any one the components could break the supply chain and jeopardise the LNG project in its entirety.

A particular entity may be concerned with only one aspect of an LNG project (eg, producing natural gas, shipping LNG or consuming regasified LNG), but that entity's long-term commercial interests cannot be considered in isolation from the other components of the supply chain. Thus, for example, while a gas producer might regard its involvement in an LNG project as ending effectively at the point where natural gas is produced and taken away for liquefaction, it should be equally interested in ensuring the effective liquefaction, sale, shipping and ultimate consumption of that natural gas, since any failure further down the supply chain may disrupt the continuing viability of the natural gas production component.

When financing the development costs of any particular component of an LNG project through third-party debt finance, the relevant lenders will also be concerned to examine the LNG project (and its attendant risks and revenue flows) as a whole.

6.2 Managing project timelines

The overall LNG project may be seen as a series of separate projects, which must come together in the proper chronological sequence if the LNG project is to be fully functional. This necessitates the appreciation of the different lead times for the completion of each project component. Increasingly, certain parts of the supply chain are being undertaken on an 'uncovered' basis (ie, with the expectation that other parts of the supply chain will subsequently fall into place). For example, this could apply to the construction of natural gas liquefaction facilities in the absence of finalised LNG sale agreements.

The construction of natural gas liquefaction, LNG regasification and regasified LNG consumption facilities should be a relatively predictable process, which can be slotted into an overall project schedule; however, environmental, political and social factors have demonstrated an ability to derail the best-laid plans. For example, the economic downturn in 2008 led to delays in a series of projects, such as Pluto 2 in Australia and Brass LNG in Nigeria, that were expected to receive final investment decisions in 2010. The construction of new LNG ships (assuming that there is no existing and uncommitted LNG shipping capacity which can be utilised) is also a predictable process, but the availability of the requisite shipyard capability could be a constraining factor. Completion of the upstream natural gas production infrastructure will require the implementation of a production plan of development whose precise nature will be unknown until there is an adequate pool of knowledge about the geological and geophysical characteristics of the underlying natural gas field.

The intended timings for the commencement and completion of each of these activities will need to be assessed and accommodated within the wider LNG project timeline. The progress of a traditional greenfield LNG project is usually measured in terms of decades, from the first conceptual economic design through to the arrival of the first cargo at the unloading port. The often-cited example of the almost glacial pace of LNG project development is the emergence of Nigeria as an LNG exporter.

The first export initiative was considered in the early 1970s and the first export cargo left Nigeria in 1999. The same remains true of some more recent projects. Australia's Gorgon gas field, which was first discovered in 1988, achieved a final investment decision in 2009, with first LNG not expected until 2014.

There is an inevitably cyclic nature to LNG project growth: in periods of increasing energy demand, there will be increased demand-side interest in securing LNG volumes and, correspondingly, increased supply-side interest in delivering those volumes. However, rising energy prices as a consequence of increasing energy demand can lead to demand destruction; in turn, this can lead to market saturation and energy price collapses, which can curb the enthusiasm of LNG sellers to invest in the development of their projects. As described above, this is ever more of a concern as the LNG industry becomes more and more exposed to the fluctuations of a liberalised energy market. The consequences of these swings of emphasis will need to be addressed through the entire project cycle.

During the time that it takes to assemble the LNG project, some political, economic, commercial or regulatory evolution will typically be underway somewhere in the envisaged supply chain; this will always have to be taken into account.

6.3 Coordinating contractual interfaces

An LNG project which is structured on the basis of the disaggregation of the various project components and the reliance on separate arm's-length agreements to manage the transition of natural gas to regasified LNG will require multiple contracts. At the heart of the enterprise will be the LNG sale and purchase agreement between the LNG seller and the LNG buyer; there will also be various agreements for upstream natural gas supply and downstream regasified LNG sales, arrangements for gas liquefaction, LNG regasification and LNG shipping, as well as a host of ancillary arrangements for infrastructure construction, operation and maintenance. When all of the necessary project consents, concessions and permits are added in, along with agreements for the financing of the various aspects of the supply chain, the total LNG project roadmap can quickly become complex to the point of being unintelligible.

The challenge of the parties to an LNG project is to ensure that the interfaces between the various contracts are as seamless and efficient as possible. At a minimum, the parties must ensure that:

- contract durations are consistent;
- rights to claim *force majeure* relief and to terminate certain contracts are applied across the supply chain; and
- costs and revenues flow properly between all project participants.

These trends are likely to continue and may become even more apparent in future LNG projects.

7. In conclusion: the future of the LNG supply value chain

Although significant evolution is undoubtedly underway in the LNG sector, the traditional components of the supply chain – and the sequence in which they occur – do not change. However, the traditional allocation of responsibility for the exercise of the components and the character of the parties participating in the performance of these components are changing. Moreover, LNG project philosophies are evolving.

Change is a constant in the options determining how LNG projects are structured. There is also continuing growth of interest in LNG projects worldwide. The evolution of the sector – across every aspect of the supply chain – will continue to be a feature. The global LNG market will continue to evolve, founded on ever-increasing levels of price transparency and a wider recognition of the opportunities for LNG arbitrage. In the shorter term, more flexible LNG sale and purchase arrangements will present unique opportunities for risk and reward. New interest in investing across the supply chain will continue to develop. With new sector entrants, unconstrained by traditional perceptions of LNG project structuring, will come new ideas on how LNG projects could be structured.

As the famous Yogi Berra once said: "The future ain't what it used to be." Prediction is difficult, but the following themes could develop:

- smaller-scale project configurations (in terms of natural gas liquefaction and LNG regasification facilities), likely fuelled by the growth of FLNG and the new flexibility this brings in allowing infrastructure to be redeployed once a field or market becomes uneconomic;
- the emergence of purely domestic LNG production and regasified LNG consumption projects in economies rich in stranded indigenous gas;
- the diversification of sales portfolios consisting of combinations of spot sales and term sales commitments as merchant sellers and even the more traditional LNG seller participants seek to protect their investment against price fluctuations in increasingly transparent energy markets;
- the arrival of new country entrants as LNG producers, exporters, importers and consumers while countries which previously dominated the industry in its nascency come to assume more of a supporting role;
- the advent of new technologies, particularly in relation to offshore liquefaction and regasification facilities, which develop and consolidate their position in the LNG industry, again introducing new participants to the industry and changing the current dynamic between existing participants; and
- countries which have traditionally assumed an export role diversifying their infrastructure portfolio such that they develop the capacity to be importers should market swings make this economic, and vice versa.

The supply chain will remain apparent as the essential contractual, physical and economic continuum, but the way in which the supply chain is constituted will always need to be looked at afresh in light of the evolution around it.

LNG sale and purchase agreements

Susan H Farmer
Holman Fenwick Willan LLP
Harry W Sullivan, Jr
ConocoPhillips

1. Introduction

Long-term LNG sale and purchase agreements are the linchpins which link upstream gas supply and liquefaction, shipping and downstream regasification and the gas sales elements of the LNG chain. Although a common group of core issues is included in every sale and purchase agreement, a model agreement 'fit for all purposes' for long-term LNG sales does not exist because of:

- the diversity of participants involved in supply and offtake;
- the different commercial models utilised (third-party sales or equity sale/affiliate offtakes); and
- the variety of gas markets that are supplied with LNG (liquid, liberalising and non-liquid or traditional).

Other factors which contribute to the development of a more diversified approach in the terms of sale and purchase agreements include:

- the strong growth of the international LNG trade and, in particular, the number of new LNG sellers and buyers;
- the dramatic shift in the overall gas supply/demand balance experienced over a short period of time; and
- regulatory changes affecting local gas markets.

Long-term LNG sale and purchase agreements with the 'traditional' Asian buyers (Japanese, South Korean and Taiwanese) have historically tended to be significantly shorter in length than comparable LNG sale and purchase agreements with other buyers. Some attribute this to the greater involvement of Western lawyers in the drafting of this second group of LNG sale and purchase agreements, adding ever greater detail to ensure that issues are comprehensively covered and dealt with. Others attribute this to the 'traditional' Asian consensus approach to dispute resolution: namely, that disputes are resolved in a spirit of mutual understanding and trust. Whatever the reason, a common set of core issues are dealt with by both groups, although issues that are dealt with in excruciating detail in the second group of LNG sale and purchase agreements may be barely touched upon in a traditional Asian LNG sale and purchase agreement. This historical difference in approach is changing with the growth in the international LNG trade and, as a result, the 'traditional' Asian LNG sale and purchase agreements are becoming lengthier.

Model agreements have been developed by various industry groups to use for 'spot' or 'short-term' LNG sales and purchases. These master LNG sales and purchase agreements models can be quite detailed and intricate, although they typically will either not address at all or address only generically some of the crucial provisions found in tailored long-term LNG sale and purchase agreements. This is a reflection of both the short-term nature of the trades contemplated to be undertaken and the master agreement contract structure, which leaves many of the key commercial terms to be agreed by the parties on a trade-by-trade basis in the confirmation notice. Master agreements are discussed in greater detail in a later chapter.

The aim of this chapter is to provide an overall survey of some of the most significant commercial and legal issues in long-term LNG sale and purchase agreements within the context of current LNG market dynamics.

2. Parties

While it is axiomatic that there must be a seller and buyer in an LNG sale and purchase agreement, the identity of the seller and the buyer and the particulars of the supply source(s) and destination market(s) are crucial in determining the individual structure of a long-term LNG sale and purchase agreement and whether it is commercially viable.

The project development structure selected by the sponsors of an LNG development project will determine whether the seller of LNG is:

- the upstream gas developer or one of its wholly owned affiliates;
- a 'project company' generally comprised of and owned by the upstream gas developers; or
- an 'aggregator'.

The buyer may be:

- a large utility or industrial company, buying LNG primarily for its own uses;
- a marketing company or natural gas aggregator, buying LNG to on-sell into a given market; or
- as the spot market for LNG evolves, a trading company buying LNG to arbitrage the LNG to other buyers based on market conditions.

In all cases the technical competence and creditworthiness of the seller to bear its performance and financial obligations under a long-term LNG sale and purchase agreement, and the creditworthiness of the buyer to bear the very significant financial obligations associated with performance and payment under a long-term LNG sale and purchase agreement, are of critical importance.

On the seller's side, the costs of developing the LNG production facilities are substantial and will often be financed on an equity and third-party project finance basis by the LNG production facilities owning entity, which may or may not be the LNG seller. The financial institutions providing the project finance will carefully scrutinise the creditworthiness and technical capabilities of the entire chain of participants, from the upstream gas supplier to the terminal developer and finally the LNG buyer. If any participant in the chain is an undercapitalised entity, the financial institutions are likely to require security, which may include completion, performance and/or payment guarantees from the creditworthy parent company of any undercapitalised/non-creditworthy entity or, where the parent companies do not meet the financial institutions' creditworthiness threshold (as evidenced by an acceptable credit rating from a recognised credit rating agency), security in the form of a bank guarantee or letter of credit from an entity, including another financial institution, that does meet the creditworthiness threshold.

A guarantor may be required to be an additional party to the long-term LNG sale and purchase agreement, or may execute a separate guarantee agreement.

3. Conditions precedent

Contracts with a large number of conditions precedent are correspondingly uncertain. Therefore, it is generally viewed as desirable to minimise, to the extent possible, the number of conditions precedent and the time period within which they must be satisfied. If all major contracts relating to the project can be signed simultaneously, conditions precedent can be greatly reduced or eliminated. However, due to the interrelated nature of the contracts in such projects (especially where parties are project financing), it is common practice to fix the central LNG sale and purchase agreement terms by executing the agreement first. The subsidiary agreements (eg, ship charters, engineering, procurement and construction (EPC) contracts, and gas supply and sales) can then be finalised as conditions precedent.

Other common conditions precedent included in LNG sale and purchase agreements are key governmental approvals required by either party, the reaching of the final investment decision and/or finalisation of financing arrangements if either party is developing/project financing new infrastructure.

It is advisable to limit, to the extent possible, the number of conditions precedent consisting of commercial arrangements to be made by one of the contracting parties with third parties, as well as to limit the time allowed for their fulfilment. The existence of such conditions precedent increases the uncertainty that the LNG sale and purchase agreement will become effective and may create a commercial option to cancel the contract, should market conditions change after the LNG sale and purchase agreement is signed.

Buyers and sellers face a major challenge during negotiations where both parties are building new infrastructure and trying to coordinate the attendant timing uncertainties of obtaining the necessary governmental approvals and finalising project finance arrangements, among other things.

The parties should also carefully think through subsidiary issues, such as whether a condition precedent can be waived by the party responsible for its satisfaction, whether approval of waiver by both parties is required and the standard of conduct required of a party to satisfy the conditions precedent applicable to it.

Where conditions precedent are included in a long-term LNG sale and purchase agreement, the time allowed for their satisfaction and the consequences for failure to satisfy (or waive, where waiver is provided for) them must be considered. The time allowed for satisfaction of conditions precedent needs to be reasonable in light of the circumstances of the transaction. Typically, a long-term LNG sale and purchase agreement will contain a 'sunset date' by which the conditions precedent must be satisfied (or waived, where waiver is provided for). If any of the conditions precedent are not satisfied or waived by the sunset date, then either:

- the long-term LNG sale and purchase agreement will automatically terminate and neither party will have any further rights or obligations under the agreement; or
- either party (or in some cases, only the 'other party' – that is, the party which has no outstanding conditions precedent to satisfy/waive) may, by written notice prior to the satisfaction or waiver of the conditions precedent, terminate the long-term LNG sale and purchase agreement, after which neither party will have any further rights or obligations under the agreement.

There is no obvious reason to prefer one approach to the other, as should the parties wish to allow more time to satisfy a condition precedent, they can always agree to extend the satisfaction period; and if one or both do not, the party wishing to terminate may give notice immediately after the sunset date. Some practitioners believe that the requirement that a party give notice to effect a termination will act as an encouragement for the parties to maintain the agreement in effect while the responsible party/ies are still working in good faith to satisfy one or more outstanding conditions precedent.

The agreement will also state what provisions of the long-term LNG sale and purchase agreement will be operative and effective upon signature regardless of whether the conditions precedent are ever satisfied, which typically includes the provisions covering satisfaction of conditions precedent, confidentiality, anti-corruption, governing law and dispute resolution.

4. Quantities

The quantity clause sets out the LNG quantity obligations of the seller (ie, obligation to make available for delivery at its facilities (if the buyer is providing for the LNG shipping – namely, free on board (FOB) as per Incoterms®[1]) or to deliver to the buyer's facilities (if the seller is providing for the LNG shipping – namely, delivery at terminal (DAT) or delivery at place (DAP), as per Incoterms® 2010[2]). These obligations include:

- the base annual quantity (ACQ) obligations;
- the amount of any downward quantity flexibility (for the buyer and, in certain cases, the seller);
- any requirement for the buyer to take delivery of ('make good') the quantities previously exercised as downward flexibility;
- the buyer's obligation to take or pay for a minimum quantity of LNG;
- the parties' obligations with respect to *force majeure* make-up quantities;

1 The International Chamber of Commerce's rules on the use of domestic and international trade terms. Incoterms® were created in 1936 and the most recent publication is Incoterms® 2010.

2 The former delivery ex-ship (DES) rule was deleted in Incoterms® 2010. The current rule is either DAT or DAP, although parties can continue to use DES, provided that the long-term LNG sale and purchase agreement refers to Incoterms® 2000.

- the parties' rights and obligations, if any, with respect to quantities in excess of the ACQ – namely excess quantities; and
- the buyer's right or the seller's obligation to take or supply make-up quantities of LNG for any take-or-pay payments made by the buyer.

The prevailing market fundamentals (ie, whether it is a buyer's or a seller's market) will determine to a large degree the level of flexibility that the seller is willing to give to the buyer in terms of its firm offtake obligations. Because the seller has made a large investment in the upstream and liquefaction facilities (and, in the case of DAT or DAP sales, in shipping), and because of the relatively limited (at least until recently) potential market for spot sales of any LNG that the buyer has chosen not to take in any year, the seller usually prefers to limit the buyer's ability to use downward quantity flexibility to reduce its take-or-pay obligation for any year and, where such downward quantity flexibility has been allowed, to require the buyer to take equivalent quantities at a later date as 'make good'. The seller's position will also be influenced by whether it is a FOB or a DAT/DAP sale: the seller will feel more exposed where, as a FOB seller, it must rely on excess shipping capacity being available to dispose of any untaken quantities in a year when the buyer elects to exercise downward quantity flexibility. During the early decades of the LNG trade, the buyer's annual downward flexibility entitlement was limited to 5% (or, at most, 10%) and the buyer was required to use 'reasonable endeavours' to make good the quantity of any downward flexibility exercised in later years.

For a brief period during the buyer's market from the late 1990s to 2002/2003, buyers were able to achieve annual downward quantity flexibility entitlements of 15% (sometimes more in the initial contract years) with no make-good obligation and, in some cases, the right to carry forward overtakes as credits against subsequent years' take-or-pay obligations, as is traditional in pipeline gas sales contracts. As the market shifted to a seller's market since 2003, there has been a return to very limited downward quantity flexibilities, reasonable endeavours and firm make-good obligations for buyers; downward quantity flexibility rights for sellers have even been introduced. It is still the norm to have a life of contract cap on the buyer's exercise of downward quantity flexibility which is less than the sum of the annual allowances (generally, in the range of 50% to 100% ACQ). Typically, this cap is a cumulative downward quantity flexibility limit, so that make-good quantities taken by the buyer allow the buyer to free up quantities under the cap for additional future downward quantity flexibility.

While downward quantity flexibility is not contained in every long-term LNG sale and purchase agreement, when it is included the parties need to be clear as to whether there is a firm obligation on the buyer to make good the downward quantity flexibility exercised during the term of the agreement or after the term, or whether the obligation to make good exercised downward quantity flexibility terminates at the end of the term. Typically, the buyer may request only make-good quantities which are in addition to its minimum annual take quantity for the year, and the seller's obligation to make available make-good quantities requested by the buyer in any year is typically limited to using its reasonable endeavours to do so.

The annual take-or-pay quantity and make-up rights in respect of take-or-pay payments are other issues that take up significant negotiation time. Buyers purchasing LNG for non-liquid gas markets are understandably concerned about the possibility of having to make take-or-pay payments and the uncertainty as to when they will be able to obtain the make-up cargoes to which they are entitled. The level of the take-or-pay commitment is a heavily negotiated issue and, in the authors' experience, generally ranges from 100% of the adjusted ACQ to 70% of the adjusted ACQ; but it can of course be any level agreed by the parties. Except during the brief period of the buyer's market, sellers have been unwilling to guarantee the availability of make-up quantities during the contract year in which buyers may wish to take their make-up entitlement, as this would imply that sellers must reserve capacity for buyers whenever they have an outstanding make-up entitlement. One might believe that buyers have every incentive to take delivery of a cargo of LNG for which payment was made in advance as soon as possible; however, many sellers insist on the inclusion of a requirement that the buyer take their make-up entitlement within a limited period of time (eg, three to five years) after the relevant take-or-pay payment was made or within a limited period (eg, six to 12 months) after the end of the normal term of the long-term LNG sale and purchase agreement. As buyers may request delivery of make-up cargoes only in years during which they can take the quantity in addition to their adjusted ACQ, buyers are often concerned that they may pay significant sums for LNG that is never delivered (or is delivered five, 10 or 12 years after payment was made).

Take or pay is being dispensed with in some long-term LNG sale and purchase agreements, particularly those involving parties selling from or buying into portfolio supplies. In its place, the standard contract law of damages applies, and similar issues to those discussed at section 7 below arise.

Other take-or-pay related issues are:

- whether interest should be paid on take-or-pay payments until make-up cargos are taken (generally, no interest is paid);
- whether make-up is calculated on the quantities for which a take-or-pay payment has been made or on the value of the take-or-pay payment (generally on a quantity basis);
- whether the make-up cargo should be delivered free of charge or whether the price differential between the take-or-pay price paid and the contract price at the time of delivery of the make-up LNG should be paid by the buyer or refunded to the buyer; and
- the length of any period allowed after expiry of the contract term for the buyer to take delivery of any outstanding make-up and whether the buyer should bear the seller's operating costs for the continued operation of the LNG facilities after the end of the term of the agreement (assuming that the facilities are being continued to be operated solely to deliver the outstanding make-up to buyer).

However, after over 30 years of international LNG trade, there are only a very limited number of reports of take-or-pay invoices having actually been sent and the few instances that are known relate to exceptional situations (eg, the collapse of the Dabhol LNG import project).

The unprecedented growth of the international LNG trade (in particular, the number of import terminals and buyers) has facilitated the resolution of certain issues. For example, a buyer developing a new gas market will typically want a period of two to four years to build or ramp up its ACQ obligations to the full 'plateau' level. Due to the spot market growth and the interest of established LNG buyers and new traders in entering into short-term sale agreements, it is now easier for sellers to accommodate new buyers' build-up period requirements.

4.1 Start date

The start date is the date on which the parties' actual performance obligations – that is, the seller's obligations to make available or deliver, and the buyer's obligation to take or pay for if not taken – commence. Where one or both parties are building facilities to enable them to perform these obligations, the long-term LNG sale and purchase agreement effective date (ie, the date on which the conditions precedent have been satisfied or waived) will typically be a date which is several years prior to the start date. The party (or parties) that is building facilities which are required to be in place in order for it to be able to perform its obligations to deliver or take is at risk to the extent that the start date mechanism does not take into account its construction/commissioning timetable.

The start date can be:

- a negotiated fixed and certain date. This alternative entails a degree of risk if the seller or buyer is building new facilities, but may be acceptable under the circumstances and if acceptable mitigation measures are included (eg, the right to supply/deliver cargoes from/to alternative facilities in the case of delay in completion of the new facilities); or
- a date determined through a window narrowing mechanism established by written notices from the party building the new facilities to the other party at pre-determined intervals.

A window narrowing mechanism must address what happens if one or more written notices are not given. The usual default utilised is that, absent a written notice, the window shall be the latest possible date(s) in the previous window. It is also important to clarify whether and what types of *force majeure* events affecting the party that is constructing new facilities will delay the start date.

On the start date, the deliver and take or pay obligations are binding regardless of whether the new build facilities have been commissioned and are operational.

5. Duration

Long-term LNG sale and purchase agreements tend to have fairly long terms – generally 20 or 25 years in length – reflecting the substantial investments required for upstream and LNG production facilities, LNG shipping and LNG receiving terminals.

Traditionally, the obligations to deliver and take end at the expiry of the contract term, without a declining production 'tail' as often seen in pipeline gas sales arrangements. LNG shipping considerations require that deliveries be made in full cargo lots. Consequently, if there were a production tail, there would be unutilised shipping capacity. This does not mean that a stair-stepped tail tied to shipping capacity could not be negotiated for the end of a contract term.

As the world LNG trade has expanded and matured, we are seeing the advent of a variety of 'mid-term' deals from three to 15 years, reflecting the fact that many of the major players are either supplying from a portfolio of supply sources and diversified shipping fleets or buying for a portfolio of consumer destinations with diversified shipping fleets.

6. *Force majeure*

In contrast to most civil code legal systems, the two most commonly used laws in long-term LNG sale and purchase agreements – English law and the laws of New York – do not incorporate principles governing *force majeure* relief. Thus, long-term LNG sale and purchase agreements must detail the types of event which will entitle the parties to relief from the parties' obligations if a *force majeure* event occurs.

Over time, *force majeure* clauses in long-term LNG sale and purchase agreements for traditional non-liquid gas markets have reached a substantially common formulation and will generally include the following main elements:

- a definition of the types of event or circumstance that qualify as a '*force majeure*', typically along the lines of "an event or circumstance which prevents, impedes or delays a party's performance of its obligations which is not reasonably within the party's control, acting as a Reasonable and Prudent Operator";
- a list of the obligations, such as the obligation to pay money, that will not be relieved by *force majeure*;
- a list of illustrative events, which usually includes:
  - natural events (eg, fire, floods, droughts, hurricanes and epidemics);
  - war;
  - embargoes;
  - terrorism;
  - strikes and other industrial disturbances; and
  - acts of government;[3]
- references to events affecting specified facilities (eg, liquefaction and import terminals, upstream supply facilities and, in certain cases, downstream facilities) and ships that are necessary for the parties to perform their obligations; and
- a list of specific events which will not constitute *force majeure*, which often consists of sub-categories of some of the aforementioned illustrative events, or other situations such as a drop in the market demand or market price for natural gas in the area of the receiving terminal.

---

3 When the seller or the buyer is a government-owned or controlled entity, careful consideration should be given to whether government action is considered a *force majeure*.

Areas that tend to be the subject of considerable negotiation include the following:

- whether the parties are entitled to relief as a result of acts of their respective host country government;
- whether the seller is entitled to relief as a result of reservoir failure and, if so, what is the exact definition of such event;
- whether the buyer is entitled to relief as a result of events affecting facilities downstream of the import terminal (eg, transportation pipelines and customer facilities);
- the extent to which *force majeure* relief is available when, on the buyer's side, more than one receiving terminal is available, on the seller's side, more than one supply terminal is available and when alternative or substitute LNG shipping is available;
- how to apportion *force majeure* relief among all buyers, in the case of a *force majeure* impacting on the seller, and among all suppliers, in the case of a *force majeure* impacting on the buyer; and
- the definition of the standard of care applying to *force majeure* in relation to LNG shipping.

A major concern for buyers, sellers and bankers is the extent to which the long-term LNG sale and purchase agreement is 'back to back' with the seller's upstream natural gas supply contracts and the buyer's downstream natural gas sales contracts. Other measures to mitigate risk (eg, insurance) will need to be considered to the extent that either party is unable to obtain identical *force majeure* relief terms along its respective contract chain.

LNG importers into the US and UK liquid gas markets are facing this problem, as gas sales in these markets are generally concluded on the basis of standard contracts, which typically have a very narrow definition of '*force majeure*'. LNG importers will be left to bear significant uncovered risk, unless they either manage to persuade gas customers to agree to the expanded *force majeure* provisions to which long-term LNG sellers are used to dealing with or can persuade the LNG supplier to accept the narrower *force majeure* relief in the onshore natural gas sale and purchase agreement.

A final issue to consider regarding *force majeure* is whether a long-term *force majeure* (eg, 24 consecutive months) should give rise to a right by the party not affected by *force majeure* to terminate the long-term LNG sale and purchase agreement. This should be carefully considered in light of financing requirements and other available markets and supplies.

7. Seller's liability for delay or shortfall

The way in which long-term LNG sale and purchase agreements address the seller's liability in the event that the LNG cargo is delivered later than scheduled – or not at all – varies widely. Both English and New York law make a distinction between direct damages and consequential damages; long-term LNG sale and purchase agreements traditionally hold the seller harmless from liability for the buyer's consequential damages suffered as a result of the seller's breach of its obligations (except in case of wilful misconduct on the part of the seller).

Until recently, most Asian sale and purchase agreements did not go into detail on the subject of damages and relied on the application of the governing laws to determine where the line between direct and consequential damages was drawn.

However, buyers – in particular, those importing LNG into non-liquid gas markets – need to consider whether they are adequately protected when relying on the application of the law of contract damages alone, especially where the sale and purchase agreement contains the typical relief against the parties' liability for consequential losses. Sellers do not have the same degree of concern over this issue because of the operation of the take-or-pay clause. Increasingly, sale and purchase agreements incorporate language that clearly establishes the seller's liability for late and missed cargoes: first, by clearly stating that a breach occurs, either on a cargo-by-cargo or annual basis, and second, by establishing the seller's liability for liquidated damages or clearly defining the type of loss that the buyer is likely to suffer as a result of the seller's breach of its delivery obligations.

The ways in which sale and purchase agreements deal with the seller's liability for failure to deliver include the following:

- Incorporation of provisions which expressly quantify or define the seller's liability for failure to deliver – for example:
  - the sale and purchase agreement prescribes a fixed amount (eg, a certain percentage of the contract price) payable as liquidated damages in the event of the seller's delayed or missed delivery. The percentage agreed to by the parties is supposed to represent a genuine pre-estimate of the damages likely to be incurred as a result of the late or non-delivery, and as such there is no universally agreed figure to be used across all trades; in our experience it is likely to fall within the 10% to 50% range. The advantage of agreeing to liquidated damages is that it provides certainty for the parties; on the other hand, the actual damage suffered by the buyer in any particular instance is likely to be different from the liquidated amount agreed, with the risk that the provision may be subject to challenge as constituting a penalty;
  - the sale and purchase agreement defines in detail how the seller's liability will be calculated (ie, the direct damages that the buyer can claim). The seller's liability for a delayed or missed delivery could be delineated as the cost of unwind transactions (eg, in the case of a US sale, the purchase of offsetting New York Mercantile Exchange (NYMEX) contracts, the basis differential and the cost of unused import terminal capacity). A further consideration would be whether specifically to exclude the seller's liability for any losses that the buyer sustained during the operation of hedging arrangements entered into in connection with the LNG purchase. This exclusion should be acceptable to buyers, provided that the basis for calculating their loss takes into account the prompt NYMEX gas purchase price and that buyers can purchase gas to replace the volumes needed to unwind the hedge; and
  - for non-liquid gas market deliveries, the sale and purchase agreement may specify broad principles as to how the seller's liability will be

determined. For example, it may establish a mitigation process requirement for the buyer (ie, the buyer must attempt to secure physical volumes of LNG, natural gas or – if applicable – alternative fuels elsewhere, or must take reasonable steps to mitigate its exposure to spot market risk); the agreement may also state that certain types of loss should be included or excluded. In circumstances where LNG buyers will either use the LNG itself or sell regasified LNG to natural gas customers in a market where there is no other readily available supply of natural gas and end users cannot utilise alternative fuels, it is important for the sale and purchase agreement to establish the type of damage that buyers or end users are likely to suffer as a result of delivery failure, as the damage is likely to be of the sort that might otherwise be considered as consequential losses (eg, damages for defaulting on contractual supply obligations for gas or electricity supply and penalties imposed for exceeding emission limits if buyers or end users are forced to generate coal-fired electricity rather than gas-fired electricity). Other issues that may be addressed include whether demurrage should be credited against damages payable for late delivery and whether to include requirements for buyers to provide verification of the steps taken and costs incurred.

In the absence of express provisions addressing these issues, the seller will be liable for the buyer's losses based on general contract damages principles (ie, the seller must compensate the buyer for losses resulting directly from the breach, which, in the case of a breach of sale of goods contracts, represents the cost of buying substitute goods). In the case of LNG sales into non-liquid gas markets, it may be impossible to obtain alternative supplies of LNG at all or at the time when the substitute supplies are required.

- Incorporation of provisions limiting the seller's liability (eg, limitation to an amount which may be determined on the basis of a percentage of the annual contract volume multiplied by the contract price) – on this basis, the seller's liability would be determined either in accordance with the law of damages or under express provisions, but the seller's aggregate liability in any year would be subject to a cap.
- Incorporation of an express provision delineating the exact circumstances in which a seller would be in breach of its delivery obligations – that is, when the seller's failure to deliver a cargo by the scheduled delivery date would constitute a breach. The traditional approach in long-term LNG sale and purchase agreement (especially with Asian buyers) has been not to address the issue of default on a cargo-by-cargo basis, but to consider the delivery obligation as annual. Arguably, this approach may be acceptable in long-term arrangements with dedicated shipping where buyer and seller are inextricably bound by a mutually interdependent relationship. However, this approach may not be appropriate when dealing with a shorter-term contract involving non-dedicated shipping and a lower number of cargos. Ways to approach this issue include the following:
  - If the seller expects to be unable to deliver a cargo in accordance with the

annual programme, it must give as much notice as possible to the buyer. Parties must then use all reasonable endeavours to reschedule delivery to a date which is convenient for both seller and buyer;

- If the buyer demonstrates that, although physically possible, rescheduling will cause it to incur specific costs (eg, purchase of offsetting NYMEX contracts, rescheduling of unloading port capacity and additional shipping costs) and notifies the costs to the seller, the seller can decide to bear some or all of these costs and to reschedule the cargo, or to default on the delivery and bear the buyer's losses and not reschedule the cargo; and
- If the parties do not reach agreement and the seller does not complete loading (FOB) or unloading (DAT/DAP) of the vessel within the agreed number of hours or days after the scheduled loading or unloading date (crediting delays in loading or unloading time for which the buyer or the seller is responsible), the seller is deemed to have failed to deliver (and is liable for the buyer's losses as discussed above).

Occasional incidences of seller default in making cargoes available are generally not included as a reason for the buyer to terminate a long-term LNG sale and purchase agreement, although sometimes prolonged/significant defaults or wilful defaults will give rise to a buyer termination right.

8. Price

Unlike crude oil, LNG, like natural gas, is not priced on a unified regional or international market basis, although, with the growth of regional gas markets and international LNG trade, worldwide gas prices are beginning to converge somewhat. Typically, as regional gas markets develop, natural gas pricing mechanisms move from a supply-based cost-plus formula to a demand-based alternative product price-based escalation formula and, eventually, to a spot or futures market gas price market index. Until very recently, most LNG was sold to Asian markets where there was little or no domestic or regional pipeline gas supply, or into European markets where it competed with pipeline supply sold on base prices escalated with the prices of a basket of competing fuels. With the advent of significant LNG imports into the United States and the United Kingdom, LNG is now priced in relation to liquid natural gas market indices. One can also witness the influence of arbitrage between various markets, with cargoes of LNG being diverted from the originally intended destination market to higher-priced markets, which, over time, will lead to a certain convergence of world gas prices (eg, US-destined LNG cargoes being diverted to higher-priced natural gas markets in Europe and Asia).

The unprecedented volatility and uncertainty in crude oil, natural gas and LNG pricing seen in the previous decade is continuing. The opening of many new gas markets to LNG imports and shale gas development in the United States and elsewhere are having a significant impact on the international LNG market. With new LNG import terminals opening in many new markets (eg, China, Poland, Thailand, Kuwait, Mexico, Chile, Brazil, South Africa, Indonesia), LNG suppliers view these new areas as a potential market when negotiating with buyers in other regions. Gas markets in Europe are going through a process of liberalisation, which, in the United Kingdom, has resulted in a gas market that is very similar to that of the United States. Moreover, the liberalisation of various Asian gas markets is either proposed or underway to various extents. A problem faced by buyers and sellers (to a certain extent in Europe and, particularly, in Asia) is that natural gas pricing is established through confidential contracts and pricing mechanisms, and individual cargo prices are not publicly available (ie, the market is not transparent). Most LNG is still sold on a long-term basis of 20 years or more; this means that long-term LNG sale and purchase agreements which were entered into many years ago or over a long period of time (during which different pricing structures may have prevailed) may still be in effect, which can result in a wide range of prices being paid for LNG in the same country or region.

The United States has a large liquid and transparent natural gas market and natural gas prices are determined purely by natural gas supply and demand. Henry Hub (a point at which 16 intra and inter-state natural gas pipelines meet in southern Louisiana) is the main pricing point for natural gas markets in the United States and is the delivery point for NYMEX natural gas futures contract. Approximately 1 billion cubic feet of gas pass through the Henry Hub each day. Futures contracts for Henry Hub delivery can be traded up to 72 months ahead, although liquidity drops sharply beyond 12 to 18 months. A 'basis' market has developed, which tracks the pricing differentials between Henry Hub and other natural gas market hubs in the United States and Canada. NYMEX also offers a series of basis swap futures contracts which are quoted as price differentials between around 30 natural gas hubs and the Henry Hub. The differential may be positive or negative; there can be significant regional differentials with the Henry Hub price, particularly in undersupplied markets during periods of high demand. LNG sold into the United States is typically sold on the basis that the seller gets an agreed percentage of the closing price of the Henry Hub or other traded gas hub price for the relevant delivery month. In this case, the differential between the percentage of the base price received by the seller and the price at which the buyer can sell regasified LNG into the market must cover the buyer's cost of regasification and transportation to the customer, plus the buyer's profit. Another way to sell LNG in the United States is for the seller and the buyer to agree on a percentage split of the net gas sale proceeds actually received by the LNG buyer or gas marketer after deducting the buyer's costs to transport the gas to the customer. In either case, the buyer and the seller can identify arbitrage opportunities to divert cargoes when the futures market price to secure cover gas in the United States is significantly lower than the price achievable in other markets. This differential must cover any increased transportation or other costs incurred as a result of diversion.

In Europe, natural gas prices have traditionally been – and, to a large extent, continue to be – set by a formula which establishes a base price that is escalated on a periodic basis (eg, quarterly) by reference to a basket of alternative liquid fuels (typically, fuel and gas oil and, in certain cases, a coal or electricity price index). The mix and respective proportions of the basket components aim to reflect actual competition with gas in the market. Continental European long-term gas or LNG contracts normally include a provision which allows the price formula to be reviewed at intervals with the intent of keeping the gas price competitive. These price review clauses are discussed in more detail below.

Liberalisation and the move towards an integrated and competitive EU electricity and gas market are leading to changes to gas contracting and pricing in Europe. The United Kingdom has already developed a well-established gas spot market with gas sold with reference to the national balancing point or other gas index. The national balancing point, commonly referred to as the 'NBP', is a virtual trading location for the sale and purchase and exchange of UK natural gas. It is the pricing and delivery point for the ICE (Intercontinental Exchange) natural gas futures contract. It is the most liquid gas trading point in Europe. Gas at the NBP trades in pence per therm. It is similar in concept to Henry Hub in the United States – but differs in that it is not an actual physical location.

Unlike continental European trading hubs such as Zeebrugge and the Dutch title transfer facility, trades made at the NBP need not be balanced, and there is no fixed penalty for being out of balance. Instead, shippers out of balance at the end of the day are automatically balanced through the 'cash-out' procedure, whereby the shipper is automatically made to buy or sell the required quantity of gas to balance its position at the marginal system buy or sell price for that day. This cash-out process is not considered to be a penalty in the same way as those imposed on shippers in continental markets, because the cash-out prices are often very close to the spot price. As a result of this daily market liquidity, the Unite Kingdom's NBP is frequently used to balance a shipper's position on the continent by way of the Bacton-Zeebrugge interconnector.

Asian LNG is generally priced at a percentage of the Japan Crude-Oil Cocktail (JCC) – which represents the weighted average price of crude oil imported into Japan – plus a relatively small constant. Historically, some Asian buyers have been able to introduce price caps or 'S-curves' into their pricing mechanisms, which protect them against very high JCC prices (and, in return, protect sellers against very low JCC prices). Depending on whether there is a buyer's market or a seller's market, Asian long-term LNG sale and purchase agreements contain pricing mechanisms with a relatively higher or lower linkage to the JCC and, in some cases, no JCC caps or S-curves.

Sale and purchase agreements with Japanese and Korean buyers are generally viewed as incorporating the principle of price review or reopener for extraordinary circumstances, whether this is set out expressly in writing and whether it is subject to final resolution by a third party in the event of the parties' inability to agree.

9. Price review clauses

Price review or price reopener clauses are a feature of many, but not all, long-term (10 years or more) pipeline gas and LNG sales and purchase agreements. This terminology is sometimes used very broadly to include the following clauses, which differ from classic price review or reopener clauses:

- the 'most favoured customer' clause, under which the seller will match better pricing terms achieved by any other buyer buying gas or LNG for delivery

into a specified market; and
- the 'market out' clause, which provides that, if the price goes below an agreed level and is forecast (eg, on the basis of the relevant gas futures market or crude oil indexation) to remain below that level for an agreed period, the seller (or, depending on the pricing mechanism, the buyer) may terminate the sale and purchase agreement on agreed advance notice.

Either of these clauses may be proposed by a party to a long-term LNG sale and purchase agreement negotiation, either on its own or in combination with a classic price review or reopener clause.

Price review clauses have been a feature of continental European long-term pipeline gas sale contracts throughout the industry's history. Until the opening of national gas markets to competition as a result of the EU Gas Directives, gas was generally sold by suppliers to national or regional monopoly gas buyers, which faced competition from sellers of alternative fuels, but not from other gas marketers. These buyers wanted to protect themselves against a gas price that could make gas uncompetitive against alternative fuels (to which customers could switch if the gas price was too high). The traditional wisdom in the industry was that buyers took the 'volume risk' and sellers took the 'price risk'. Typically, the parties would be given from three to five chances during a 20-year contract term to invoke the price review procedure.

An example of the type of language commonly used in these clauses was as follows:

If...economic circumstances in the [buyer's market]...have substantially changed as compared to that expected when entering into the contract for reasons beyond the parties' control....and the contract price...does not reflect the value of natural gas in the [buyer's market...].

These clauses also typically contained a provision along the following lines:

The Contract Price as revised...shall in any event allow the buyer to market the LNG...in competition with all competing sources...of energy...in the buyer's market...and on the basis that the buyer utilises sound marketing practices and efficient operations...and that such revised Contract Price shall allow the buyer to achieve a reasonable rate of return on the sale of regasified LNG derived therefrom...

The reaction of many – if not most – common law lawyers to this type of clause is one of extreme discomfort. When the authors first encountered these clauses in the early 1990s, European gas commercial managers made assurances that such clauses were regularly invoked, resulting in lengthy review discussions which ended in either mutually agreed revisions to the price mechanism or agreement to leave the existing price mechanism in place. They also made assurances that there was no instance where the review process had ended with one party referring the matter to arbitration.

With LNG suppliers and gas buyers from the early 1990s facing the impending liberalisation of the EU gas market, the question of how to approach the traditional price review or reopener clause in new long-term contracts became a major challenge in negotiations. Although almost all major industry players realised that significant changes were required, there was no consensus on what those changes should be. Nevertheless, a number of long-term LNG and pipeline gas supply contracts entered into in the mid to late 1990s with major European gas buyers incorporated only slightly modified versions of the traditional clause. Unsurprisingly, when the first reviews were invoked under these contracts in the early 2000s, the parties found that their interests diverged more widely than in the past. Over the past few years, a number of price reviews (with potentially very large sums at stake) have been referred to arbitral panels for resolution.

If the traditional price review clause no longer works, what should replace it? This issue has been the subject of much discussion at gas conferences over the past few years. Most lawyers (at least, most English and New York qualified lawyers) view price review provisions with great trepidation for the following reasons:

- They allow the most important commercial element of sale and purchase agreements to be reopened, typically upon the occurrence of uncertain future events which are difficult to define.
- They generally provide that the revised price terms shall reflect a 'market price' benchmark which, in the non-liquid and non-transparent gas markets outside the United States, the United Kingdom and Europe, would require the parties to have access to the confidential terms of all relevant sale and purchase agreements and to the results of any price renegotiations which have been implemented in these sale and purchase agreements.
- Depending on the benchmark market that must be considered to determine the revised contract price mechanism, if significant regional differences exist in geographically separate gas market prices, the revised price mechanism may result in the performance of the sale and purchase agreement becoming uneconomic or 'sub-economic' for one or even both parties.
- If the pricing mechanism of the sale and purchase agreement includes a floor or ceiling price, S-curve formula or annual price change volatility dampening mechanism which has operated to the benefit of one party, either by design or due to unexpected changes in the energy market (ie, where the seller has sold LNG during a certain period at a higher price or the buyer has bought LNG at a lower price than the market price which would otherwise be applicable in the absence of the adjustment mechanism), should the reopener benchmark to be applied when the disadvantaged party invokes the review process just match the current market or also redress the accrued imbalance to some extent?

Despite the misgivings of common law lawyers, price review or reopener clauses still appear to be viewed as an essential part of most long-term LNG sale and purchase agreements supplying non-liquid gas markets. Many – if not all – LNG suppliers and buyers enter negotiations with a firm intention to include a price review or reopener clause, without appreciating how difficult it will be to reach agreement on the terms of the clause, particularly if they want the clause to be enforceable.

The following key elements must be addressed when negotiating a price review provision:

- The trigger event or criterion entitling a party to invoke the price review procedure must be defined – typically, this will be defined as a change of circumstances beyond the control of the parties, with limitations on the time of invocation throughout the contract term.
- The elements of the price mechanism which are subject to review must be defined – these elements may include one or more of the following:
  - base price;
  - indexation (degree of indexation and index used);
  - floor price;
  - ceiling price; and
  - inflection points of the S-curve formula.
- If a requesting party has satisfied the trigger event or criterion, which benchmark should be applied to determine the revised price mechanism? This has become a challenge for LNG buyers and sellers, as their view of the relevant market is likely to vary significantly.
- In a non-transparent market, how can the parties access the necessary information and use data that is almost always confidential in price review discussions or dispute resolution process?
- If the parties cannot agree on a revised price mechanism, can the matter be submitted to a third-party expert or arbitrator for resolution? Many LNG contracts contain 'meet and discuss' price review clauses which do not allow for referral of the matter to a third-party expert or arbitrator for resolution in the event that the parties cannot agree.

Without the ability to refer a disagreement on a revised price mechanism to a third-party expert or abitrator, the parties will be left without recourse, unless a specific recourse is otherwise specified. For example, one price review or reopener clause provides that if the parties are unable to agree on the revised price mechanism, the seller has the right, upon written notice, to terminate the long-term LNG sale and purchase agreement.

10. Specifications and off-specification LNG

The quality specifications of the LNG to be delivered under the long-term LNG sale and purchase agreement will be specified in detail, generally in an annex to the agreement. The minimum quality specifications, generally specified as a range for various factors, are often a matter of some negotiation where the buyer is attempting to meet the quality requirements of its facilities or its customers. One important point to be clear on is whether the quality specification is an 'as loaded' specification (applicable for FOB sales) or an 'as unloaded' specification (applicable for DAT/DAP sales).

The long-term LNG sale and purchase agreement will provide procedures for determining the quality of the LNG delivered. It will also generally include provisions that require the seller to notify the buyer when the seller obtains information that a particular LNG cargo is or will be outside of the quality specifications. The buyer traditionally has the remedy of rejecting LNG which is outside of specifications. However, this right, if unfettered, can lead to unnecessary waste in a situation where the buyer may suffer little or no harm from the off-specification LNG. Most long-term LNG sale and purchase agreements will put some limits on the buyer's right to reject LNG which is outside of specification, provided that the buyer can take the off-specification LNG without the risk of damages to persons, property or the environment. Where the buyer does not reject off-specification LNG, the buyer will generally be entitled to be reimbursed for direct and verifiable actual out-of-pocket costs incurred in receiving, treating and/or disposing of the LNG (or the natural gas regasified therefrom) as a result of the LNG being outside of the quality specifications. Sellers often attempt to limit their exposure for such reimbursement by applying a cap (generally tied to a percentage of the price of the off-specification LNG – for example, 15% or 25%) to their liability.

11. Shipping and LNG vessels

The long-term LNG sale and purchase agreement will contained detailed provisions on the LNG vessels to be utilised under the agreement, including LNG vessel specifications, vetting requirements, matters related to LNG vessel and terminal compatibility, responsibility for port charges and tug charges and compliance with terminal requirements.

Provisions are also generally included to address the use of alternative LNG vessels and/or substitute LNG vessels.

12. Other legal and financial issues

The traditional approach to dispute resolution, governing law, contract language and financing requirements in long-term LNG sale and purchase agreements is worth mentioning.

12.1 Governing law

While historically there are a few examples of other governing laws being chosen for long-term LNG sale and purchase agreements (eg, Japan, Indonesia), the substantial costs of LNG projects in today's market, and the resultant greater need for project financing, have resulted in the need for a well-established and certain commercial law as the governing law for long-term LNG sale and purchase agreements (almost universally the law of a common law jurisdiction, namely England or New York).

12.2 Dispute resolution

Dispute resolution in LNG sale and purchase agreements is almost universally achieved through international arbitration (eg, at the International Chamber of Commerce, the United Nations Commission on International Trade Law or the London Court of International Arbitration). The parties agree on an arbitration venue in a neutral location (often London or Singapore). The parties often agree that disputes of a technical nature (eg, measurement and certain shipping issues) will be referred to a single expert or panel of experts.

12.3 Contract language

English is by far the most common language used in LNG sale and purchase agreements. However, where one of the counterparties originates from a country whose official language is not English, the party in question often requires that the parties agree on an official translated version of the contract (in particular, in the Japanese market). It then becomes essential to establish which version will take precedence in the event of a difference in interpretation between the two versions. Lenders generally require that the English version should prevail in the event of any differences.

12.4 Assignment of contract for financing purposes

In the event that one or both parties is project financing the construction of infrastructure necessary for the performance of the LNG sale and purchase agreement, such party or parties will require that provisions be incorporated in the agreement to allow them to assign their LNG rights under the agreement to financiers to secure financing and oblige the counterparty to enter into a direct agreement with the financing party's lenders (which gives the lenders certain step-in rights and obligations in case of default of the financed party under the agreement).

13. Evolving LNG market issues

Certain issues and developments, which were not a part of traditional LNG purchase negotiations or trading environment in the past, are becoming the focus of much attention in the current market.

13.1 Contracting with aggregators

The LNG aggregator is typically a separate trading subsidiary of one of the major international oil and gas companies that purchases LNG from affiliated entities, which themselves purchase the LNG from both non-affiliated and affiliated LNG producer entities and then on-sell the LNG on a short, medium and long-term basis both to their own marketing affiliates and to third parties. The affiliate buying LNG for on-sale to the aggregator must have the contractual rights and practical ability to divert cargoes from an agreed nominal base destination. Typically, the LNG sale and purchase agreements supplying the affiliate buyer will agree a price applicable to the nominal base destination and allow for a split of any additional net proceeds received as a result of a diversion to a higher price market.[4] The combination of flexible supply sources, shipping fleet and capacity at import terminals in liquid gas markets such as the United States and the United Kingdom enables the aggregator to maximise revenues and minimise shipping costs.

LNG aggregator supply entities are generally relatively minimally capitalised entities which are contractually a step or two removed from the actual LNG producer/supplier. The security that a traditional buyer derived from contracting directly with the LNG producer/supplier which would give warranties on the underlying gas reserves, agree to maintain ownership and to operate the liquefaction

4 Again, anti-competition issues must be studied carefully in light of the jurisdictions involved.

facilities for the contract term, undertake to allocate excess capacity among its customers on a *pro rata* basis, dedicate ships to the contract's performance where the sale was on a DAT/DAP basis and so on are not available to a buyer from an LNG aggregator. A buyer from an LNG aggregator is exposing itself to the overall risk of the LNG aggregator's LNG trading business model, whereas a buyer from a traditional LNG producer has a much easier to quantify risk profile – that is, presumably the traditional producer will continue to perform so long as it is selling LNG for more than the cost of producing it.

Some of the issues which are likely to require a different approach from the traditional when contracting with an LNG aggregator seller include the following.

(a) *Dedicated reserves/*force majeure

Notwithstanding that the LNG aggregator will stress the advantages of its access to a portfolio of supplies, when entering into a long-term sale, the aggregator will almost certainly be looking to a specific supply contract as the base case to supply its customer, combined with flexibility to substitute alternative supply sources. Although the aggregator will be seeking *force majeure* relief for events affecting the designated supply source, it will be reluctant or unable to offer the traditional warranties on reserves and allocation of supply in the event of shortfalls. The circumstances in which the seller will be able to rely on *force majeure* events affecting supply sources and LNG ships need to be carefully considered and drafted.

(b) *Flexibility of supply source/cargo size*

To optimise profitability, the aggregator wants to maintain flexibility after the development of the annual delivery programme to vary supply sources and cargo size/LNG vessels utilised. The consequences of the potentially significant variations to the quantity and quality of LNG delivered from time to time, as compared to what was scheduled, need to be carefully addressed. The aggregator will also want the right to additional supply sources after entry into the sale and purchase agreement. The buyer will need to consider carefully what mechanisms need to be incorporated into the sale and purchase agreement regarding approval of new LNG supply sources.

(c) *Take or pay versus damages*

With the increasing liquidity of the LNG market and the aggregator's access to import terminal capacity in liquid gas markets, the traditional drivers for including take or pay mechanisms in sales contracts are disappearing. Aggregators are likely to find it inconvenient to have to accommodate the make-up supply obligations which accompany take or pay and thus may prefer a standard damages approach rather than take or pay.

(d) *Seller default*

Particularly where a buyer is situated in an illiquid gas market, challenges arise in arriving at an agreed description of the damages payable in the event of a late delivered or missed cargo. Assuming that the parties can agree on either a good description of the damages recoverable or a liquidated damages mechanism, the buyer is then likely to be particularly concerned when buying from an aggregator that the aggregator may be tempted to divert one of 'its' cargoes to a higher-priced destination if the extra revenue available to it from such a sale exceeds the damages payable by it to the buyer under its contract. Accordingly, the parties will need to consider how to structure the contract to disincentivise 'commercial default'.

13.2 Reserves issues for LNG sourced from coal bed methane and shale gas

Long-term LNG sale and purchase agreements have traditionally been supplied from designated conventional gas reservoirs. Buyers, which were extremely concerned about security of supply, typically insisted that the seller provide certification from a third-party expert such as DeGolyer & McNaughton that the reservoirs held sufficient gas to supply the entire contract term. Buyers would often also ask that sellers provide undertakings not to 'over-sell' the reservoirs. Buyers also took comfort from the fact that the majority of the sellers' expenditure would be made 'upfront' in drilling wells and installing gas gathering and liquefaction facilities prior to start-up of deliveries, with relatively modest capital costs (for a limited number of additional wells and perhaps compression equipment and facilities) generally being required thereafter.

The planned coal bed methane to LNG projects in Queensland, Australia represent a massive increase in scale from the current regional coal bed methane business. It has been estimated that the number of wells required to be drilled in 2015 to supply the new LNG projects under development will be four to five times the number drilled in 2010. Buyers of LNG dependent on coal bed methane as feedstock also need to be aware of other characteristics of coal bed methane production, including the following:

- While the 'average' offshore conventional natural gas well can deliver up to 200 to-300 million standard cubic feet (MMscf) a day, a typical coal bed methane well delivers around 0.3MMscf/day.
- Coal bed methane wells generally have a shorter production life (five to 15 years) than offshore conventional gas wells.
- Due to the limited scale and duration of coal bed methane production to date, estimates of recoverable reserves vary widely.
- Significant environmental issues may arise from the fracturing techniques required to produce the gas and from the significant quantities of water required to be injected or disposed of during the development and production of coal bed methane.

From a buyer's perspective, buying LNG reliant on coal bed methane feedstock presents a very different reserves and seller performance risk profile from buying LNG produced from conventional gas reservoirs.

In the United States, several LNG import terminals are proposing to install liquefaction facilities to enable them to take advantage of the recent growth in the US gas supply resulting from the development of shale gas reserves. Similar issues to those discussed in relation to coal bed methane also apply to shale gas.

13.3 New terminal models

As mentioned in the previous section, a number of US LNG import terminals are pursuing projects to install liquefaction facilities so that the terminals can operate as both LNG importers and exporters. The Singapore LNG import terminal, currently under construction, is designed to be able to load as well as unload cargoes, although it does not include liquefaction facilities. Clearly, the introduction of terminals with 'bi-directional' capabilities will offer commercial opportunities as well as introducing complexities in the interpretation of standard long-term LNG sale and purchase agreement terminology, particularly with regard to the source of LNG feed gas. The commercial structure of prospective LNG sales from the US terminal projects has yet to emerge, but it would seem that any sales contracts which are agreed will need to address many of the traditional long-term LNG sale and purchase agreement issues in an innovative way.

13.4 Destination flexibility

Traditionally, restrictions on alternative cargo destinations have taken the form of an absolute prohibition against the buyer taking a cargo to any destination other than the buyer's facilities (as defined in the relevant sale and purchase agreement) without prior consent of the seller. The opposite extreme would be a situation where the buyer could take the cargo anywhere without the consent or even knowledge of the seller.

Long-term LNG sale and purchase agreements were usually structured on a 'tramline' basis: cargoes, whether sold on a FOB or a DAT/DAP basis, were restricted to delivery at the buyer's named import terminal. For a variety of reasons, buyers (in particular, in FOB sales) are now increasingly negotiating to have the right to divert cargoes to other destinations, either because being able to trade LNG cargoes in the world market is an inherent part of their business model or as a way to mitigate high take-or-pay obligations.

Traditionally, sellers have strongly argued that such restrictions are reasonable for two main reasons:

- LNG pricing has been (and, to a large extent, continues to be) very regionalised and *ad hoc*; and
- It would be unreasonable for sellers to find themselves competing for a spot LNG cargo sale with their own long-term customers.

For sales to traditional, non-liquid gas markets, it is still the norm to specify the cargo destination and to allow diversions only in limited amounts and circumstances. Buyers may be able to achieve limited diversion rights (particularly in FOB contracts), which allow them to divert cargoes if they experience operational difficulties (which may include lower-than-expected demand for regasified LNG in their import terminal end-user market). Additional restrictions may be included in any provision allowing the buyer to divert cargoes; for example, minimum price criteria may be required for any cargo resale or there may be a requirement that any sale should not cause other customers of the seller to exercise their downward flexibility rights. Other compromise positions may appoint the seller as the buyer's agent to arrange the sale of the cargo in return for an arrangement fee. Alternatively, the seller may act as seller of the cargo, remitting to the buyer only the equivalent of the sums that the latter would have paid for the cargo under its take-or-pay obligations; this prevents the buyer from profiting from the sale.

EU competition law restricts the ability of contracting parties to agree to limit cargo destinations within the European Union, particularly in FOB sales. Buyers are in a weaker position when arguing for diversion rights in DES contracts, as the seller's ships will have to make voyages to destinations which may be further away than the buyer's facility and may involve increased unloading port charges or risks, among other things.

Increasingly, sale and purchase agreements with buyers in liquid gas markets include cargo diversion clauses, which provide for cargoes to be diverted to higher-priced gas markets at the election of the buyer or the seller upon agreed advance notice. The buyer and the seller agree to split the net incremental revenue resulting from the diversion after deducting the cost of the buyer's 'cover' gas, any increased shipping costs incurred and, in certain cases, a charge for the buyer's sunk import terminal costs.

13.5 Seller's step-in rights

LNG suppliers negotiating with some of the emerging LNG buyers have encountered difficulties in finding buyers with the credit capacity or regulatory approvals to enter into large long-term LNG sale and purchase agreements. LNG sellers are increasingly looking to address these concerns by incorporating provisions in their LNG sale and purchase agreements which provide that, in the event of significant performance default on the part of the buyer, the seller will be able to 'step in' to the buyer's import terminal use and pipeline transportation contracts. A number of difficult commercial, legal, tax and regulatory issues must be investigated and addressed in order to make these step-in rights effective.

13.6 Access to LNG shipping spare capacity

LNG counterparties which are not responsible for transportation under the long-term LNG sale and purchase agreement often seek to include provisions which allow them access to the LNG ships in the following circumstances:

- default of the transporting party under the long-term LNG sale and purchase agreement;
- where shipping capacity is not being used by the transporting party due to its exercise of downward flexibility; or
- during any initial build-up period of the LNG sale and purchase agreement.

14. Conclusion

Long-term LNG sale and purchase agreements are heavily negotiated and tailored contracts between diverse sellers and buyers. Although increasing volumes are being traded on a spot and short to mid-term basis, it is anticipated that the business will continue to rely on long-term LNG sales to underpin the majority of the volume traded for some time. Thus, practitioners will continue to have to address the various contract clauses which have been discussed in this chapter. Continued evolution of the market will require that long-term LNG sale and purchase agreements continue to evolve to adapt to the growing international LNG market.

LNG trading

Anthony Patten
Allens Arthur Robinson
Philip Thomson
Ashurst

1. Introduction

This chapter covers the issues relating to the buying and selling of LNG other than on a long-term, high-volume, take-or-pay basis.

It examines why the global LNG market is witnessing significant growth in LNG trading. We consider what the different trading models are and how the documentation issues which arise on spot and short-term sale arrangements can best be addressed. While the risks relating to a spot or short-term contract are essentially the same as for a conventional long-term sale and purchase agreement, we consider how these risks can be treated in the context of a short-term or spot trade. We also examine some swap structures and the issues specific to LNG swaps. Finally, we offer some thoughts on how the spot and short-term market for LNG might grow and what systemic constraints there are to limit that growth.

Our conclusion remains that, while the spot and short-term market will probably continue to grow both absolutely and as a percentage of global LNG traded, it is unlikely to displace the long-term, high-volume take-or-pay contracting model as the dominant model used in the industry.

2. The market

2.1 The traditional chain

The LNG industry has traditionally been structured around a series of 'chains'. The physical links of each chain consist of the gas producing, transportation and liquefaction infrastructure in the exporting country and the storage, regasification and transportation infrastructure in the importing country, as well as the LNG tankers to transport the LNG from the loading port in the exporting country to the unloading port in the importing country. The links in the chain are connected through contractual arrangements which, ideally, clearly allocate risk to the appropriate party and facilitate the limited recourse financing of some or all of the infrastructure comprising the chain. Under such a structure the LNG supplier and offtaker typically sign up to long-term, high-value supply and offtake obligations, with correspondingly severe penalties for failure to perform. As a result, they have little incentive to pursue more flexible supply or procurement contracting strategies.

The traditional 'chain'-based model has in recent years ceased to be the only business model used in the industry as a result of two interrelated trends: the growth of market players acting as traders or aggregators and the increasing liquidity of the global LNG market.

2.2 The rise of the aggregators

As the LNG industry has grown, a number of market players (principally among the international oil companies) have established portfolios of liquefaction, regasification and shipping assets. These market players are increasingly active as aggregators. What distinguishes aggregators is their contractual flexibility to take advantage of opportunities created by virtue of their own asset portfolio. This flexibility can create value by, for example, helping to meet different seasonal fluctuations for LNG demand in different markets.

For example, an aggregator will wish to have the contractual right to divert a cargo that was intended for one regasification terminal in one market to a more profitable market while at the same time minimising its exposure in the first market by sourcing energy from elsewhere. That replacement energy may be in the form of:

- a replacement LNG cargo, whether sourced from elsewhere within its portfolio or from another LNG supplier;
- replacement natural gas purchased in the market from which it had diverted LNG; or
- an alternative fuel, such as naphtha.

Aggregators have generally been more active in the Atlantic/Mediterranean Basin than in the longer-established Asian market (although the emergence of the US shale gas industry has clearly and significantly reduced US LNG demand). This may reflect the fact that, in Asia, LNG's role in the energy mix has been to help compensate for the lack of domestic energy sources in the East Asian industrialised economies. Given that the industrialised economies of Europe and North America do have alternative sources of supply (eg, pipeline gas from Russia or gas from the North Sea in the case of Northwest Europe), more attention has been placed in the Atlantic Basin on developing a flexible business model for LNG that allows LNG to compete with other energy sources. This can be contrasted with Japan, where pipeline gas does not exist as an alternative to LNG importation.

2.3 Why increasing liquidity?

The global LNG industry has grown significantly in recent years. According to the US Energy Information Agency, for example, the volume of LNG traded grew from 4993.03 billion cubic feet (bcf) in 2000 to 8,688.42 bcf in 2009.[1] With growth have come new market entrants and hence liquidity. The evidence suggests that this trend is set to continue.

In the Atlantic/Mediterranean Basin, new liquefaction capacity is being or has been added in Egypt, Equatorial Guinea, Nigeria, Norway, Angola and Trinidad and Tobago. New regasification capacity has been, or is being, added in the United Kingdom, the Netherlands, France, Italy and elsewhere. In addition, where long-term

---

1 http://www.eia.gov/emeu/international/gaslngimports.html.

contracts are approaching expiration, suppliers are now less likely automatically to renew these contracts, preferring to pursue more flexible marketing strategies. At the same time, new market players – and new types of market player – are emerging.

The market has grown despite a significant increase in LNG production costs. Despite significant reductions in capital costs before the early 2000s, with liquefaction costs falling to $200 per ton by 2001,[2] costs have soared over the past decade. Although costs can vary widely, the data suggests that construction costs for new liquefaction capacity have more than tripled, with certain 'problem trains' quoted at $1,200 per ton and above.[3]

Capital costs have also risen further along the LNG chain. Construction costs for LNG tankers increased by 40% to 50% between 2003 and 2008, while the data suggests that the cost of regasification facilities and receiving terminals increased by more than 50% over a similar period.[4]

Capital costs have increased as a result of:

- the increased costs of raw materials such as nickel and steel;
- a shortage of experienced workers and contractors (eg, in Australia, where numerous LNG projects are in development); and
- longer delivery times for key equipment.

In a climate of increasing concern over security of energy supply among energy importers, LNG has the advantage over pipeline gas of permitting imports from alternative sources should the original source cease to be available (whether as a result of reserve depletion, *force majeure* or deliberate suspension or diversion of supplies).

It is important to emphasise that the global LNG market is not yet a fully liquid market (as is the case for oil). LNG prices in the United States, European and Asia-Pacific markets remain divergent, so creating arbitrage opportunities for LNG buyers and sellers. As discussed below, it is unlikely that the LNG market will achieve such liquidity in the foreseeable future. Liquefaction projects are not being structured on the basis of sales into the spot market on a long-term basis (although some sponsors are willing to commit to construction costs without the LNG offtake being fully contracted).

These two trends are interrelated. Aggregators depend on a liquid market for LNG. Conversely, the development of aggregators – market players with asset portfolios – has stimulated increasing liquidity in the market.

Increasing market liquidity has in turn facilitated the emergence of a spot and short-term market for LNG.

3. Growth of spot and short-term LNG market

In these circumstances, the short-term LNG market (ie, spot contracts and contracts of less than four years) has grown from virtually zero before 1990 to 16.3% of the

2 "Understanding Today's Global LNG Business", Energy Business Reports (Atlanta, United States).
3 http://www.eia.gov/oiaf/aeo/otheranalysis/aeo_2008analysispapers/lnggc.html.
4 *Ibid.*

world LNG trade in 2009 and 18.9% in 2010.[5] In 2010, 727 cargoes of LNG were traded under short-term contracts. The leading short-term sellers were Qatar, Trinidad and Tobago, Nigeria, Egypt and Russia. Major short-term importers were Japan, Korea, Spain, the United Kingdom and the United States.[6]

4. Different trading models

Given these changes in the LNG market, there are, in addition to the 'chain' model described above, three different types of trading model:

- short-term contracts (ie, up to about five years) – for example, to cover a mismatch in the commissioning dates for liquefaction and regasification terminals that will eventually be linked by a long-term supply contract;
- spot cargoes – for example, a cargo that becomes available as a result of the originally intended regasification terminal being affected by some type of short-term *force majeure* incident; and
- swaps (strictly speaking, a separate topic) – where two buyers or two sellers agree to swap cargoes.

Spot/short-term cargoes can become available for a number of reasons, such as:

- default by a proposed long-term buyer (eg, Dabhol);
- earlier than expected commissioning (as was the case with Egypt LNG); or
- de-bottlenecking, giving rise to increased production capacity (subject to any applicable club rules or other priority rights for base load offtakers).

5. Documentation for spot and short-term cargoes

5.1 Master sales agreements and confirmation notices

While many short-term and spot LNG supply contracts are drafted on a bespoke basis, this is clearly not the best approach in terms of maximising speed of execution and minimising legal cost and management time.

A better approach is to have in place a master sales agreement between the seller and buyer, setting out the standard terms intended to apply to any sale between those parties, and then a confirmation notice which sets out, on a sale-by-sale basis, the terms to apply to that particular sale (eg, price, quantity, quality, name of tanker to be used, applicable demurrage rate, delivery date and downstream market).

As well as settling these practical matters, the parties should, before entering a confirmation notice, check that the risk allocation set out in the master sales agreement is compatible with the upstream and downstream sales, purchases and infrastructure. The seller will not wish there to be a mismatch between its contractual positions with the upstream project and with the buyer. Likewise, the buyer will not wish there to be a mismatch between its contractual positions with the seller and the downstream market. In addition, the parties should check that there are no adverse

5 See http://www.giignl.org/fileadmin/user_upload/pdf/A_PUBLIC_INFORMATION/Publications/GNL_2010.pdf.

6 *Ibid.*

tax consequences arising from the transfer of title in the relevant jurisdiction. Although the use of a confirmation notice enables the counterparties to execute a spot cargo trade in a short timeframe (as the market now demands), the need to confirm risk allocation and the different characteristics of particular markets (eg, the Atlantic Basin compared to the Asia-Pacific) have meant that confirmation notices still need careful attention to take account of the context of the particular trade. As difficult issues may need to be resolved in the confirmation notice (eg, the liability regime or the point at which title to LNG is to be transferred), a degree of negotiation is usually still required.

The shipping provisions illustrate how the master sales agreement and the confirmation notice fit together. Long-term sale and purchase agreements contain extensive provisions relating to port facilities, loading or unloading procedures, LNG tankers and the shipping schedule. This is because this is the area of the transaction where the practical interface between buyer and seller is greatest. In a spot or short-term contract, most of these provisions can be placed in the master sales agreement and will therefore apply to every trade entered into between the parties. If this is the case, the confirmation notice need deal only with those issues which are specific to the particular trade in question, such as the name of the LNG tanker or tankers which will be used on that trade and any peculiarities of the loading and unloading ports.

The form of confirmation notice is generally attached as a schedule to the master sales agreement (although this form can be amended on a sale-by-sale basis).

The master sales agreement and the confirmation notice together form the binding contract for the sale and purchase of LNG. It will be important to include in the master sales agreement language making clear that, until the confirmation notice is signed by both parties, there is no such contract and that, on execution of the master sales agreement, there is no subsequent obligation on either party to enter into any confirmation notices.

In addition, it will be sensible to have language in the confirmation notice stating that if there is an inconsistency between the terms of the two documents, the confirmation notice will prevail. This gives the parties the flexibility to agree in the confirmation notice on a sale-by-sale basis amendments to the standard position set out in the master sales agreement without requiring the master sales agreement to be amended. This could be useful if, for example, the quality specification for a particular trade into a particular downstream market varies from any minimum quality specification set out in the master sales agreement.

5.2 Reciprocal agreements

Some master sales agreements and confirmation notices are drafted so that they can be used on transactions where either party may be the buyer or the seller, and for both ex-ship and free on board (FOB) transactions, with the specifics being detailed on a transaction-by-transaction basis in the relevant confirmation notice. While this may lead to a slightly longer document than would otherwise be the case, the fact that each party can be the buyer or the seller tends to make them fairly well balanced.

More generally, given the increased number of market players and the increased likelihood of such players acting as both buyers and sellers of LNG, negotiation of master sales agreements has, in our experience, become less partisan and the trend is now towards relatively neutral, balanced agreements which in time will help to facilitate LNG industry uniformity.

5.3 Standard forms

There is currently no market standard form of master sales agreement equivalent to the International Swaps and Derivatives Association arrangements used in the commodity, currency and interest rate hedge markets.

As with long-term sale and purchase agreements, each party tends to have its own standard form which it would prefer to use. The negotiation of a master sales agreement and form of confirmation notice can therefore involve a 'battle of the forms'.

The Association of International Petroleum Negotiators has developed a model LNG master sales agreement and confirmation notice, which was published in 2009 and which its sponsors intended would become recognised as the industry-wide standard for all the regional markets. There is little evidence supporting widespread use of this model, or of a separate model developed by the European Federation of Energy Traders and published in 2010. Instead, most major industry players still prefer to use their own contract forms.

As a rule of thumb, our experience is that a structure based on a master sales agreement and confirmation notice is used for contracts of up to about one year. For a longer-term contract, the parties tend to want to use a more bespoke structure.

6. Risk allocation

The risks relating to a short-term or spot contract are essentially the same as for a conventional long-term sale and purchase agreement. Likewise, a similar allocation of risk between buyer and seller ought to deliver a robust LNG supply arrangement.

However, it should be possible to achieve a simpler solution to many of the issues on a short-term or spot contract than on an equivalent long-term contract. Exactly how simple will depend in large measure on the duration of the contract. For example, the contractual provisions relating to the spot sale of a cargo on the water and for delivery within a few hours or days will be simpler than those for a short-term contract covering several cargoes or for a spot cargo for delivery several weeks or months ahead. For simplicity, in the analysis that follows, we generally refer to spot cargoes.

6.1 Price

It should be possible to have a very simple pricing clause for the spot sale FOB of an LNG cargo that is already on the water. For spot cargoes sold in advance of loading, the price will ordinarily be set by reference to an index, although sometimes (albeit rarely) they are sold at a fixed price. Counterparties to fixed price deals will then often seek to hedge their position.

On a longer-term, but nonetheless short-term, contract, pricing will usually be set by reference to an index (eg, the JCC Index if for delivery into Asia).

Some long-term supply arrangements contain price review clauses, enabling one or both parties to reopen the price where the contractual price varies significantly from the market price of energy in the downstream market. This protects the seller from the risk of having to sell into that market for significantly less than the market will bear. It also protects the buyer from the risk of having to take LNG at a price that is uncompetitive in its downstream market. Clearly, the risk of the market price moving in such a way as to cause this risk to crystallise over a short-term arrangement is remote. For this reason, the parties should not need a price reopener.

6.2 Volume and destination flexibility

On a spot cargo trade, the concept of take or pay is in commercial terms redundant because the buyer will either take the cargo (and therefore all the contracted volume) or not. However, in structural terms, it can be helpful to state that the contracted volume comprising the cargo is the take-or-pay volume, and so set the liquidated damages payable by the buyer should it fail to take that cargo. As an alternative structure used in some spot cargo contracts, the buyer that fails to take the cargo is entitled to receive the net proceeds of any sale of the LNG cargo to a third party (provided that the buyer has paid the applicable contracted price for the LNG not taken).

A recent development in some spot cargo contracts has been the introduction of a stipulated 'unwind' mechanism whereby, on the giving of notice by the seller, the parties will agree that the cargo trade will not proceed. A buyer would expect to receive a specified sum in exchange for the exercise by the seller of the right to 'unwind' the trade (eg, payment for any regasification charges and/or a sum to allow the buyer to obtain replacement gas in order to satisfy its downstream customers). The seller can then divert the LNG cargo to a more profitable customer and/or destination. Given the uncertainty created by this type of provision, a buyer of LNG would expect to obtain a price reduction to reflect the increased risk of non-delivery. Regular 'unwinding' of LNG sales by a particular seller might lead financial regulators to treat the underlying LNG contract as a financial instrument rather than a physical contract for delivery, and accordingly require the seller in some cases to be regulated by relevant financial regulatory authorities (as occurs in relation to the trading of other commodities such as oil). This is a particular consideration for investment banks and commodity traders which have developed LNG trading platforms.

A short-term contract for numerous LNG cargoes will normally contain a traditional take-or-pay provision. The buyer is still likely to want the maximum achievable make-up rights to mitigate any take-or-pay liability that arises. Clearly, the shorter the term, the less likely it will be that the buyer will be able to negotiate any make-up rights. Spot LNG sales usually provide for a net proceeds mechanism as an alternative to make-up rights.

On a short-term contract (eg, a contract for incremental supply over the peak demand period), the buyer may need less volume flexibility than on a long-term supply contract, where downstream demand may be more seasonal.

Likewise, the buyer under a short-term contract may be confident in its ability to forecast at the outset where it will want to take the LNG and therefore need less, if any, destination flexibility.

6.3 Termination

Long-term sale and purchase agreements typically give the buyer termination rights where the seller persistently fails to ship LNG. Exercising this right allows the buyer to source replacement LNG from elsewhere without risking a take-or-pay liability should the original suppliers resume supply. In the case of a spot contract for a single cargo, the termination rights can be much simpler because the contract is 'live' only for the period starting when the confirmation notice is signed until the cargo is delivered and paid for. In these circumstances, termination rights may be limited to insolvency affecting either party or the failure by the buyer to provide/maintain any credit support (eg, a letter of credit supporting the payment obligation).

As the spot and short-term LNG market matures, one issue that may attract more attention than it has to date relates to cross-default. This may be an issue if it is likely that at any time the parties may have two or more trades on foot and documented under one master sales agreement, but several confirmation notices. In these circumstances each party may need to consider whether it should be able to terminate the contract created by the master sales agreement and one confirmation notice on breach or the occurrence of prolonged *force majeure* affecting the contract created by the master sales agreement and the other confirmation notice. Much will depend on the gravity of the circumstances giving rise to the termination right and whether the default on one trade interferes or is likely to interfere with performance of the other. It may also be appropriate to provide for different treatment for prolonged *force majeure* and breach, given that the former is likely to be outside the control of the affected party. Cross-default may be a particular issue where multiple trades are documented under several confirmation notices, but are in fact all part of the same commercial transaction. For example, if neither party has all of the shipping capacity required for the transaction as a whole, the parties may agree to structure the transaction partly as an FOB sale and partly as an ex-ship sale. The structure would likely require two confirmation notices (one for the FOB sales and one for the ex-ship sales), underpinned by a single master sales agreement. In this case, a party may want robust cross-default rights so as to be able to terminate all parts of the wider transaction should the other party default.

In addition to the termination of the contracts created by the master sales agreement and the confirmation notices, the parties may wish to be able to terminate the master sales agreement itself. In legal terms, the termination of the master sales agreement is not, in the absence of any confirmation notices, a drastic remedy because the obligations contained in it (when no confirmation notice is in place) are boilerplate in nature (eg, confidentiality). However, it may give the practical advantage of ending the entire relationship with the counterparty (which may be important, for example, for reputational reasons).

It is worth emphasising that, given the lack of liquidity to date in the spot and short-term LNG market, cross-default has tended not to be the subject of detailed discussion. However, as the market deepens and the scope for more complex transactions increases, parties may wish to consider the area in more detail. A default under the contract created by a confirmation notice and a master sales agreement may have implications elsewhere in the project structure – for example, under a credit agreement.

7. Swaps

7.1 Swap structures

Many short-term and spot trades are transacted in the context of a wider swap transaction including two or more cargoes contracted on a long-term take-or-pay basis. A common swap arrangement is illustrated below.

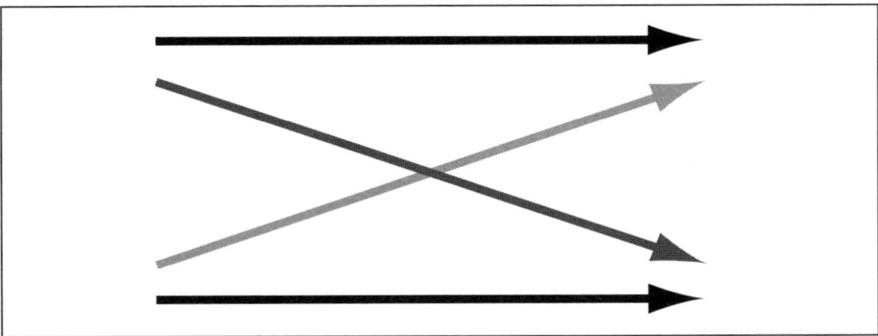

In this case, suppose that Seller 1 is located in Trinidad and Tobago and is contracted to sell LNG to Buyer 1 in Spain. Seller 2 is located in Algeria and is contracted to sell LNG to Buyer 2 in the United States. Since both of these trades involve considerable shipping distances (and therefore associated cost), it makes good sense for the parties to swap their cargoes so that Seller 1 in Trinidad and Tobago delivers to Buyer 2 in the United States and Seller 2 in Algeria delivers to Buyer 1 in Spain. The reduction in costs is the upside created by the swap that can then be shared among the parties as they may agree.

Another common swap structure is illustrated below.

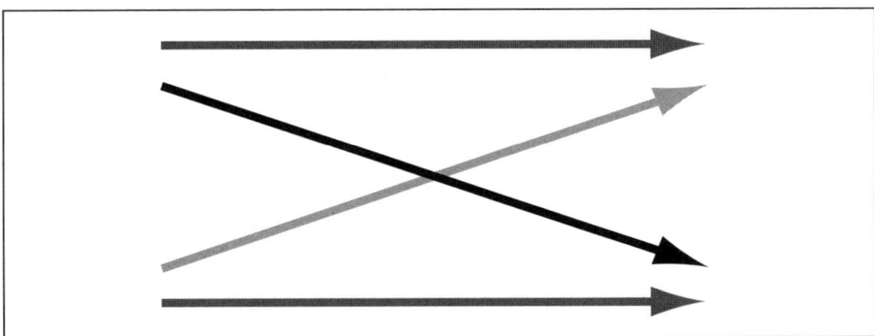

Suppose here that Seller 1 is contracted to deliver LNG to Buyer 1 in January.

Seller 2 is contracted to deliver LNG to Buyer 2 in February. However, as a result of scheduling mismatches, Buyer 1's regasification terminal is due to shut for general maintenance in January and Buyer 2's in February. It therefore makes sense for the parties to swap the cargoes so that Buyer 2 will take in January the cargo scheduled to be delivered to Buyer 1 and Buyer 1 will take in February the cargo scheduled for delivery to Buyer 2. The swap saves both buyers' liability to their respective sellers under the take-or-pay regime. This saving is the upside created by the swap.

7.2 Upside sharing

If two buyers or two sellers want to swap cargoes that were originally intended to be purchased or delivered on a long-term take-or-pay basis, it will normally be necessary to obtain the respective counterparty's consent under the original long-term sale and purchase arrangements. Depending on the intended swap structure, this consent could be required under the shipping provisions, the destination clause and/or the scheduling provisions. The parties whose consent is required will typically wish to share in the potential upside that the swap transaction will generate without being exposed to the downside. How the upside is shared will be a matter for negotiation between the four parties.

The negotiations on the sharing of the upside will be simpler if the two underlying long-term sale and purchase arrangements provide for similar risk allocation. If this is the case, the parties will not need to price into the upside sharing discussions the extra risk they are being asked to assume.

Nonetheless, how the upside is shared will often be the subject of some debate. There is no single correct solution. In order to simplify the transaction as a whole and minimise the role of the parties whose consent is received, the parties proposing the swap may find it easier to offer a share of the upside by way of a fixed sum (eg, by way of adjustment to the cargo prices) rather than by means of a percentage-based structure. This approach also means that there is no need for the consenting party to verify the upside created by the swap (which can be practically difficult and can raise issues of confidentiality).

At the same time as settling the commercial terms of the swap, the parties will need to verify that the onshore infrastructure at the various terminals and the shipping and the LNG tankers that are proposed to be used are mutually compatible. Likewise, the two cargoes to be swapped will need to be of consistent quality.

Care should be taken to ensure that any arrangement to share, between buyer and seller, any upside generated by diverting LNG does not offend applicable competition law. For example, the Directorate General – Competition of the European Commission takes the view that, in respect of LNG sales into or within the European Union, upside sharing may constitute a disincentive for the buyer (which would be obliged to share profit with the seller) to sell the LNG to other customers or to a different destination. If upside sharing provisions do constitute a disincentive, they may be seen to be an unlawful restriction of competition.

7.3 Non-performance

A tricky issue that will arise relates to the risk of non-performance. Consider the first swap structure illustrated above. If Seller 1 agrees as part of the swap transaction to deliver a cargo originally intended for Buyer 1 to Buyer 2, does Seller 1 or Buyer 1 take the risk that Buyer 2 fails to take that cargo? If Seller 1 assumes the risk, it will have to seek and recover the loss suffered directly from Buyer 2. If Buyer 1 takes the risk, it will likely have to protect Seller 1 from the loss suffered and recover its liability to Seller 1 from Buyer 2. This issue will be a factor in negotiations between Seller 1 and Buyer 1 on the sharing of the upside generated by the swap. The risk of Buyer 2 failing to perform is likely to be more acceptable to either Seller 1 or Buyer 1 if the risk profiles of Buyer 1 and Buyer 2 are broadly the same under the original two contracts, and if Buyer 2 represents a robust counterparty credit which is at least as good as Buyer 1.

Parties to swaps are also sometimes taking short positions, increasing the risk of non-performance. For example, a seller may agree with a buyer to divert an LNG cargo to a buyer in another market, on the basis that the original buyer would be 'resupplied' with another LNG cargo, sourced elsewhere and transported by a vessel other than that nominated in the original contract. Ideally, the seller would not agree to divert the original cargo without having first secured the 'resupply' cargo. As the market matures, it can be expected that sellers will assume a degree of resupply risk.

8. The future

The LNG spot market is some way from being a genuine spot market (in the sense of being a real-time commodity market for the prompt sale and delivery of a good or service). The market still operates on a timescale of weeks or months ahead.

The industry consensus is that the LNG spot/short-term market will likely continue to grow both absolutely and as a percentage of the global energy market. While opinions vary, the general view is that LNG spot and short-term sales will not comprise more than between 15% and 20% of the global LNG market. In particular, it seems unlikely that spot/short-term sales will displace the long-term, high-volume take-or-pay contracting model as the dominant contracting model in the industry. A number of factors constrain the development of a liquid market for LNG, and, therefore, the supply of LNG on a spot/short-term basis. These include:

- the fact that the ability to arrange short-term and spot sales depends on having, or being able to source at short notice, available regasification capacity giving access to the relevant downstream market;
- lack of standardisation of LNG quality specifications;
- difficulties in achieving required levels of programming flexibility, especially where programming remains primarily driven by sellers;
- lack of standardisation of terminal and tanker design to ensure maximum interoperability; and
- lack of transparent price reporting (given customary confidentiality provisions, in LNG sale and purchase arrangements and the absence of published sources (eg, Platts) for the price of LNG).

LNG shipping

David Gardner
Energy Transact LLP

1. Introduction

1.1 The history of LNG shipping

Two events made possible the transportation of natural gas by sea. The first was the process of liquefaction of natural gas; the second the successful design of containment systems in which to store the liquid gas. LNG carriers require a special containment system as normal steel cannot withstand the low temperature at which the liquid gas is carried. Although the cargo is insulated, a small amount of cargo is allowed to 'boil off' or vaporise in order to keep the cargo at a constant temperature, and this boil-off gas is traditionally used as a fuel to power the vessel.

The first commercial LNG carrier delivered into service was the Methane Pioneer, which was converted in 1959 from an existing cargo vessel, with the first commercial cargoes delivered in the mid-1960s. The Methane Pioneer had rectangular cargo tanks made of a special aluminium to withstand the cold cargo and its internal holds were fitted with balsa wood faced with maple and oak plywood to provide insulation. The vessel was built to transport LNG from Lake Charles, Louisiana, to Canvey Island, United Kingdom, and was partially owned by the British Gas Council, the forerunner of British Gas.[1] The Methane Pioneer was followed in 1962 by the Methane Princess and the Methane Progress, both of which were built by the Vickers Yard in the United Kingdom with tanks of a prismatic design. The early LNG tankers built during the 1960s and 1970s were generally constructed at European shipyards, with membrane containments systems and limited cargo capacity.[2]

The next generation of LNG carriers were built in the 1980s and were required primarily to service new LNG projects in Abu Dhabi, Indonesia and Australia, transporting the gas to the emerging markets of Japan and, to a lesser extent, Korea and Taiwan.[3] By 1990 there were approximately 70 LNG tankers in service, and until the mid-1990s the industry remained essentially a point-to-point trade with the great majority of employed vessels dedicated to specific projects and trades.

The wider changes in the LNG industry from the late 1990s and during the first decade of this century spurred dramatic changes to the LNG shipping industry. These included a series of technical innovations intended to reduce the cost of LNG

1 Michael Tusiani, *The Petroleum Shipping Industry*, Vol 1.
2 A large ship would be in the order of 75,000 cubic metres.
3 Michael Tusiani, *The Petroleum Shipping Industry*, Vol 1.

transportation and to bring the technology employed in the LNG shipping industry into line with other sectors of the maritime industry. In addition, at the end of the 1990s the entry into the international market of the Korean shipyards, and their aggressive pricing aided by the fall in the value of the Korean won against the US dollar, drove down shipyard prices from around $250 million in the mid-1990s to around $160 million by the end of that decade. This reduction in the cost of a newbuilding LNG carrier, coupled with a huge expansion in demand for LNG ships as investment was sanctioned in new projects, led to a doubling of the world's fleet and the introduction into the industry of a whole generation of new LNG owners, including many of the established tanker owners.

The last few years, by contrast, have been a period of relative stagnation for the LNG shipping industry. Between June 2008 and January 2010, no new orders for LNG vessels were placed at any shipyard worldwide, reflecting the decline in the world economy and a tailing off of new LNG projects. However, at the time of writing, the LNG shipping market is again tight, with daily rates for steam turbine vessels in the region of $80,000 at the start of 2011. Moreover, a further expansion to the world fleet is generally anticipated to cope with increased global LNG demand and the likely increases in transportation distances, and these considerations have led to a recent flurry in orders for newbuilding LNG carriers.

1.2 The LNG shipping industry today

By comparison with oil, the transportation of LNG remains relatively small, in terms of both volume and value. In 2005 the LNG trade accounted for about 142 million tonnes,[4] whereas the volume of oil carried was estimated to be 2.308 billion tonnes.[5] Oil transportation is part of a mature commodity market where there are a large number of tanker owners and operators carrying substantial volumes of cargo under a variety of contracts for spot cargoes, period trading contracts and long-term charter contracts or contracts of affreightment. Oil is an intensely competitive market dominated by movements in the oil price. As such, operators and vessels in the market come and go in conjunction with fluctuations in the freight market. Second-hand sales and scrappings are an inherent feature of the market. The shipping market is highly fragmented, with independent shipping companies[6] owning approximately two-thirds of the tankers above 200,000 dead weight tonnes between 1976 and 1995, with oil companies and government-owned or controlled entities accounting for the remaining one-third,[7] and with no single owner holding more than a small percentage of the overall fleet. In addition, the trend during the 1970s, 1980s and 1990s was for the oil companies to obtain less and less of their tonnage on long or medium-term charter, so that by 1995, approximately half of the world's very large crude carrier fleet traded on the spot market.[8]

The profile of the LNG shipping industry is very different. The LNG shipping

---

4 *Lloyds List* May 31 2006

5 Intertanko Annual Review 2005.

6 Including those companies that are effectively owned and run by one person.

7 Michael Tusiani, *The Petroleum Shipping Industry*, Vol 1.

8 Tanker Market Report of September 15 1995, Intertanko Oslo.

market remains divided between the fleets controlled by specific projects and those controlled by the large gas companies that use their portfolio of vessels to trade worldwide. The role of the independent shipowner, though increasing, remains small. Second-hand sales and scrappings are rare, and few LNG carriers trade spot for any prolonged period of time.

1.3 Technical innovations

There have been three principal innovations in the design of LNG tankers in recent years:

- a significant increase in cargo capacity;
- the introduction of new propulsion systems; and
- further innovations in the cargo containments systems.

Whereas the largest LNG carriers ordered during the 1990s were generally 138,000 cubic metres (m^3) or 145,000m^3, a new generation of larger Q-Flex and Q-Max vessels with cargo capacities of 215,000m^3 and 240,000m^3 was developed during the Rasgas and Qatargas tenders during the early part of the first decade of this century. While these ships offer the obvious advantages of economies of scale and lower-cost transportation, many current LNG receiving terminals cannot accommodate vessels of that size and draught, restricting their trading scope. In general, however, the trend towards increased size has continued, with vessels of cargo capacity in the order of 170,000m^3 increasingly common.

Linked with the development of larger ships has been a change within the industry to other forms of propulsion, in particular diesel electric engines. Until the start of this century, almost all LNG tankers operated with steam turbine propulsion. Although reliable and with low maintenance costs, steam turbine vessels have relatively poor fuel consumption and have been discarded by most other shipping sectors, making it increasingly difficult to find engineers experienced in their operation and maintenance. In addition, the design changes to the vessels needed to accommodate the increased cargo capacity and, in particular, the widening of the aft body hull form have led to a re-examination of the propulsion systems, and the use of twin shaft lines in order to maintain efficiency. The use of diesel electric and slow speed diesel engines is now common, with many new LNG carriers also having onboard reliquefaction facilities capable of capturing and reliquefying boil-off from the cargo tanks.

The cargo containment systems have also undergone radical change over the last 10 years. Until the start of the last decade, the great majority[9] of LNG carriers used the Kvaerner Moss Spherical System, which utilises spherical tanks made from aluminium alloy or 9% nickel steel heavily insulated. Today, the majority of newbuilding LNG carriers employ the membrane containment system currently manufactured under licence from Gas Transport Technigaz (GTT). GTT is the result of a merger between Gas Transport and Technigaz, each of which had its own membrane system. The merged company has also developed a new membrane tank containment system, 'Combined System 1' (CS1), which comprises a primary

---

9 Two tankers used the structural prismatic design developed by the IHI shipyard in Japan.

membrane made from a nickel-steel alloy known as Invar, a Triplex secondary membrane and a reinforced polyurethane foam insulation. This new system can be extensively pre-fabricated and the assembly process has been rationalised, enabling the shipyards to achieve significant savings of time in one of the most complex parts of the vessel construction.

This switch reflects both the lower operating costs of vessels with the GTT membrane system[10] and the fact that most LNG tankers today are constructed in Korea, whose yards have specialised from the outset in membrane containment systems.

1.4 Floating LNG

There is one further area of increasing innovation in LNG production and transportation. The difficulty in obtaining environmental consents for the construction of onshore LNG import terminals – particularly acute in the United States – has led to the increasing development of offshore receiving terminals. There are many different technologies available, but the two principal alternatives are the use of either an LNG regasification vessel or a floating storage and regasification unit (FSRU). An LNG regasification vessel is usually a purpose-built LNG carrier with an onboard regasification facility capable of mooring against a turret loading buoy and discharging into a subsea pipeline. The second alternative is where the FSRU that receives LNG from a conventional LNG tanker, regasifies this and transports it to shore again via a subsea pipeline.

The development in particular of FSRUs has provided an alternative use for older tonnage coming off long-term charter contracts. Those vessels using the Kvaerner Moss containment system give rise to less sloshing issues and are particularly suited to conversion into FSRUs. The contracts of charter for such units tend to be based on their oil industry equivalents, the floating storage unit (FSU) or the floating production storage and offtake unit (FPSO), with the owner warranting the regasification or send-out capabilities of the unit. Also increasingly in prospect is the development of the LNG FPSO, capable of liquefying gas produced offshore and exporting direct to offtake LNG tankers. These FPSOs are likely to follow the approach adopted in the oil industry where they comprise a combination of newbuilding units and converted tankers procured on both lease and operate (ie, charter) and engineering, procurement, construction and installation terms. The converted units in particular offer, through lower construction costs and shorter lead-in times, significant benefits for the development of previously marginal fields.

1.5 Safety issues

Historically, the perception is that LNG is a dangerous, high-risk cargo, but in reality there have been very few incidents associated with either LNG or the vessels used to transport it. The Society of International Gas Tanker and Terminal Operators' Confidential Incident Report of LNG problems since 2001 records that on average

10 The Suez Canal charges of a Kvaerner Moss design vessel are significantly greater than a GTT membrane system vessel due to the void spaces below the spheres. The membrane vessels generally also have a superior boil-off performance.

there are between three and seven problems involving LNG vessels each year. While it is true that some of the more recent problems regarding containment systems have worrying features,[11] most of the problems have been of a relatively minor nature or unconnected with any specific problem relating to the carriage of LNG. As yet there have been no pollution incidents involving an escape of LNG and no significant breach of cargo containment in around 40,000 voyages.

It remains to be seen whether this relatively low claims record remains the case as the markets expand and the spot market plays an increasingly prominent role. Concerns have been expressed that the increase in use of spot markets may result in operators becoming more risk averse in order to meet time pressures. There are also increasing concerns that the rapid expansion in fleet size is causing a shortage of skilled personnel. Problems may also be experienced in ensuring compatibility between the vessels currently being built and existing terminals and existing vessels and new terminals being built.

2. Owning LNG carriers

Side by side with the process of expansion and technical innovation have been dramatic changes in the terms on which LNG carriers are purchased and operated. A gas company can acquire the use of a LNG carrier in two ways: by purchasing and owning the vessel or by chartering it under a time, voyage or bareboat charter or other contract of affreightment.

2.1 Ownership issues

Every ship sailing in international waters must be registered with and fly the flag of an individual state. A ship that does not possess this 'nationality' enjoys no protection under international law and will be unable to enter ports or engage in lawful trade. The most obvious expressions of a ship's nationality are to register and/or fly the flag of a particular country. By and large, all nations are free to determine the conditions that a shipowner must satisfy before it is entitled to register its ship on that country's registry and/or to fly the flag of that country.

In choosing a flag for their vessel, shipowners must consider a range of factors, including:

- any restrictions of eligible shipowners (some flags do not accept foreign companies as the owner of the ship, or require the foreign company to register locally as a 'foreign maritime entity');
- any extra port dues which may be charged on foreign flag ships in some countries;
- crew and manning requirements, which may vary from one flag to another;
- the acceptability of the flag (which will determine the law of the mortgage and thereby many aspects of mortgage registration/enforcement requirements) to potential financiers;
- the cost of registration and operation (many onshore flag will have more stringent technical and security requirements);

11 It should be stressed that this is in terms of the size of claim rather than any safety concern.

- tonnage tax options (for onshore flags) and tax leasing opportunities;
- convenience of ship and mortgage registration (some registries require provisional and later permanent registration involving a translation of the mortgage deed and registration documents from English to a local language); and
- the reputation of the flag in the marketplace.

Generally, flag states fall into one of two group: the onshore flag states and the offshore 'flags of convenience'. Most onshore European flags have traditionally been more expensive for the owner than offshore flags, and this has led to a steady move away from the onshore flags in favour of offshore jurisdictions. In recent years, however, many European ship registries have adopted tonnage tax regimes in order to compete and retain their commercial shipping fleets, with some success. When this can be combined with tax leasing structures, it can be very attractive to owners of high-value ships such as LNG carriers.

2.2 The sale and purchase of LNG tankers

(a) Generally

As with vessels in other sectors of the shipping industry, LNG carriers can be acquired second hand or direct from a construction yard as newbuilding tankers. Second-hand vessels are usually bought and sold on terms based on the Norwegian Sale Form 89 or the later 93 form, though in practice the sale of second-hand LNG vessels remains exceptional.

The prospective owner looking to acquire an LNG carrier will therefore generally approach one of the leading shipyards in the Far East to seek to agree terms and timing for the delivery of a newbuilding. The limited role played by European shipyards in the construction of LNG carriers has now declined almost to the point of extinction, and the principal construction yards are now located in South Korea – particularly at the facilities of Samsung Heavy Industries, Daewoo Shipbuilding & Marine Engineering and Hyundai Heavy Industries – and to a lesser extent in Japan, with the leading Chinese yards also attracting increasing orders, often for projects linked to the import of LNG into mainland China.

(b) The shipbuilding contract

There is no standard form shipbuilding contract for LNG carriers and, as in the tanker and dry cargo sectors, the form of shipbuilding contract concluded with a Korean, Chinese or Japanese yard will generally be based (albeit sometimes loosely) on the standard form contract prepared by the Shipbuilders Association of Japan (the SAJ form). In the LNG industry, however, the trend in recent years has been for the charterer to negotiate and agree the terms of the shipbuilding contract and technical specification directly with the shipyards, and this has led to a degree of departure from these standard forms and variance in the form and terms agreed in LNG shipbuilding contracts.

Design: Generally, the shipyards have been willing to develop and adapt their own 'in-house' designs to accommodate the changing requirements of owners and charterers over the last decade, and to retain contractual responsibility for this design. This assumption of contractual responsibility will usually extend even to the more complex features of the vessel, including the propulsion and cargo containment systems, and sometimes to specific innovations such as the re-liquefaction plant. Generally, the builder will retain title to the vessel during construction and until delivery, with the consequence that the rights of the buyer (and charterer), in the event that the builder defaults, will usually be limited to the rights to terminate the shipbuilding contract and to require the builder to refund any instalments of the contract price paid by the buyer prior to this termination.

Contract price and delivery: Even where the the dock and berth space required to construct an LNG carrier is booked several years in advance, LNG shipbuilding contracts are almost invariably concluded on a fixed-price basis, with the contract price calculated in US dollars and payable in instalments before and on delivery of the vessel. Although the price is fixed, it will be capable of adjustment to reflect any modifications required by the buyer and sometimes by the charterer. The proportion of the price payable before delivery will depend in part on the respective bargaining positions of the buyer and builder, but also on the choice of shipyard, with some builders more willing than others to agree flexible terms.

In addition to agreeing to build the vessel for a fixed price, the builder will usually agree to deliver the vessel at the shipyard on or before a fixed future date, though this date will usually be capable of extension by any 'permissible delays' including *force majeure*, modifications to the vessel's specification,[12] delays by the buyer in the performance of its obligations and bad weather delaying sea or gas trials.

This delivery obligation gives rise to difficulties in LNG projects in at least two respects. First, although the builder will usually be obliged to pay liquidated damages if delivery of the vessel is late, these damages will rarely be sufficient to compensate both the owner for its additional costs of supervision and financing during the delay and the charterer which is unable to sell or purchase its intended production. Second, the delivery record of the Korean yards has in recent years been rather more reliable than the delivery record of the various LNG projects, leaving many charterers obliged to pay hire for ships for which they have no immediate use. For this reason, many shipowners and gas companies have sought from the builder the right to delay delivery of the vessel to accommodate a limited degree of delay in the project. Where this is achievable, it can provide a useful flexibility for the charterer, which would otherwise be obliged to sub-let the vessel to try to recover some part of the hire payable under the long-term time charter.

Supervision of contracting and subcontracting: The buyer under an LNG shipbuilding contract will almost invariably obtain, both for itself and for any

---

12 These will include both modifications introduced by the buyer and statutory modification arising from any changes in the regulatory regime with which the vessel is required to comply.

charterer of the vessel, wide-ranging rights to inspect the vessel and all its component parts as these are fabricated, assembled and installed both at the builder's shipyard and at the premises of the builder's subcontractors. The contract should also include a provision obliging the builder to rectify any aspect of the construction, materials or workmanship that the supervision team identifies as defective.

Given the ever-present need to control costs, it is perhaps unsurprising that both the Korean and Japanese shipyards should subcontract a substantial proportion of their work, with an increasing tendency, for example, for large items of steelwork to be fabricated in China. From the buyer's perspective, it is critical to its ability to control the quality standards achieved by the builder that it be entitled to approve the subcontracting of any significant portion of the construction work. It is also usual to include within the technical specification a detailed 'makers' list' identifying the agreed manufacturers to which the builder will subcontract the supply of certain key items of equipment and machinery.

Performance criteria: In addition, to any specific requirements set out in the technical specification, the shipbuilding contract will usually contain a description of the vessel, including its intended length, breadth, depth, tonnage, cargo capacity and draught. In addition, the builder will typically guarantee certain key aspects of the vessel's performance including, for an LNG carrier, its speed, fuel consumption, cargo capacity, draught and rate of boil-off. If the vessel fails to meet these guaranteed performance standards, the contract will usually provide for payment of liquidated damages by the builder and, if the deficiency is excessive, for the buyer to be entitled to terminate the shipbuilding contract.

The speed of the vessel is usually measured during sea and gas trials over an agreed distance and under defined sea and weather conditions set out in the technical specification, with the fuel consumption of the vessel's main engine usually measured at a testbed facility in the workshop of the engine manufacturer. As such, the builder's warranty will usually extend only to the performance of the engine under these conditions, and the builder does not guarantee that the performance of the engine when installed in the vessel will be the same as its performance during shop trials. This gives rise to a potential exposure for the buyer/owner, which will be required to guarantee the speed and fuel consumption of the vessel under actual voyage conditions. Finally, for both practical and technical reasons, the level of boil-off will not usually be measured until after the vessel has been delivered and commenced trading.

The condition of the vessel on delivery: The physical condition of the vessel, and in particular its readiness on delivery to commence employment immediately, are obviously matters of the greatest concern for any buyer. However, given the technical sophistication of a modern LNG carrier, it is in practice almost impossible for any shipbuilder to ensure that the vessel complies entirely at delivery with the requirements of a lengthy and complex technical specification.

In practice, therefore, a line will need to be drawn between defects and shortcomings which are critical to the vessel's operating capabilities (and which must usually be rectified before delivery), and those which can be resolved by the shipbuilder after delivery has taken place.

Post-delivery rights: Following delivery, the builder will generally exclude liability for any defects or deficiencies in the vessel, other than the obligation to repair defects arising under the builder's warranty. This is perhaps the area of greatest divergence between LNG shipbuilding contracts and those in use in other shipping sectors. In most tanker and dry cargo shipbuilding contracts, the builder will usually offer a single warranty encompassing any defects to the vessel arising within 12 months of delivery. In most LNG shipbuilding contracts, the degree of technical innovation in recent years – coupled with the significant increase in the production capacity of the leading LNG yards, and certain well-publicised problems – has led buyers to seek extensions to the builder's warranty obligations to match more closely the obligations that the owner will be asked to assume under a time charter. The form of the extensions varies considerably, but will often include both a longer primary warranty period and further extensions covering specific systems within the vessel, including the cargo containment and coatings systems.

2.3 Financing the acquisition of LNG tankers

(a) Sources of funds

The sources of finance available for a shipowner will obviously depend on its market strength, but will generally include debt finance, lease finance and capital markets. Both debt and lease finance involve the repayment of the sums advanced by instalments (repayment or rental instalments), but there is a significant difference in the security enjoyed by the financial institution, in that in the case of a lease, the lessor will be the registered owner of the ship, whereas in the case of a loan, the bank will be secured (among other things) by a mortgage over the ship.

(b) Credit perspective

There are essentially three credit perspectives for the ship financier to take into account when financing the acquisition of an LNG carrier newbuild:

- asset finance, where the financier looks purely to the vessel itself and, most significantly, its value;
- corporate finance, where the financier looks at the strength of the borrower and takes a view as to whether to finance the purchase of the vessel based primarily (but not exclusively) on that strength; and
- project finance, where the financier looks for a guaranteed income stream from the use of the vessel over a period of years in the context of a particular project or to fulfil a particular contract.

Although some LNG carriers have been financed directly on the back of a corporate balance sheet, the majority have been financed through a combination of asset and project-based finance. The limited spot market for LNG tankers, and the lack of historic data for the revenues obtainable from this, mean that financiers will look most closely at the income stream that the vessel is anticipated to generate under the long-term project charter. The financier will want to ensure that these contracts are concluded with a creditworthy charterer and may require the project sponsors to provide further credit support if the financier has concerns as to the financial viability of the charterer. In addition, a financier will be keen that the term of the vessel's employment (and consequential income stream) correspond to or be longer than the term of the finance facility.

Despite the fact that the LNG vessel is often referred to as the 'pipeline' for an LNG project, it is uncommon for the vessel to be financed as part of the project, and there will thus be a different set of banks acting as vessel lenders from those financing the project itself. In such circumstances, the relative positions of the two sets of banks will need to be regulated by means of a 'coordination agreement', which will deal with default and other situations in order to ensure the continuous availability of the vessel to the project. Such agreements can become quite complex, covering the ability of the project lenders to cure defaults by the project company to ensure that the vessel remains available under the charter, the vessel lenders to cure defaults by the owner to ensure that hire remains payable to reduce the debt, and the project company or lenders, in certain circumstances, to cure defaults by the owner.

(c) *Principal risks*

In addition to the creditworthiness of the charterer, two further principal risks are likely to be of concern to the financier of an LNG carrier: the pre-delivery risk and the political and country risks.

Where pre-delivery financing is required, the lender will be concerned to ensure that its exposure to any overruns in the costs of building the vessel is limited. The lender will also be concerned to ensure that the delivery arrangements under the shipbuilding contract match those under the time charter, and in particular that the vessel is delivered within the window permitted under the charter. The financiers will be concerned to limit the risks associated with the vessel's intended trade, and in particular the proposed loading and discharge terminals. These risks can include the selection of an unusual flag, or simply concerns as to the vessel's ability to load and discharge freely or a change in law affecting either the project company or the charter or the payment of hire.

2.4 Insurance

The owner is responsible for insuring its vessel. The two principal forms of insurance are hull and machinery (H&M) cover and protection and indemnity (P&I) insurance. H&M policies provide cover for loss or damage which the vessel sustains due to perils of sea, fire, explosion, theft, piracy, collision with or damage to port installations, cargo accidents and natural disasters. The policy can also be extended to provide cover for engine damage and latent defects (but not the cost of rectifying the defect), crew negligence and negligence by repairers or charterers. Under most policies, the vessel's H&M underwriters will provide cover for three-quarters of any damage caused to another vessel as a result of a collision (the other one-quarter of any liability is covered by the vessel's P&I cover). The policy will also cover the owner's liability to contribute to general average. The H&M policy will not normally give cover for loss or damage arising out of the risks of war or terrorism, and for this reason it is necessary for the owner to obtain separate cover for these risks from war risks underwriters.

P&I insurance provides the owner with indemnities for third-party liabilities such as cargo damage, damage to fixed and floating objects (including terminals), pollution, wreck removal and crew claims. Where a vessel is required to trade to a designated war risk area, the charterer will usually be required to pay for the cost of specific war risk insurance. In addition, LNG carriers trading to the United States are required to have a certificate of financial responsibility (effectively additional pollution coverage) issued by the P&I clubs, and those trading to Japan are required to take out social responsibility insurance against loss and damage caused by the vessel.

The charterer will also generally take out charterer's liability cover, which will provide cover in respect of the liabilities that it assumes as charterer of a vessel. This will include:

- loss of or damage to the vessel by reason of the vessel being ordered to an unsafe port or due to some characteristic of the cargo;
- liabilities that the charterer may have for death of or injury to the crew; and
- third-party liabilities such as liability for pollution caused by the vessel and wreck removal.

This cover is available either through P&I clubs or from specialist underwriters in the market.

3. Charters and contracts of affreightment

Two types of basic charter contract are in use in the shipping industry: the time charter and the voyage charter. Under a time charter, the vessel is hired for a specific period of time (eg, 12 months), with the charterer paying a daily rate of hire for the vessel. The shipowner provides the vessel with the crew, stores and provisions, ready to load cargo and proceed on a voyage. The charterer pays for the expenditure directly resulting from compliance with its instructions, such as bunkers (fuel costs and a significant item), port charges and the costs of loading and discharging the cargo.

Under a voyage charter, the vessel is hired for a specific voyage or series of voyages. The owner pays all of the operating costs of the ship, including bunkers, canal and port charges, pilotage, towage and ship's agency. In return, the charterer typically pays freight either by way of a lump-sum amount or by reference to the quantity of cargo actually shipped. The contract of affreightment constitutes a variation to the voyage charter under which the shipowner agrees to transport a quantity of cargo in shipments of size and timing nominated by the charterer. The shipowner is free (subject to compliance with any requirements of the loading and discharge ports) to nominate the vessel it will provide to transport the cargo, and will usually be responsible for voyage costs (usually on voyage charter terms), and will again be paid freight calculated typically by reference to the quantity of cargo shipped.

3.1 Time charters

Although the voyage charter is in many respects a more flexible format for short-term trading arrangements, the overwhelming majority of LNG shipping contracts are made on time charter terms. There is also in ShellLNGtime 1, a standard form of charter contract developed for use in the LNG industry.

3.2 Delivery

The owner is obliged to deliver the vessel at the agreed delivery port, and in the agreed condition within an agreed delivery window. The charter will usually include specific provisions setting out the required condition of the vessel at delivery, but it should as a minimum be seaworthy, with a full complement of crew and with all necessary trading certificates on board to perform the required charter service. In addition, the vessel will usually be required to have a valid Ship Inspection Report Programme inspection and safety management systems, and to be vetted and approved by the proposed ports of loading and discharge. The owner will usually be required to pay liquidated damages if the vessel is not delivered within the required window, and the charterer will usually be entitled to cancel the charter if delivery is delayed beyond the agreed cancelling date.

3.3 Charter period and charter hire

The owner will let the vessel to the charterer for a fixed period, though sometimes with options to extend. The charterer is obliged to pay hire, usually calculated at a daily rate, from delivery of the vessel throughout the period of the charter, provided that the vessel is available for service. The charterer is obliged to pay hire irrespective of whether the charterer can use the vessel for the purposes originally envisaged and irrespective of the occurrence of unexpected and unforeseen events. The charter will rarely allow the charterer to terminate early for convenience, though some flexibility can be obtained through the judicious use of option periods.

3.4 Trading limits

The charterer will usually be permitted to direct the vessel to any part of the world, subject to trading limits imposed by the vessel's H&M insurers. However, only limited liquefaction and regasification facilities are available worldwide, and the most important constraint on the trading range of an LNG tanker is the extent to which it is physically compatible with the available loading and discharge terminals and otherwise acceptable to the authorities at those terminals. In this regard, the charter will usually contain a list of 'primary' terminals with which the owner will warrant the vessel is compatible and to which the vessel can trade. Where the vessel has not visited the terminal either at all or in recent years, this can be surprisingly difficult to establish, making terminal compliance one of the more complicated areas in charter negotiations.

There are certain additional constraints – principally areas that are unsafe by reason either of war, piracy or terrorist activity, or of ice. It is also usual in most time charters for the charterer either to warrant that the ports and berths to which the vessel is ordered are safe or at least to agree to exercise due diligence to ensure that any port or berth which the vessel is ordered is safe.

Finally, the increasing use of offshore receiving terminals, and the potential for the development of LNG FPSOs as offshore export terminals, give rise to one further practical and contractual constraint. Although ship-to-ship transfer, typically for lightering purposes, is frequently employed in the oil industry, the ship-to-ship transfer of LNG is still relatively unusual. The International Chamber of Shipping Oil Companies International Marine Forum has produced a Ship-to-Ship Transfer Guide applicable to the LNG industry, but many charters still permit such transfers to be made only in emergencies or where the interests of the safety of the vessel, crew or environment so require. This constraint does not prevent the use of offshore terminals where discharge can be conducted using the vessel's normal systems, but may restrict the type of ship that can be employed for such purposes and affect the overall cost of transportation.

3.5 Operation and maintenance

The standards to which the vessel must be maintained are one of the most critical issues for any charterer under a long-term time charter. In the LNG industry, the lack of a readily available pool of vessels, the potentially hazardous nature of the cargo, the scheduling requirements imposed on the charterer and the fluctuations in the price of the LNG in delivery markets all emphasise the importance to the charterer of obtaining safe and reliable transportation. The balance to be struck here is one of the most crucial elements of a long-term LNG time charter.

Under the ShellLNGTime form, the owner warrants the condition of the vessel both at delivery and throughout the charter period. It is not easy, however, to reconcile this with the later obligation imposed on the owner to exercise due diligence to maintain the condition of the vessel, and most long term LNG charters adopt a compromise approach. In addition the vessel will be off-hire whenever it fails inspection by a port authority or terminal or by a major charterer and this failure prevents the normal commercial operations of the vessel.

3.6 Breakdown and off-hire

If the vessel is unable to perform the required service, whether by reason of mechanical breakdown, deficiency of crew or otherwise, the charterer will usually be entitled to put the vessel off-hire – that is, the charterer is not obliged to pay hire during any periods when the vessel is unable to perform the required service. The owner will also usually be required to pay any fuel, cooldown, boil-off and port charges incurred by the vessel during a period of off-hire. However, it does not follow, simply because the vessel is off-hire, that the owner is in breach of the terms of the charter. The owner does not warrant the continuous availability of the vessel and the charterer's remedy where the vessel is unavailable for service will usually be only that hire is not payable for this period.

The calculation of the period of off-hire can also give rise to difficulties. Generally, LNG charters provide that the vessel will be off-hire only to the extent that the vessel is unable to perform the service then required and only for the 'time lost' as a consequence. If, for example, the cargo discharge systems are not functioning correctly, the vessel will remain on-hire until it is ordered to discharge, as it is only at that point that it will be unable to perform the service then required. In addition, the vessel will be entitled to credit for service performed and distance made good during any period of off-hire. For reasons considered in more detail below, many LNG charterers allow significant margins in the loading and discharging schedules, and the vessel will often remain able to meet its intended target even after a minor breakdown, thereby limiting the extent of any off-hire period.

3.7 Performance of the vessel

The owner will normally guarantee certain aspects of the performance of the vessel, including its speed, fuel consumption and the rate at which LNG in the cargo tanks 'boils off' (ie, vaporises).[13] For the charterer, however, this is only a limited remedy. If the vessel is unable to make the guaranteed speed, the charterer will be entitled to deduct the value of the time lost[14] measured at the daily rate of hire. Importantly, the owner does not warrant that the vessel will arrive at the next loading or discharge port in accordance with its schedule.

Two performance-related issues typically give rise to difficulties in LNG charters. The first is the period over which performance is to be measured. It is unusual in oil charters for performance on a long-term basis to be measured over any period less than one year, unless the charterer is willing to pay a bonus to the owner in circumstances where the owner exceeds the warranted performance. Most LNG tankers still trade on a 'milk run' between a single loading and discharge port and, in these circumstances, the charterer will gain little advantage from any improvement in the ship's performance, but may be obliged to pay demurrage where the ship fails to meet its loading and discharge windows. As a consequence, the charterer will often seek to measure performance over a single voyage,[15] but will be reluctant to offer a bonus where the vessel exceeds the performance guarantees.

The second difficulty lies in determining those periods, if any, that should be excluded from the overall calculation. Generally speaking, performance[16] will be measured on each voyage from pilot station to pilot station. The calculation should plainly exclude any periods where the vessel is off-hire, but most LNG charters also exclude periods during which the ship cannot ordinarily be expected to meet its performance warranties. These exclusions typically include periods of adverse weather, poor visibility, congestion and time spent saving life.

3.8 Long-term charters

Although most long-term charters contain the essential features described above, they give rise to different considerations.

---

13 A modern LNG tanker typically 'loses' around 0.1% of cargo each day when laden.

14 That is, the difference between the time that would have been needed to complete the voyage if the guaranteed speed had been achieved and the time taken at the actual speed.

15 Or even, in some charters where passage through the Suez Canal is required, a voyage 'segment'.

16 Boil-off where the relevant measurements are those taken at the opening and closing of the custody transfer measurement is the obvious exception to this.

(a) ***Construction phase***

Most long-term charters are for newbuilding vessels to be constructed at an agreed shipyard. The charter will therefore usually address the rights of the charterer during the construction phase, including the right for the charterer to send its own technical team to the shipyard to supervise the construction of the vessel, to modify the specification of the vessel, to be present during sea and gas trials and to provide a technical acceptance of the vessel prior to delivery. In addition, the delivery mechanism described above will usually be modified to reflect the permissible delays in construction that will extend the delivery date under the shipbuilding contract, and sometimes the place of delivery. There is here an area of considerable difficulty, where the objectives of the owner, its lender and the charterer will diverge. Both the owner and its lenders will wish to deliver the vessel at the builder's shipyard immediately on and 'back to back' with delivery under the shipbuilding contract. From the charterer's perspective, however, it would like to take delivery of the vessel on arrival at the first load port and ideally in a cooled down and ready to load condition.

(b) ***Project* force majeure**

Time charters do not generally contain 'conventional' *force majeure* clauses. They contain an exceptions clause that will usually give the owner some protection against loss and damage arising from collisions, errors in navigation, accidents at sea and certain aspects of vessel breakdown. The clause may also contain some exceptions that apply to the charterer as well, but these will usually be limited in scope.

In particular, the events that constitute *force majeure* under a gas sale and purchase agreement will not excuse the charterer from performance under a time charter, and in particular will not excuse the charterer's obligation to pay hire, which arise irrespective of its ability to use the vessel. In recent years this principle has been modified to some extent in long-term LNG charters, with the owners agreeing a right for the charterer to terminate the charter early and for convenience in the event of a prolonged *force majeure* at the project production facilities, or sometimes at the intended receiving facilities. These modifications remain limited in scope, however, and are always a topic of concern for the owner's lenders.

(c) ***Default and remedies***

Most standard form time charters do not contain a specific clause identifying the various events that will constitute default by either party, preferring instead to treat specific piecemeal events as defaults, and often limiting these to failure by the charterer to pay hire. However, a long-term charter is too complex an undertaking for this approach to be retained, particularly as the respective strengths and capabilities of the parties will change over the charter period. The LNG charter will therefore usually contain an extensive series of additional defaults, including at least insolvency, material breach and failure by the owner to perform certain key obligations under the charter, including obligations in relation to the ownership, classification, flagging and insurance of the vessel. In addition – and controversially – many long-term LNG charters seek to measure the owner's performance, giving the charterer the right to terminate if the number of days of off-hire exceeds an agreed target.

From the charterer's perspective, the right to terminate is at best an unsatisfactory remedy. The task of obtaining a replacement vessel of an equivalent size, age and specification operated by a satisfactory owner will rarely be straightforward, and the charterer's claim in damages against the defaulting owner will not necessarily cover the additional cost of such a replacement vessel.

For this reason, the charter will usually contain a right on default by the owner for the charterer to convert the time charter into a bareboat charter for the balance of the charter period. Under a bareboat charter, the charterer is responsible for the operation and maintenance of the vessel, thereby providing it with the means (usually by appointing an independent ship manager) to improve the reliability and performance of the vessel. Although the charterer can secure a greater degree of practical control over the vessel by this means, it will assume responsibility for payment of the operating costs and may assume a greater degree of liability for any loss or damage caused as a consequence of the operation of the vessel.

(d) Security and financing issues

In any long-term charter, the security provided by the charterer is critical to the owner's ability to obtain financing for the construction of the vessel. This is one of the most complex areas of negotiation in any long-term time charter. The owner will often be a single vessel-owning company, albeit a subsidiary of a significant ship-owning group. The charterer will usually be either a subsidiary of an integrated gas company, national utility or (where the LNG is sold ex-ship) one of the project companies for the LNG project.

The evaluation of the creditworthiness of the charterer's guarantor is, from the owner's perspective, perhaps the most critical single issue in the charter. Where the LNG is being sold ex-ship, the sponsors of the LNG project have tended in recent years to offer one of the project companies as the charterer, but have often declined to provide any form of guarantee themselves for the obligations of the charterer. In these circumstances, although other forms of access to the project income may be available, the owner will frequently have to evaluate for itself the likely long-term success of the LNG project in order to decide whether to enter into the charter contract.

4. Sale and purchase agreement

Generally, LNG traded under short-term contracts will be bought and sold under a master sale and purchase agreement, with a confirmation notice (or equivalent) setting out the terms of the individual trade. The terms of such master sales agreement are dealt with elsewhere in this book, but below are outlined some of the issues that arise in tying the shipping contracts into these sale arrangements.

4.1 Loading and delivery

Where the sale is effected on a delivered ex-ship basis, the seller will be responsible for delivering the LNG to the buyer within the narrow window provided in the confirmation notice. If the vessel does not arrive within this window, the buyer will usually be entitled to refuse delivery of the cargo and the seller will be obliged to compensate the buyer for at least the direct losses it has incurred by reason of the seller's failure. Similarly, the buyer under a free on board sale will be required to provide its vessel in a ready to load condition within the specified delivery window and, if it is unable to do so, to pay the seller for the LNG it agreed to purchase.

The timing and condition of the vessel on arrival at the loading or discharge port are therefore often of critical importance. However, as described above, most time charters do not oblige the vessel, other than on first loading, to arrive at the load port within a specified window. By contrast, voyage charters will generally contain a provision giving an expected arrival date at the load port, and in such case the owner undertakes to proceed to the load port with convenient dispatch and to give regular notices of the vessel's expected readiness for delivery. The effect of these provisions is that the vessel must commence the voyage to the load port in sufficient time so it is reasonably certain that, proceeding normally, it will arrive by the cancelling date.

It is not sufficient for the vessel simply to arrive; it must also be ready to load. Following arrival, the master will be required to tender a notice of readiness, the document that triggers not only delivery under the charter, but also commencement of the time allowed to the charterer for loading the cargo. While a number of requirements must be met before a valid notice of readiness can be tendered, the most important of these is that the vessel be in a physical and legal state of readiness to load. In particular, its tanks must be cooled down to the relevant temperature to permit loading. If the vessel is not ready and the cancellation date passes, the charterer has the option to cancel the charter.

4.2 Laytime and demurrage

The sales agreement will generally permit the terminal an agreed period of 'allowed laytime' during which to load or discharge the cargo, and require the party responsible for the terminal to pay demurrage (a form of liquidated damages) to the party responsible for the shipping where this period is exceeded other than for reasons attributable to the vessel. The charterer will remain obliged to pay hire under a time charter during any period of delay in loading or discharging, and such delay will usually result in the payment of demurrage or an adjustment to freight under a voyage charter. The payment of demurrage under the sales agreement is therefore intended to compensate the charterer for these costs.

However, some LNG sales agreements also require the party responsible for the shipping to compensate the terminal for any delays in loading or discharging beyond the allowed laytime where these are caused by the vessel. Although the owner under a time charter will usually warrant the ability of the vessel to load or discharge a cargo of LNG within an agreed period, and agree to compensate the charterer for any additional time required, these warranties are not usually back to back with the sales agreement and can give rise to additional costs and expense for the charterer.

4.3 Terminal arrangements

Under most charter contracts, the charterer is obliged to make arrangements at both the loading and discharge ports to secure berth space for the vessel and to arrange tugs and pilots (though often on behalf of the owner). Under the sales agreement, the party responsible for providing terminal facilities will supply this berth space, though the right of the vessel to load or discharge will depend on timely arrival within its allotted window. The party supplying the shipping will generally be responsible for ensuring that the vessel is compatible with the terminal specifications as advised to it, and for ensuring that the vessels signs and complies with the terminal conditions of use.

This latter requirement has given rise to some difficulties in recent years.[17] The reason for this lies in the pooling arrangements of the P&I clubs that insure LNG carriers against loss and damage caused to third parties, including loading and discharging terminals. Each of the 13 P&I clubs that form the International Group of P&I Clubs pool risks in excess of $6 million under a common sharing arrangement. To ensure that only risks underwritten on a common basis form part of these pooling arrangements, certain risks – for example, nuclear or war risks – are excluded altogether, while others can be covered only to the extent that they fall within established guidelines. The guidance used in recent years permits the clubs to underwrite contractual indemnities, provided that the vessel owner is in part at fault in the loss, or loss is allocated on 'knock for knock' terms and either the national or international limitations on liability remain in force or the conditions of use contain a separate but acceptable limit to the vessel's liability. There has been a trend in recent years for LNG terminals to seek to impose onerous liabilities on the shipowner (including unlimited liability and/or no fault liability for damage to the terminal and other vessels), and that can lead to the P&I clubs refusing cover or imposing high additional premiums.

The terms of the conditions of use and the extent to which these and any indemnities contained in the sale agreement are compatible with the charter and the owner's P&I cover are therefore critical issues for any charterer.

4.4 Consequential loss

As set out above, a buyer or seller that is unable to provide a vessel at the loading or discharge port within the required window will often be exposed to significant losses under the sales agreement. However, these losses will rarely be recoverable by the charterer from the shipowner, even if the owner is in breach of the terms of the time charter. The owner will often seek to exclude such claims expressly, but even where the charter does not prohibit recovery of such 'consequential' loss, such claims are unlikely to be regarded as foreseeable and to succeed. It is no doubt for this reason that any breakdown of the LNG carrier will usually be treated as *force majeure* under the sales agreement.

4.5 Loss and damage to cargo

Although the charter will set out the terms on which the shipowner makes the vessel

---

17 These risks were explained in more detail by Nigel Carden of Thomas Miller at a talk given in Houston in 2006.

available to the charterer, the contract between the owner of the cargo and the shipper will often be set out not in the charter, but in the bill of lading. Although in the LNG industry the charterer of the vessel and the owner of the cargo will often be part of the same group, they will frequently be different companies.

The charterer will generally be responsible for preparing and issuing the bill of lading on behalf of the master and owner. The bill of lading is the principal document of title to the LNG being transported and, unless the charterer is also the cargo owner, forms a direct contract of carriage between the cargo owner and the vessel owner.

Historically, bills of lading issued in the LNG industry have been non-negotiable – that is, the consignee and destination are named in the bill. The master will generally be obliged to deliver the cargo to the person and to the destination named in the bill of lading, and such bills therefore act as a constraint on the trading of LNG. In recent years there has been an increasing trend towards the use of negotiable bills in which the consignee is left blank or marked as 'to bearer' or 'to order'. Such bills permit a further sale of the cargo after loading, thereby increasing trading flexibility.

Where the charterer instructs the master to discharge the cargo at a destination other than that named on the bill of lading or to discharge the cargo other than to the holder of the bill of lading, it will generally be required to provide an indemnity to the owner against any claims that may thereby result.

As yet, there have been few cargo claims involving LNG voyages, but it is likely that as short-term trading expands, the number of claims may increase. These claims are likely to fall into three main categories:

- claims for short delivery caused either by inaccuracies in the bill of lading or by losses arising during the voyage due to leakage or excessive boil-off;
- claims that the LNG is off spec – that is, does not meet the contractual specification; and
- claims for late delivery.

Where there is a loss of or damage to the cargo during the voyage, then in most circumstances the owner will face a cargo claim under the bill of lading which may or may not incorporate the terms of the underlying voyage charter. The claim will arise under the voyage charter directly only where either the charterer owns the cargo or the charterer deals with the cargo claim under the sales contract and then needs to seek an indemnity for this claim from the owner. For this reason, it is important to ensure that the bill of lading, the voyage charter and the relevant sales contract have the same or similar schemes for dealing with cargo claims and that, as far as possible, all of these contracts have the same governing law and jurisdiction clauses. This is best and most commonly achieved in the context of a charter by ensuring that either the Hague or the Hague-Visby Rules are validly incorporated into both the voyage charter and the bill of lading.

The owner of the cargo will also usually take out insurance against fortuitous loss of or damage to the cargo caused during the voyage. This cover will provide an indemnity where cargo is lost or damaged due to all types of risks. However, it will not provide cover against losses attributable to the inherent nature of the cargo or the voyage. Thus, boil-off losses occurring during the course of the voyage will not be recoverable unless it can be shown that the loss was due to fortuitous operation of a peril insured under the terms of the policy.

5. Conclusion

The terms on which LNG ships are purchased, chartered and financed have become increasingly competitive in recent years. The leading producers and buyers have led the drive to reduce the cost of LNG transportation through innovations in the design of LNG tankers, often negotiating these changes directly with the shipyards. The emphasis on performance has also been felt in the terms of the long-term contracts of employment where the entry into the market of new shipowners has led to increasing competitiveness, with shipowners willing to adopt a greater measure of the operating risk.

In addition, although many aspects of the contracting arrangements remain specific to the LNG industry, shipping contracts are now increasingly flexible, with short-term time charters, negotiable bills of lading and time charter trips (ie, the hire of a vessel on a time charter basis to perform a specific voyage or series of voyages) increasingly common. This process is likely to continue and indeed accelerate in the coming years.

Coal bed methane for LNG

Daniel Gosewisch
Queensland Treasury Corporation
Toby Hewitt
Dart Energy Limited

1. Introduction

This chapter discusses the use of coal bed methane (CBM) as feedstock for LNG projects.

We examine the role of both LNG and CBM in the global energy mix and look at the status of various CBM to LNG projects sited in Queensland, Australia. The technical and commercial dynamics of CBM extraction differ materially from their equivalent in conventional natural gas exploration and production. We consider how this impacts on the structuring of CBM to LNG projects, how the main project risks may be allocated between the project and how these challenges are being or might be addressed. Finally, we consider whether the progress made in Queensland is repeatable in other geographies.

2. Commercial background

2.1 The role of LNG in the global energy mix

Much has already been written on the remarkable growth of the LNG trade in the past 30 years or so. Natural gas now represents about 21% of global energy consumption and LNG represents about 7.5% of that. It is widely accepted that natural gas has a key role to play in meeting burgeoning long-term global energy demand and as a relatively environmentally friendly 'bridging fuel' for the transition between coal-fired power generation and renewables-based power generation. On average, gas-fired power produces 60% less carbon dioxide emissions than black coal-fired power. LNG enables the connection of major natural gas producers with major and often far-flung markets. However, to support a commercially viable LNG project and bring economies of scale, very large natural gas resources must be available in the country of origin. For consuming nations, LNG brings enhanced security of supply by virtue of not having to rely on pipeline gas sourced from neighbouring countries with which good neighbourly relations are not always enjoyed.

2.2 The role of CBM in the global energy mix

Notwithstanding that the first commercial CBM project began in 1977 in the San Juan Basin, western United States, the CBM industry is seen as being relatively young. First commercial production was in 1999 in Australia and 2002 in Canada. In 2000, CBM contributed about 7.5% to US natural gas consumption, although it does not yet contribute significantly to global consumption. In 2010 CBM accounted for more than 70% of domestic gas consumption in Queensland, Australia. However, it is the potential for CBM's future contribution to the global energy mix which excites. Global LNG demand is expected to increase from approximately 180 million tonnes per annum in 2009 to 435 million tonnes per annum in 2030. Of this total figure, up to 10% could come from the CBM to LNG projects planned in Queensland.

The world still has massive coal resources, and therefore by extension massive amounts of natural gas co-located in the coal. This gas has hitherto been 'locked up' in the coal, but advances in extraction technology – particularly in drilling and completion technology – have enabled the extraction of large quantities of CBM. For CBM developers, the relative proximity of large CBM accumulations to major population and industrial centres, together with existing gas transportation and processing infrastructure, makes for better profit margins. In places where this combination of factors is present, use of domestic CBM resources to satisfy domestic natural gas demand becomes commercially compelling and, in some cases, mandated by national legislation. In such cases CBM and other sources of 'unconventional gas' (eg, shale gas) may actually become a competitor to LNG as a supply source. In this regard, witness the contribution that CBM, shale gas and 'tight gas' make to US natural gas consumption and the accounts of spare capacity in US LNG receiving terminals. The same potential exists in the Upper Carboniferous coal basin of Northern Europe, where gas-consuming countries worry about dwindling domestic supplies of conventional natural gas and increasing reliance on Russian pipeline imports.

The eastern states of Australia are blessed with a combination of factors which make satisfaction of domestic gas demand plus sizeable LNG exports viable. The Commonwealth Scientific and Industrial Research Organisation estimates that CBM resources in Queensland and New South Wales are in excess of 250 trillion cubic feet, enough energy to power both states for 400 years at current demand. The coal deposits of the Bowen and Surat basins in Queensland are approximately 450 kilometres from the coast and pipelines connecting these deposits to LNG plants at the Port of Gladstone are in development. This is coupled with healthy LNG demand from north Asian buyers, which see Australia as a reliable supplier country.

2.3 Queensland CBM to LNG projects

The world's first major CBM to LNG projects are set to become a reality in Queensland. There are four major projects, all with liquefaction facilities to be sited on the coast at Gladstone, South East Queensland:

- GLNG – this is operated by Santos and involves joint venture partners Santos (30%), Petronas (27.5%), Total (27.5%) and KOGAS (15%). A US$16 billion, 7.8 million tonnes per annum (mtpa) liquefaction facility is planned. The project has signed a 3.5 mtpa LNG sale and purchase agreement with KOGAS and the final investment decision was announced by the partners in January 2011.
- Queensland Curtis LNG (QCLNG) – through its acquisition of Queensland Gas Company in early 2009 and its later acquisition of Pure Energy, BG Group's CBM reserves and resources in the Surat and Bowen Basins total about 17 trillion cubic feet. BG announced the final investment decision for

the first phase of the QCLNG project in October 2010. The first phase will consist of two LNG trains with a combined capacity of 8.5 mtpa. BG has entered into a number of agreements with the China National Offshore Oil Corporation (CNOOC) under which:

- CNOOC will purchase 3.6 mtpa of LNG over a 20-year period;
- CNOOC will purchase 5% of BG's interests in certain tenements in the Surat Basin;
- CNOOC and BG will jointly participate in a consortium to construct two LNG ships in China; and
- CNOOC becomes a 10% equity investor in the first LNG train.

Separately, BG has signed a 1.2 mtpa LNG sale and purchase agreement with Tokyo Gas Co, Ltd. Tokyo Gas will also take small equity stakes in certain of BG's Surat Basin tenements and in the second LNG liquefaction train.

- Australia Pacific LNG (APLNG) – Origin Energy Limited (Origin) and Conoco Phillips are the 50/50 joint owners of the LNG liquefaction plant to be sited in Gladstone. The two train LNG plant is planned to have a processing capacity of up to 18 mtpa. Origin is currently the largest producer of CBM in Australia (this is used for domestic consumption) and claims, by some margin, the largest CBM reserves base in Australia. In April 2011 APLNG and China Petroleum & Chemical Corporation (Sinopec) signed a 20-year, 4.3 mtpa LNG SPA and an agreement such that Sinopec subscribes to a 15% equity interest in APLNG, thus diluting the respective interests of Origin and Conoco Phillips to 42.5% a piece. Subsequently, APLNG announced the final investment decision for the initial train, 9.0 mtpa liquefaction facility. While APLNG operates the downstream part of the project, Origin remains the operator for the bulk of the upstream acreage and also retains some CBM acreage not part of the APLNG joint venture. APLNG also has joint venture interests in some Queensland Gas Company operated CBM tenements. This has resulted in a measure of cooperation between APLNG and QCLNG, in that BG has entered into agreements with APLNG to source CBM to help manage natural gas feedstock requirements for the start up of QCLNG.
- Arrow LNG (formerly known as Shell Australia LNG) – Arrow LNG is the fourth major CBM to LNG project to be sited in Queensland. Arrow Energy Pty Ltd (Arrow) was an Australian listed company until it was acquired in August 2010 by a privately owned 50/50 incorporated joint venture of Shell and PetroChina. Arrow proposes initially a two-train 8 mtpa LNG liquefaction facility, to be fed from Arrow's CBM reserves in the Surat and Bowen Basins. Arrow currently supplies about 20% of Queensland's gas demand. Shell and Petrochina will each take half of the LNG produced from Arrow's project.

3. CBM extraction versus conventional natural gas extraction

While CBM operations share some of the characteristics of both conventional oil and/or gas (COG) exploitation and coal mining, it is a unique industry sector at the same time. Several countries have applied – with a few adjustments – the regulatory regime for conventional oil and gas operations to the CBM sector. This does make some sense, because it is natural gas rather than coal that is being targeted for exploitation and the gathering, processing, transport and sale of natural gas in connection with CBM operations can be similar to the corresponding activities in a conventional gas development. Moreover, as with conventional gas production, it is drilling that is the primary exploration and production activity in CBM operations. However, to some extent that is where the similarities between CBM and conventional natural gas exploitation activities end. CBM operations differ significantly from conventional gas development in a number of respects.

3.1 Technical comparison

CBM wells tend to be much shallower than conventional gas wells (usually less than 1,000 metres deep) and tend to have a much shorter life. This necessitates the drilling of a commensurately greater number of wells in a more closely spaced pattern. Unlike with conventional gas developments, in which the greatest risk to the project is in the exploration phase (ie, it is usually more likely than not that a conventional gas wildcat well will not result in a commercial natural gas discovery), natural gas is, relatively speaking, omnipresent in coal beds. In consequence, CBM exploration programmes are faster and lower in cost and risk than their conventional counterparts. The difficulty with CBM is not so much finding the CBM resource, but establishing with a reasonable degree of certainty that extracting natural gas from the coal beds in question will be commercially viable. In other words, in a CBM project, the lion's share of technical risk occurs during the appraisal, development and production phases rather than in the exploration phase, and as such, relatively longer and higher cost/risk appraisal programmes are required.

This is because of how CBM is produced from the coal beds. For most coal beds, the quantity of gas held in the coal is primarily a function of coal rank and pressure, bituminous or mid-rank coal generally containing the most natural gas in place. Gas production from coal beds occurs by a three-stage process in which gas first flows from natural fractures or 'cleats' within the coal, then desorbs from the cleat surfaces and finally diffuses through the coal matrix to the cleats. As well as natural gas, coal beds contain large quantities of water. When a well is drilled into a coal bed, the pressure of the coal bed is reduced. With decreasing pressure, water and then gas are produced to the surface. In a coal reservoir, the water must be removed from the cleats before gas can effectively flow to the well. The length of the dewatering process and the magnitude of the producing rates of gas and water are controlled by the physical properties of the coal bed. No two coal beds are the same and so, depending on the factors outlined above, the volume and rate of gas production can fluctuate greatly from coal bed to coal bed.

There is clearly thus a need to understand as much as possible about the characteristics of the relevant coal reservoir before commencing production drilling. A well drilling technique is chosen to suit the particular coal formation. In Australia, the most common is a vertical well down to the target coal reservoir, although the drilling of a series of bores within and parallel to the coal bed is becoming more widespread. This surface to in-seam drilling technique is used extensively in areas of low permeability to allow gas to flow more freely. However, the downside of this methodology is that each in-seam well can target only CBM one coal seam at a time. Where the permeability of the coal is greater, so that CBM flows more easily to the well through the coal, having just one vertical well targeting multiple coal seams at different depths is possible.

The experience of pioneers of CBM exploitation in Australia is that two basic types of drilling completion techniques are employed to increase CBM flow and to improve coal permeability. The first of these is 'hydraulic fracture stimulation' – or 'fracking' – in which specially engineered fluids are pumped at high pressure into a relevant reservoir. Although fracking has been equated with concerns overseas, fracking in Australia is tightly regulated, with bans on the use of certain chemicals. The second is 'open-hole cavity completion' or 'cavitation', where the well is injected with high-pressure gas and water and is then rapidly depressurised. The resulting pressure differential between the wellbore and the surrounding coal causes a sudden CBM expansion, which leads to the coal matrix bursting and sloughing into the wellbore to the surface.

Once CBM is produced to the surface, gas and water are then separated in a separator, the gas is sent for further processing, compression and pipeline transport and the produced water is sent for storage, processing and/or disposal. The predicted large volume of produced water and the variable quality of this water make water management a key issue. The CBM operator must have an understanding of the quantities and quality of produced water, which is often rich in salts, coal fines and other constituents that render it unsuitable for many uses – hence the need for large holding or evaporation ponds. Depending on the circumstances and level of water treatment employed, produced water can also be reinjected into isolated formations, released into streams and used for irrigation. Water of potable quality may have significant value, although obviously, treating produced water to remove dissolved salts and any residual hydrocarbons requires the application of suitable technology and involves additional costs. In addition, treatment of CBM produced water would involve the storage and disposal of large quantities of salt waste.

3.2 Economic comparison

Each CBM well typically experiences a 'ramp-up' period between first production of gas and peak production of gas. During this ramp-up period, gas production increases as water production declines. The variable characteristics of coal bed reservoirs make the number of wells required to achieve a certain level of production more unpredictable, and with it the level of costs. Unlike in conventional gas developments – where capital expenditure on development is incurred upfront immediately prior to a field going into production and where a reasonably steady rate of operating expenditure over the life of the field may be assumed – CBM developments require the drilling and completion of new wells during the field production phase in order to maintain targeted levels of gas production. This can lead to 'spikes' in capital expenditure, which may not necessarily be covered by revenue from gas sales. One of the other consequences of this is that the costs profile of a CBM project is more evenly spread over the life of the projects as opposed to a conventional gas project, where most of the costs are front-loaded. The longer time taken for a CBM well to reach peak production and the generally smaller peak volumes produced mean that the profile of CBM revenues is also flatter and more spread out over the total life of the field.

On the plus side of economic comparison, coal seams tend to extend for long distances with only gradual changes in depth, gas content, thickness and other technical characteristics. Therefore, once a foundation project is established in an area, the suitability of the surrounding acreage for gradual 'step-out' field expansion can be relatively easily determined. The foundation development creates a logistical and infrastructure hub, from which expansion projects may benefit. It is this high level of comfort on the expansion prospects of existing developments which has led to the influx of investment into Queensland CBM tenements from 2007 to the present day. Investors have demonstrated a willingness to make decisions on likely future reserve availability in an area based on information gleaned from neighbouring tenements and generic understandings of the extension of the same coal seams for many kilometres beyond existing reserve boundaries.

CBM has a generally lower content of carbon dioxide and other impurities than conventionally produced natural gas, leading to generally lower field processing costs. A typical CBM raw gas specification might be more than 96% methane, with the remainder being nitrogen, carbon dioxide and small amounts of oxygen. The flipside of the coin is that, being 'drier' gas (in the absence of hydrocarbon liquids such as propane, butane or condensates), CBM has a lower gross heating value than conventionally produced natural gas, and this may cause issues downstream if CBM is commingled with conventional natural gas of different specification in a transmission pipeline. The location of CBM production with respect to gas transmission pipelines and infrastructure and similar infrastructure can also dramatically affect the economics of CBM development.

3.3 A typical upstream CBM project

A simple description of a fairly typical upstream CBM exploration and production project may be instructive. Following grant of the relevant exploration licence and any environmental approvals required at that stage, CBM exploration activities would usually comprise:

- the drilling of widely spaced exploration holes to determine depth to coal and structural complexity;
- the drilling of core holes to establish gas contents, saturations and permeabilities of coal seams;
- the drilling and testing of pilot wells to confirm completion methodology and commercial producibility of seams targeted;
- the certification of proved and probable, and proved, probable and possible reserves;
- initial marketing (perhaps including the development of a memorandum of understanding/gas sales agreement based on the volume of reserves certified); and
- discussion of the production licence/development plan with government.

The pace at which exploration moves generally depends on the attitude of the exploring company. However, this can be quite quick if the company already knows the geology and is aiming at rapid development. The process of exploration and appraisal could take anywhere from two to four years.

In the Australian context, the operating company would then submit its application for a production licence and its development plan, and it is at this stage that issues of overlapping coal tenures and environment, health and safety are addressed. In Australia, this process could take from six months to one year for a moderate development. However, it may take two years or more for major developments requiring a full environmental impact statement and commonwealth government environmental approvals. Once the production licence is issued, the company is free to proceed with development drilling and installation of gathering lines, compressors, evaporation dams and so on. Some activities may require specific environmental approvals – for example, in relation to the construction of a large dam and noise/emissions permits in relation to the operation of compressor stations. Water management often requires additional approvals, especially where the water disposal methodology employed involves beneficial use of the water resource, such as through irrigation or industrial uses. The development drilling timeframe will depend on the depth to coal, type of completion and production methodology, number of wells to be drilled and so on. A 30-well programme was completed on one project in about six months and a 75-well programme was completed in about nine months. This will depend on the number of rigs working, 12 or 24-hour operation, drilling conditions and other factors such as availability of contractors and equipment.

Many CBM developments have occurred initially on quite a small scale, with so-called 'sweet spots' being developed first to satisfy local market demands. At the same time, reservoir appraisal and production testing programmes could be carried out in neighbouring areas with a view to further incremental phased developments. Investors are keen to ensure that any gas that is produced in the initial phase of any CBM development can be commercialised quickly at minimum cost. Experience in Queensland shows that, with the right regulatory and economic conditions, it is possible to achieve first commercial gas sales within approximately two years of the time of grant of the relevant production licence. In Queensland, Arrow has also demonstrated a successful business model in developing its own small-scale power generation capacity using CBM feedstock and, later, vertically integrated its operations by acquiring full ownership of a 450 megawatt (MW) power station connected into the national electricity grid. Origin Energy was already one of the largest integrated energy companies in Australia prior to its CBM developments, but it has also expanded this portfolio to include a 600MW gas-fired power station using CBM as feedstock.

3.4 Tenure issues

From a legal perspective, a fundamental consideration for any prospective CBM to LNG developer is good legal title to the gas, or at least the right for the developer to monetise that gas. The common thread among most countries with significant CBM resources, with the exception of the United States, is that title to the CBM in situ lies with the state. However, different countries have adopted different legal frameworks pursuant to which commercial extraction of CBM may be permitted, and it is crucial for the developer to understand the advantages and drawbacks of the particular legal regime. In some countries, such as the United States, Australia and the United Kingdom, no distinction is made between natural gas situated in coal beds and natural gas situated in other rock formations. The licensing regime for onshore petroleum applies in each case. In other places, such as Indonesia, India and China, a special type of tenure has been created for CBM in the form of CBM production sharing contracts. These are separate and distinct from both coal and COG tenure, albeit that CBM production sharing contracts borrow heavily from the provisions of their COG counterparts. The relative novelty of commercial CBM extraction in certain developing countries has meant that the legal framework pertaining to CBM can be quite immature and sometimes ambiguous.

In Indonesia, India and China, currently the only certain legal means of pursuing commercial CBM extraction is through a CBM production sharing contract, although in Indonesia, the holders of existing COG production sharing contracts and coal mining tenements are each given priority rights in relation to the making of CBM production sharing contract applications in relation to prospective CBM areas covered by the existing tenement. In practical terms, however, the logical means of incentivising the holders of other types of tenure (whether coal or COG) to cooperate with a CBM developer is to include these stakeholders as equity participants in the CBM development consortium. Where, for whatever reason, a CBM production sharing contract is not available in relation to a target area, CBM developers will need to consider concluding private contractual arrangements with the holder(s) of the relevant coal mining tenement(s) providing for, among other matters, rights to a share of the revenues from coal mine methane sales. A factor to bear in mind in such cases is that the CBM developer's right to CBM sales revenue is only as good as the coal miner's underlying right to produce and commercialise that gas itself. Unfortunately, in India for example, the coal miner's legal title to the CBM contained in its coal lease has been questioned; although anecdotally, the Indian government is now moving to clarify the position.

From a practical perspective, where CBM operations are planned to co-exist with COG operations and/or coal mining operations, there will be a need for the various stakeholders to coordinate their activities with a view to avoiding undue interference with each other's activities and to optimise the exploitation of each resource. For example, CBM operators would certainly benefit from the use of any existing mine and/or COG infrastructure, and mine operators could benefit from the commercial exploitation of methane, which would otherwise have been vented or flared during coal extraction. In some cases extraction of CBM prior to coal mining may in fact be a safety-related prerequisite to the performance of underground coal mining operations. However, it is not realistic to expect that the natural commercial interests of the various players will be aligned in all cases, especially if either the coal mining or COG player does not have a participating interest in CBM extraction. At present, particularly in developing countries, it appears that there is no formal mechanism by which the various interest groups are forced to cooperate or coordinate with each other.

There is a statutory regime in Queensland which deals with these issues, in the form of the Petroleum and Gas (Production and Safety) Act 2004 and the Mineral Resources Act 1989. Here, a system of coordination has been enshrined in legislation. In practice, this leads to a negotiation process between CBM developers and coal miners, covering several elements. First, in what is generally referred to as a 'co-development agreement', commercial matters are dealt with, such as the priority of production in particular areas, allocation of operational and capital costs associated with degassing/extraction of CBM, the allocation of costs for rehabilitation and access to the relevant land and compensation for either the loss of access to a resource, delays in obtaining access to a resource or accelerated access to a resource. It also deals with the sharing of data and the interplay of the relevant production activities, such as the location of infrastructure, to enable day-to-day activities to occur. A further regulatory element deals with the relevant consents to enable various activities to occur and the process to be followed to enable the amendment of the coordination agreement. Other elements include provisions in relation to safety, disposal of coal formation water and dispute resolution. Given the high level of personal responsibilities placed on both coal site safety executives and petroleum site safety managers, extensive discussions take place to ensure that safety systems and processes are consistent and integrated across both operations. These agreements are often negotiated well before production takes place and before either a mining lease or a petroleum lease is applied for.

In Queensland, in circumstances where there is an existing overlapping petroleum lease, unless there is an approved coordination arrangement, an application for the coal mining lease will not progress and may ultimately be rejected. Similarly, where there is an existing coal mining lease which overlaps with a petroleum lease application, the petroleum lease application cannot be granted until there is an approved coordination arrangement. A coordination agreement has no effect unless it is approved by the relevant minister. In considering whether to grant such consent, it appears in summary that the minister will have regard to safety aspects, optimisation of resource exploitation and ecologically sustainable development. Given that the coal party and petroleum party must agree to the coordination arrangement between themselves before it is submitted to the minister, the commercial negotiations between those parties during the exploration phase will normally include processes for agreeing the coordination arrangement once production commences. (These agreements are not so much as an agreement to agree as an agreement to a particular future base-case plan, with a process for makes changes as the plan evolves.) Those negotiations will also cover the triggers for granting mutual consents to each other's production tenure, as a formal consent from the overlapping tenure holder is the preferred method of obtaining ministerial approval to the granting of either a petroleum production lease or coal mining lease.

Stratification agreements may also be employed in areas where separate ownership interests in CBM resources on the one hand, and COG on the other, co-exist. Fundamental to such agreements will be to define as precisely as possible the subsurface target zones for respectively CBM and COG, and to separate rights of exploitation in relation to each zone, thereby effecting the stratigraphical division of the tenement. The parties will also be obliged to minimise interference with each other's surface activities and set an order of priority in carrying out each work programme. There will also usually be a system of reciprocal indemnities under which the owner of CBM rights will indemnify the owner of COG rights in respect of losses arising out of CBM activities and visa versa. In cases where the two types of interest co-exist pursuant to the same tenement, there may be provisions covering the appointment of a tenement manager and obligations on each of the parties to do what is necessary to maintain the tenement in good standing. In Queensland at least, there is no regulatory methodology for such vertical subdivision of tenements, with the result that whichever party is listed on the government register as the tenement holder will remain liable for compliance with all legal requirements for both COG and CBM activities. For that party, it is therefore crucial to consider whether contractual remedies can appropriately cover all liabilities and responsibilities involved.

4. Supply chain: from CBM field(s) to liquefaction plant and beyond

4.1 Aggregation of reserves

According to received wisdom, about 5 trillion cubic feet of natural gas reserves are required to make the construction of a single train onshore liquefaction facility commercially viable in terms of economies of scale. Where relatively long pipelines are required to transport the CBM from the field areas to the liquefaction plant (as is the case in Queensland), a greater quantity of available reserves will be necessary to justify the project capital expenditure involved. These economic realities result in perhaps the single greatest challenge for greenfield CBM to LNG projects to overcome: the aggregation of sufficient deliverable reserves. In a conventional gas-fuelled LNG project, gas reserves will normally be sourced from two or three separate tenement areas at most, and from a similar number of discrete fields in relatively close proximity to each other. Contrast this with CBM. In a CBM-fuelled LNG project, for the technical reasons previously stated, meeting the aggregation challenge will likely require the acquisition of (or access to reserves in) a large number of different tenement areas covering tens of thousands of square kilometres, land access for the drilling of hundreds if not thousands of wells, associated gas gathering infrastructure and pipeline(s). As such, in order to ensure the delivery of sufficient quantities of CBM to the liquefaction plant, a commensurately greater number of contractual relationships will be required (see Figures 1 and 2 on pages 98 and 99), and a commensurately heavier burden in satisfying regulatory obligations will be experienced.

It is not just CBM from specifically CBM tenements which may be aggregated for the CBM to LNG project. Coal mines, and in particular underground coal mines, can be a substantial source of gas with normal CBM specification characteristics. Aggregating gas available from these mines, even where there is no coordination arrangement or regulatory requirement for cooperation, is a significant opportunity to increase resources. In particular, if coal miners are compelled either to pay tax or to acquire permits based on the intensity of their greenhouse gas emissions, they will have a significant incentive to do deals with CBM developers to take the CBM that the miners produce. For coal miners themselves, doing anything with the CBM is difficult, as they have none of their own infrastructure to monetise the gas and the quantities of CBM they produce are far more than they could use in the energy needs of their own operations. Also, as stated above, the removal of gas from coal seams in advance of coal mining can be important from a mine safety perspective.

The sheer footprint size of a major CBM development means that wells are located on land owned by perhaps hundreds of different landholders and their families. Creeks, rivers and other waterways must be crossed by roads and pipelines. Water drawn from coal seams must be used or responsibly disposed of. Community engagement is therefore a crucial part of the development process. Indigenous communities, farmers, local town groups, roads authorities and other owners of linear infrastructure (eg, power lines and water distribution pipelines) may have CBM development co-located with or proximate to their own activities. Understanding their needs, and ensuring that their legitimate expectations are met or exceeded, has just as much of an impact as any of the technical details of a CBM project. Aggregation of reserves is not just about aggregating the rights to the resources in situ; it is also about doing all those things necessary above ground to ensure that access to the reserves is not restricted or delayed.

Finally, where there are different ownership interests as between an upstream joint venture and the downstream pipelines and/or the LNG plant, several issues will arise. At a contractual level, the value creation must be fairly allocated to upstream development, transportation infrastructure and the LNG plant. It is not necessarily the case that the same level of risk appetite or return on investment will be acceptable to each owner along that chain. Transfer pricing will be a commercial negotiation, in addition to being a legal issue with respect to the calculation of government royalty. Government royalty in Queensland is a set percentage of the 'wellhead value' of gas produced, which is the market value of the gas, less processing and transportation costs such as arm's-length pipeline tariffs, arm's-length processing charges, operating costs and depreciation of capital costs of owned processing and pipeline infrastructure. Where an LNG producer owns all of the infrastructure from the wellhead through to the LNG plant and sells LNG to a related party internationally, negotiations with government are required to satisfy the regulators that the amount of royalty paid is not manipulated. Presumably, joint venture partners that are selling CBM from an upstream joint venture to their partner's LNG plant will have similar concerns to ensure that they receive a fair price. In projects with vertically integrated ownership, it will be in the interests of the project partners to push profits as far downstream as possible to minimise royalties.

Where two LNG developers with separate LNG plants share an upstream joint venture permit, the timing of production of the CBM becomes a key commercial decision if there are differences between the parties as to when they want gas to be produced. It is common for joint venture arrangements to contain no joint marketing and instead simply to refer to each party taking its proportionate share of CBM. However, unless storage options are available, it is unlikely to be an easy matter for joint venturers to take CBM at different times, due to the characteristics of CBM reservoirs. This may require some commercial flexibility to overcome. For example, acreage is shared by APLNG and BG in the Surat Basin. Due to BG's LNG plant being developed first, an agreement has been announced for APLNG to sell some of its share of CBM to BG at the beginning of the project. Speculating, this is probably because APLNG could not take or store that share prior to its own project coming on line.

In addition to the issues encountered when separate LNG project developers co-exist in the same upstream joint venture, LNG developers will have adjacent acreage targeting the coal seams which extend beyond the boundaries of one permit and into the other. In conventional oil and gas, where the reservoir can be considered as a single pool, it is common for this to lead to the use of cooperative unitisation agreements to allocate the petroleum produced. These agreements were originally used in the United States and their negotiation was motivated by mutual benefit as a jointly managed reservoir was able to be more productive. Such arrangements are generally unusual in CBM and so far are not a major feature of the significant adjacent acreage held by Queensland's CSG to LNG proponents. Although there is no question that CBM can to some extent be produced across tenure boundaries, the extent of the influence is limited compared to the full size of the reservoir. The relevant legislation in Queensland does contemplate such agreements. Where there are adjacent petroleum leases, it is not permitted to produce petroleum in one lease that is from part of the reservoir that is in the area of the adjacent lease, unless a coordination arrangement between the respective lease holders is put in place.

4.2 Basic contractual frameworks

Due to the large number of joint venture consortia which are likely to be involved in producing the upstream feedstock for a CBM to LNG project, it makes sense – certainly from any LNG buyer's perspective – that there be a single LNG seller counterparty to each LNG sale and purchase agreement with each LNG buyer. Also, having a single aggregator entity increases flexibility in sourcing the necessary quantities of feedstock from the various different supply sources. The aggregator will source feedstock gas for the liquefaction plant from the various tenements in which it has a participating interest pursuant to gas sales agreements with each consortium. These arrangements will form the primary source of supply, but sometimes it may be necessary to source additional gas from third party suppliers – for example, where the aggregator's fields are still 'ramping up' to full production. Aggregators may also consider owning or contracting underground gas storage facilities in order to be better able to cope with unexpectedly large fluctuations in demand or field deliverability. As previously stated, CBM field deliverability is more linked to dewatering of the reservoir than its conventional equivalent and CBM wells cannot simply be turned on and off to same extent as their conventional counterparts. Where a CBM well is shut in, the longer that the well is shut in the more likely it is that water will return to the coal seam. Once the well is put back into production, the previously completed process of dewatering may have to be repeated to bring the well back up to full production.

There will be a need for the upstream supplier consortia to have access to shared processing facilities or to build such infrastructure themselves. CBM produced by the upstream supplier will be gathered at an integrated processing facility. From there, services to the upstream producers will include gas processing, water treatment and power generation under tolling arrangements. For facilities owned separately from the gas field, the fee structure is likely to be made up of a fixed reservation charge and a variable charge to cover variable operating costs such as power. The owner of the integrated processing facilities will usually purchase system use gas from the upstream suppliers and also contract out operating and maintenance services for the facilities.

A main trunk pipeline will then connect the integrated processing facilities to the gate of the LNG liquefaction plant. In Figure 1, the ownership of the pipeline is separated from that of the gas. In Queensland or other places where relatively long and capital-intensive pipeline infrastructure is required, it may be sensible for the pipeline to be owned separately in order to provide greater funding flexibility. Separate ownership should also promote the sharing of capacity in main pipeline infrastructure with the costs savings and the efficiencies that this would entail. However, these factors will be subject to the need of each set of LNG project sponsors to secure sufficient pipeline capacity for their particular project. At present, there are no plans for the four major Queensland LNG projects to share pipeline capacity. Open access competition rules can apply to pipelines in Queensland under the National Gas Law, whereby a pipeline owner would be required to offer spare capacity to the market at reasonable rates. However, exemptions are also available in certain circumstances for newly constructed pipelines and foundation customers may also contract for all of the available capacity.

BG has sought and obtained a 15-year no-coverage exemption for its pipeline from the Surat Basin to Gladstone from the National Gas Law. In its application, BG argued that this exemption was appropriate as the pipeline would be economic to duplicate by someone else (as demonstrated by the proposals from the other CBM to LNG proponents to do just that), and it would not be contrary to the public interest as it provided an appropriate incentive for the major pipeline investment required. This application received ministerial approval in June 2010, with a determination being made which agreed with BG's application and held that the exemption would not impact on competition. It is expected that other CBM to LNG proponents will seek similar no-coverage exemptions for their own pipelines. Each application is assessed on its own merits and economic factors in particular may change from one pipeline to the next.

In circumstances where more than one LNG project company wishes to ship CBM down the same trunk line in a commingled stream, other matters may need to be addressed in addition to the usual matters covered in pipeline joint venture and gas transportation agreements (eg, allocation of capacity rights and liabilities, tariffs, ship or pay). There may also need to be pipeline gas balancing arrangements entered into between the shippers among themselves and between the shippers and the pipeline owner(s). These will encompass the physical balancing of the pipeline, whereby each shipper undertakes to balance inputs into and offtakes from the system and is subjected to sanctions in the case of imbalances. In the context of

Queensland, where the different LNG projects are in different stages of development and subject to different offtake requirements, there is also likely to be a proviso to the general rule that each shipper must match its actual pipeline deliveries and offtakes with its corresponding nominations. This would allow the shippers to agree between them to borrow gas from each other. This could be on a short-term basis to allow a shipper to substitute another shipper's gas to fulfil its delivery obligations to the gas buyer and thereby manage unexpected fluctuations in production and demand. On the other hand, there could be an arrangement where one shipper may contribute to another's gas sales obligations over an extended period and have its contribution repaid over a similar period. To the extent that different shippers contribute CBM into a pipeline system at different specifications (especially heating value), a gas allocation agreement is desirable to ensure that to each shipper is re-delivered a quantity and quality of gas equivalent to what it put into the system.

Figure 1: Pure aggregator model

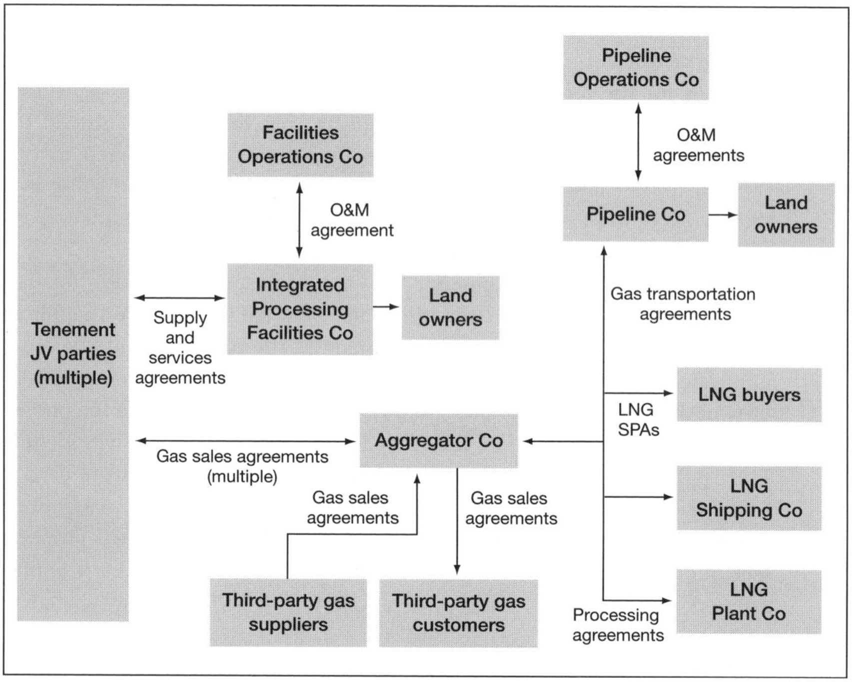

In Queensland, the major announced projects have some level of vertical integration in ownership, though not necessarily operatorship. For example, the APLNG project is operated by Origin in the upstream, but by an APLNG corporate entity in the downstream. This structure has been used to leverage the particular knowledge brought to the project by different equity owners, but means that despite ownership integration, the contractual structures are closer to the aggregator model. Santos has a similar model and the Origin and Santos own natural gas and/or CBM acreage sitting outside the acreage they share with their multinational partners, creating the potential for more arm's-lengths transactions in the future if that gas is directed towards LNG. This compares with QGC and Arrow, which are integrated corporate entities, owning and operating most of their own acreage and building their own pipelines and LNG plants. Both do, however, have some acreage subject to third-party ownership, including tenements that they share with each other.

Even for a wholly integrated CBM to LNG project, there are likely to remain a number of contractual relationships which are unavoidable. The onshore portion of the project will need to deal with land holders of all kinds, including not just rural properties, but also indigenous owners, public authorities and coal miners. Equally, all Queensland LNG projects are currently planned for the Port of Gladstone, which is operated by a government-owned corporation. Further, by using contractual relationships with third parties, various options are available for dealing with ramp-up timing differences. Some major proponents have announced plans to use depleted conventional reserves held by third parties as gas storage or gas swapping options to achieve a similar outcomes. Others have domestic power generation businesses or domestic gas supply businesses which can act as a sink for gas during ramp-up.

Figure 2: Integrated proponent CBM to LNG gas flow model, with some third-party gas supplies, gas usage, gas storage and gas swapping options

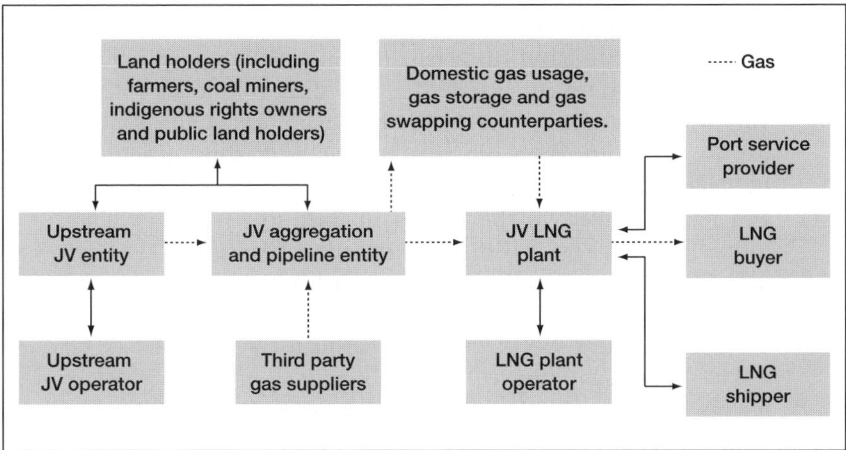

4.3 CBM gas sales agreements

The gas sales arrangements between the upstream supplier consortia and the company which aggregates the CBM for the purposes of liquefaction and on-sale to LNG buyers are clearly a fundamental element in the supply chain. There will inevitably be a tension between the aggregator company's need to secure reliable and steady supplies of gas to meet its LNG sale and purchase agreement commitments on the one hand, and the upstream providers' need to build flexibility into their delivery commitments due to the intrinsic technical differences between CBM and conventional gas extraction (see above) on the other. The aggregator company will seek to make the upstream gas supply arrangements and the LNG SPAs 'back to back' as far as possible, but as we will see, there is limited appetite from LNG buyers under long-term contracts to compromise on security of supply, given the high prices they are paying for the gas.

The allocation of *force majeure* risk was demonstrated to be a particularly important clause when considering a gas sales agreement from an upstream supplier of CBM in Queensland, due to widespread flooding experienced across the state between December 2010 and February 2011. It was not just the immediate rainfall event and flooding which impacted on CBM operators, but also the filling of water storage dams with rainwater, which then impinged capacity to store produced water from CBM operations. Queensland gas supplies were constrained at the time and this led to significantly higher electricity prices for a period during which gas-fired powered stations had reduced CBM supply.

CBM wells are known as 'low turn-down' wells. This means that the production of gas from each well cannot be significantly reduced or shut in without risking the long-term loss of that well as a gas producer. In conventional gas sales agreements, the buyer will generally have the ability to vary its gas take on any given day. This gives rise to the question of what the CBM producer can do with the gas from a low turn-down well when the buyer reduces gas takes on a particular day. Without the ability to dispose of unsold production through gas storage or spot sales/gas swaps, the excess gas is generally flared. Flaring is clearly wasteful of the CBM resources and environmentally unfriendly. A number of 'gas sinks' may be available for the disposal of gas excess to the buyer's daily requirements, but if these are not available then the CBM producer must either seek to impose daily take-or-pay obligations in the gas sales agreement or accept the risk of a certain unsold output. Nominations flexibility day to day is therefore generally limited, with daily quantities rarely having flexibility below 80% or even 90% of full volumes. Equally, upside nominations are also tightly limited, as the field is unlikely to have much ability to increase quantities. Thus, a buyer with a large quantity of banked gas as a result of a previous failure to take may find that recovering that banked gas later can be done only in limited quantities per day.

In many other respects, the closely negotiated areas of CBM gas sales agreements will be the same as in conventional gas sales agreements. In case of reserves risk, the buyer will usually wish the seller to have an obligation to supply which is not linked to a particular source, thus leaving it up to the seller to meet contractually committed quantities from whatever sources are available. In this case the reserves risk will sit with the seller, subject to any *force majeure* provisions relating to the geology of the CBM field. There will be a ramp-up period while production builds up to peak levels during the dewatering phase. During this period, if the buyer wishes to take ramp-up gas, the seller may ask the buyer to commit to take whatever is produced on a day-ahead nomination basis.

4.4 CBM LNG sale and purchase agreements

In basic terms, a CBM LNG sale and purchase agreement should not be materially different from a conventional LNG sale and purchase agreement. The critical features of quantities, timing and specification remain the same. However, the nature of CBM is such that the negotiated end point with respect to each of those features may be substantially different. Specification will naturally be different as a result of the different chemical composition of the CBM produced, which is almost pure methane.

On quantities and timing, the nature of a CBM field production profile feeds into the design of the LNG sale and purchase agreement. An average CBM well will not reach peak production for perhaps 12 months from being drilled, as the coal seam is being dewatered, and then production will reduce down to a fraction of the peak over the following two or three years. As mentioned above, this means that large numbers of wells must be drilled over the life of the project. In the life of a major LNG plant, thousands of wells will be involved. If the assumption is that a CBM well produces most of its gas over a period of approximately four years, new wells must be continually drilled to keep the LNG plant running at capacity. However, it is generally either technically, financially or commercially impossible to drill four times as many wells in the first year as are then needed for each year after that. In most markets, the drilling rigs required would simply not be available. Various options are available to deal with this. One is for the buyer simply to agree to taking fewer contracted cargoes in the first few years and building up to a plateau after several years of fewer cargoes. Alternatively, the ramp-up may take place during LNG plant construction, if another use or storage option for the gas is available during that time. In practice, a mixture of the two is likely, with the LNG SPA supply obligation starting at some point during ramp-up. This ramp-up period favours an LNG plant design with the ability to run efficiently while still liquefying a lower percentage of full plant capacity than a conventional LNG plant – although anecdotally, designing LNG plants to run at less than 70% capacity is difficult to achieve from a technical perspective.

For long-term LNG buyers, certainty of supply into the future is a critical issue, which extends from the LNG plant further upstream. In LNG markets, security of supply is a major factor in concluding agreements and buyers may seek a variety of mechanisms to satisfy themselves that the supply will be available when needed, including imposing conditions precedent relating to the upstream development. *Force majeure* clauses in upstream CBM contracts were mentioned above, but passing this risk though to the buyer in an LNG sale and purchase agreement is not guaranteed. Changes in law and changes in government policy that impose additional costs on CBM production will not normally be grounds for a *force majeure* claim under an LNG sale and purchase agreement, although a change which prevents production or a substantial part of production may be. It is also not unusual for upstream CBM contracts to contain some level of flow through to pricing in relation to changes in taxes or other compliance costs, while the ability to adjust pricing under an LNG sale and purchase agreement where such costs are incurred upstream would be unusual. Any difference in terms may need to be actively managed by the aggregator of CBM, such as through having multiple supply options or the ability to switch gas between domestic and LNG offtakers, or by using insurance and derivative products such as firm options to acquire spot gas at short notice from third parties. Similarly, damages regimes for upstream CBM contracts are likely to focus on daily shortfalls, which may be modest if short-lived. However, from an LNG perspective, a shortfall in supply may result in a missed shipment, and the resultant liability for that could be much more substantial.

4.5 Financing

Although there are now three CBM to LNG plants in Queensland progressing beyond their respective final investment decisions, and plans for more, there has been limited activity so far in relation to project financing. Those projects where final investment decisions have been reached have funded their programmes through equity or corporate level debt. Most projects have secured offtake LNG sale and purchase agreements to provide revenue certainty, although Arrow's shareholders Shell and PetroChina have indicated that they will take the LNG produced directly into their own portfolios. For the two Australian companies that have remained independent through their CBM development, their funding needs have been met in large part by selling down a percentage of the project to multinational oil and gas companies or LNG buyers and allocating the sale proceeds to development. The Australian companies that have been acquired by multinationals outright have since used shareholder sourced funds for their development.

However, this is not to say that a project financing model is not available. More than one Queensland project has sought to appoint advisers to progress project financing. For the LNG plant itself and the transmission pipelines, from a financing perspective there is no substantial difference between a CBM project and any other LNG project worldwide.

With the upstream CBM fields, however, project financiers will need to understand some unique features of CBM and how this feeds in to the financing model being employed. Unlike other conventional projects banked on a model of proved reserves, the nature of CBM is that investments are generally made on the basis of proved and probable reserves. In the Queensland domestic gas market, project financiers for power stations have accepted this proved and provable model, so it is not expected to be a barrier to the larger financing required for an LNG project. Another difference is that the thousands of wells that CBM involves are not all drilled in the one place and not all drilled at the beginning of the project. Unlike a typical project-financed project, where major drilling and infrastructure development activity stops at commissioning of the LNG plant, in a CBM project new wells will be drilled over most of the project life continuously. New pipelines and infrastructure will be needed in that period to connect the wells to the transmission pipelines. Therefore, it will not be the case that all project assets will be in place, and therefore available for securitisation, at the beginning of the project. Neither will it be the case that the firm contracted prices will be available to insert into an economic model. The same goes for consents from third parties such as land holders and coal miners, and approvals from government. If drilling and development of surface facilities in a particular area are not planned to take place until 15 or 20 years after project commencement, the level of certainty in relation to those matters required by financiers at financial close must be realistic.

The phenomenon of reserves-based lending facilities, sometimes referred to as 'accordion facilities', has now been around for a while in the COG context. Reserves-based lending can be used as a method of corporate debt finance which involves a lending facility under which the amount of available funds expands as additional reserves are added across a company's portfolio (and, of course, contracts as funds are drawn down or assets within the portfolio are sold). Repayment of the debt comes from revenue derived through the sale of a field or a field's production. Given the greater uncertainties around the timing and scale of future funding requirements associated with CBM development projects, corporate reserves-based lending facilities may become more prominent in the CBM world, as there is no requirement to fund any particular project and the timing of drawdown is left flexible within the overall tenor of the facility. In some cases, more progressive lenders may even be prepared to lend against contingent resources and make funding available prior to execution of gas offtake arrangements. With CBM for LNG projects, it is perhaps conceivable that the upstream element of the supply chain could be funded using a reserves-based lending facility, with the downstream elements funded using a more traditional project finance-based structure, although this would add further complexity to what is likely to be an already complex network of contractual relationships.

5. The future of CBM for LNG outside Queensland?

A particular set of geological, market and regulatory conditions has made massive investment into Queensland's CBM to LNG projects viable. In this final section, we consider whether these successes are repeatable elsewhere.

In relation to new 'greenfield' CBM to LNG projects (ie, cases where CBM feedstock is used as the foundation for the construction of new LNG liquefaction facilities), the obvious place to look is southwards across the state boundary to New South Wales (NSW), where a very similar set of circumstances to Queensland exists. First, there are potentially very large volumes of gassy coals located in a north-south belt (running approximately parallel with the Eastern seaboard), stretching from south of Sydney through the Hunter Valley region to the Gunnedah basin in the northeast of the state. In the absence of new local sources of supply, there is expected to be yawning gap between gas supply and demand in NSW over the coming decades. Hence, local gas demand (primarily for power generation and other industrial uses) has the potential to provide the initial market for early stage CBM pilot and development programmes and underpin investment in increasing the reserves base incrementally over time. In time, it is expected that the quantity of CBM reserves in NSW will be such that a sizeable surplus is left over for export outside the state – enough potentially to support greenfield LNG export facilities in NSW and long-term LNG sale and purchase agreements with north Asian buyers. Given the relative infancy of the CBM sector in NSW, the regulatory framework governing CBM extraction is somewhat less sophisticated than its counterpart in Queensland (the regulation of onshore petroleum activities is done at a state level in Australia). This is particularly the case with respect to the resolution of overlapping rights between coal and petroleum tenement holders. However, this is unlikely to remain the case for long, and in general the current regulatory regime in NSW is unlikely to constitute a significant hurdle to a CBM-based LNG export project. At this point, the main hurdles seem to be commercial and political. The commercial hurdle to an NSW-based LNG export project is that NSW CBM reserves are a natural option for diversifying supply sources available to the major LNG projects in Gladstone, and a new pipeline is proposed linking the Hunter Valley and the Gunnedah basin with Southeast Queensland – the Queensland Hunter Gas Pipeline. The political hurdle is clearly the high-profile opposition among certain sections of the public to CBM development being experienced in Australia currently. Clearly, further work needs to be done by the CBM industry to allay perceptions of risk and persuade opinion formers that the economic and indeed environmental benefits far outweigh the potential risks involved. At the time of writing, the NSW state government is reviewing the regulatory framework applying to CBM and the results are keenly awaited.

Outside Australia, Indonesia could be the 'next cab off the rank' in terms of realising a CBM for LNG project; the obvious candidate comes in the form of the existing LNG liquefaction facilities at Bontang in East Kalimantan. Due to the depletion of conventional gas supplies, a large proportion of Bontang LNG's processing capacity lies idle. The coal fields of East and Southeast Kalimantan are widely viewed to have great CBM potential, as evidenced by the strong competition for CBM acreage in that area. The objective is clearly to utilise spare capacity in Bontang LNG to process CBM into LNG. While the challenge of gas aggregation in Kalimantan is less acute than in Queensland (because LNG liquefaction plant and significant pipeline infrastructure have already been constructed), one issue will be the extent to and terms on which new producers may access existing infrastructure. It would be fair to say that Indonesia's regulatory framework around third-party access is a good deal less sophisticated than Australia's. The general weight of government bureaucracy and slow speed of decision making (particularly in the areas of procurement, plan of development approval and gas marketing) also have an impact on the pace of CBM development in general. However, the government appears to be recognising that CBM development is subject to a different set of economic and technical parameters from COG, including the step-by-step approach to proving up reserves. An example of this was the government's recent decision to permit CBM developers to commercialise gas produced from pilot programmes, in advance of full field development approval. Also, significant progress has been made on the ground, with the Dart Energy-operated Sangatta West project producing Indonesia's first CBM in April 2011. All things being equal, it seems just a matter of time before CBM is used as feedstock for LNG production in Indonesia, although initial production is likely to be used for local power generation purposes. The question that LNG producers in Indonesia are all grappling with is ultimately how much gas will be permitted to be exported, given the so-called 'domestic market' obligation which has now been written into both COG and CBM production sharing contracts. Indonesia's archipelagic geography means that, paradoxically, its major domestic markets of Java and Sumatra experience major gas supply shortfalls, and yet Indonesia currently exports all of its LNG production from East Kalimantan and West Papua (Tangguh LNG). At present, Indonesia simply does not have the capability to supply its own domestic market with Indonesian-sourced LNG, but this may change if domestic LNG receiving projects get traction.

In most other places in the world (with the possible exception of North America), geology, geography and demography conspire to make it unlikely that CBM will be used as feedstock for a major greenfield CBM for LNG project in the foreseeable future. It so happens that the world's major coal (and therefore CBM) resources are commonly located in countries where the shortfall between domestic supply and demand is currently at its greatest. In these places (eg, Northern Europe, China, India, Central and Southeast Asia and Southern Africa), the beauty of CBM lies in its strategic, environmental and commercial value: strategic in that it reduces reliance on external sources of supply; environmental in that it provides a viable and cleaner alternative to coal-fired power generation; and commercial in that transportation costs are vastly reduced. This is not to say that no CBM for LNG development will occur outside Australia and Indonesia, and shorter-term, niche opportunities may arise. For example, moveable floating natural gas liquefaction facilities could potentially be used on a short-term basis as a gas sink during the ramp-up phase of a major CBM production project and LNG cargoes produced in this way sold on a spot basis.

6. Conclusion

The technical challenges in relation to the development of a CBM to LNG project are reasonably well known, and the facilities involved as individual elements are all familiar and tested technologies. However, as this industry is evolving, some unique issues are emerging, with inventive commercial and legal solutions. Assuming that a suite of agreements for a conventional LNG project would be suitable to a CBM to LNG project would be dangerous, as the physical constraints of CBM development do create new commercial requirements.

From a whole of project perspective, being primarily onshore and occupying fairly large areas of land means that the end-to-end supply chain for LNG from wellhead to ship is extensive. This zone of activity means that CBM projects have more exposure to local conditions than conventional projects. As CBM requires thousands of wells spread across a large region, a stable and workable framework for the drilling, water disposal, processing and pipeline infrastructure to be built in harmony with existing mining, industrial, rural and residential communities is essential. For investors, financiers and buyers of LNG, it is this regulatory regime which can provide certainty that will underpin the investment required. Even in a country with a large land mass as sparsely populated as Australia, the regulatory environment continues to evolve so as to find the right balance to allow CBM to LNG projects to go ahead. With hundreds or even thousands of adjacent or overlapping owners of property and infrastructure, having the methodology to allocate rights efficiently so that the critical connecting infrastructure can be built is a key factor in the success of the industry.

On the other hand, the large acreage required to develop a CBM to LNG project is a repeatable business model, where cost efficiencies should be able to be extracted from undertaking similar work over wide regions. As mentioned, and leaving aside specialist completion techniques, the technology involved is relatively straightforward for gas field development, involving smaller drilling rigs and shallower wells than conventional gas. While commercialisation of reserves must still be properly assessed, the major benefit of CBM is the relatively lower risk process of reserve certification for field expansions where similar coal seams extend for many hundreds of kilometres. With appropriate geology and technology, single wells can target multiple coal seams to maximise CBM production and lower unit costs. It is this large resource accessible through low-cost technology that has seen CBM to LNG become a viable and attractive unconventional alternative.

This chapter draws on material from Peter Godfrey, Tan Ee and Toby Hewitt, "Coal Bed Methane Development in Indonesia: Golden Opportunity or Impossible Dream?" *(2010) 28 JERL (International Bar Association).*

Floating LNG

Matthew Griffiths
Royal Dutch Shell plc

1. Introduction

In understanding the role of floating LNG (FLNG), it helps to take a step back and understand the drivers in the wider natural gas business. Global demand for energy could triple from its 2000 level by 2050 if emerging economies follow historical patterns of development. The key drivers for this are a rising global population and prosperity growth as emerging nations enter their most energy-intensive phase of economic development.

To keep pace with increasing demand, the world will need to invest heavily in the entire mix of energy sources. At the same time, there is an urgent need to address greenhouse gas emissions. Therefore, many governments, buyers and energy companies are increasingly focusing on natural gas, for the following reasons:

- It is abundant – the International Energy Agency (IEA) estimates that total recoverable resources could sustain today's production levels for over 250 years. Those reserves are widely dispersed throughout the world, with significant reserves in the Russian Federation, Iran, Qatar and Australia. As well as ongoing conventional natural gas extraction, new methods of extracting natural gas are being introduced, including the extraction of 'unconventional gas' (ie, tight gas, shale gas and coal bed methane, where gas is trapped in low porosity and/or low permeability rocks). These new methods of extraction will significantly increase those reserves and their global distribution (eg, in North America, Australia and China).
- It is also an affordable fuel for power generation, given its energy efficiency compared to, say, coal plants; the lower capital costs per megawatt installed of gas-fuelled power plants over other types of power plant, such as coal, nuclear and wind; and the fact that gas plants are the quickest type of plant to construct. Gas is also competitive against alternative fuels in other applications such as residential use, industry and transport.
- It is environmentally acceptable, as gas is the cleanest burning of the fossil fuels and can therefore help to reduce carbon dioxide, particulates and other harmful gas emissions when replacing, say, coal.

For these and other reasons, the IEA currently estimates that gas use will rise by more than 50% on 2010 and account for over 25% of world energy demand by 2035, and believes that we may well be entering a "golden age of gas".

By liquefying that natural gas in multiple locations and shipping it around the world to multiple destinations, many governments and buyers gain additional security and diversity of energy supplies, especially where they are heavily reliant on pipeline gas supplies from or via neighbouring countries. LNG now accounts for a significant share of global gas trade and there is strong growth in LNG shipments.

So, in order to ensure that the production and trading of natural gas is being maximised and optimised, there needs to be a technically and commercially attractive solution to the extraction of natural gas which, in the past, would not have been produced. That is exactly where the FLNG concept fits in.

2. Background and specific projects

FLNG is a new and innovative concept which allows for a significant amount of the LNG infrastructure that would previously have been placed onshore to be placed offshore.

2.1 Conventional LNG infrastructure

The LNG industry has traditionally operated on the basis of the following infrastructure (set out in the order in which the gas molecules flow):

- onshore or offshore upstream natural gas production and processing facilities;
- onshore or offshore natural gas pipelines;
- onshore LNG liquefaction/export facilities which receive, process and liquefy natural gas and store LNG, and also provide LNG loading facilities;
- LNG carriers;
- onshore LNG regasification/import facilities which provide LNG unloading facilities and store and regasify LNG and send out natural gas; and
- onshore natural gas pipelines to end users.

2.2 Upstream FLNG infrastructure

However, the FLNG concept, as it relates to upstream facilities, significantly alters the infrastructure upstream of the LNG loading facilities, as follows:

- offshore upstream natural gas production facilities;
- offshore natural gas flowlines and flexible risers from the seabed to the FLNG facility;
- offshore FLNG facility, which is moored above or near the natural gas production facilities and receives, processes and liquefies natural gas into LNG (and maybe other byproducts such as LPG and condensate), stores LNG and provides LNG loading facilities; and
- downstream of the LNG loading facilities, the same infrastructure as for a conventional LNG project.

FLNG facilities have several advantages over conventional land based infrastructure:

- Remote offshore gas reserves (also known as stranded gas) can be economically produced, avoiding the need for long subsea pipelines and other infrastructure. For example, Australia has a lot of gas that is classified as stranded (about 140 trillion cubic feet, according to the Commonwealth Scientific and Industrial Research Organisation in 2008), so there will be many other opportunities for FLNG in Australia, particularly in remote areas far from shore.

- Smaller offshore gas reserves can be economically produced on a standalone basis or clustered via tiebacks. A larger FLNG facility or more than one FLNG facility can be used on larger fields.
- As a result of its offshore location, there may be less impact from an environmental, social or health perspective. For example, no near-shore works (ie, dredging, jetty construction or pipeline laying) or onshore construction or associated works (ie, land clearing, roads, laydown areas) are needed. In many cases the FLNG facility will not be visible from shore. Also, overall, far less materials are used. The location of the FLNG facility offshore will, of course, need to take into account marine plants and animals.
- It is possible to move from a field discovery to final investment decision, and then on to the start of operations, in a shorter period of time, allowing for earlier production and accelerated revenues. A modular construction process and controlled shipyard environment should assist greatly with construction times.
- Capital expenditure may be lower, due to the potential for cost efficiency through repeatability gains where there is a generic design concept, with adaptations for each particular project. In addition, as most of the construction activities will take place in an established shipyard used to sourcing labour and materials, costs should be more controllable. Finally, the lack of offshore pipelines and near/onshore facilities will reduce costs and there is no need to purchase or lease waterfront land for facilities.
- As long as it is properly maintained during its operational phase, it will be possible to re-use the FLNG facility in a different location. The design life for topsides can be as much as 25 years, and 50 years for the hull, and maintenance and refurbishment can extend the FLNG facility's life.
- Larger FLNG facilities can remain on station during harsh sea and wind conditions, due to their size and ability to weather vane.
- Due to their completely offshore location, they may be more secure and less prone to interference than an onshore LNG facility.

However, certain issues may need to be considered in relation to FLNG facilities:

- There may be concerns from host governments that constructing the FLNG facility in, perhaps, another country will reduce local employment opportunities during construction and the development of local skills and supporting industries. However, local jobs will still be created during the operational phase (as the operation of an FLNG facility is not dissimilar to the operation of a conventional LNG liquefaction facility). Moreover, there will be significant additional tax revenue for the host government, and also significant expenditure on goods and services in the host country in both the construction and operations phases.

- Host governments may also be concerned that an offshore solution is detrimental to any domestic gas supply ambitions. However, if LNG regasification facilities are in place in the host country, then the FLNG facility may be able to provide LNG to the host country over a relatively short shipping distance. Alternatively, commercial arrangements for gas supply from existing onshore facilities could be entered into in order to fulfil any domestic gas supply obligations.
- Finally, there may be general concerns over the safety, reliability and operational availability of FLNG facilities, given the novelty of this new technological solution. However, these key considerations would need to be addressed upfront by all FLNG providers (ie, developers of an upstream FLNG facility concept) in gaining acceptability from customers for their particular design concepts.

An example of such an upstream FLNG project is the Prelude FLNG Project, which will be located approximately 200 kilometres offshore of northwest Australia. The Prelude FLNG Project, which is owned and operated by Shell, is the world's first FLNG project to make a final investment decision to proceed, in May 2011, and will be operational in the second half of this decade. It will be constructed in Samsung Heavy Industries' shipyard in Geoje Island, South Korea, where there is a sufficiently large dry dock to build the FLNG facility. The total production will be 5.3 million tonnes per annum (MTPA) of hydrocarbon liquids, comprising 3.6 MTPA of LNG (easily enough to satisfy Hong Kong's annual natural gas needs) and the remainder condensate and LPG. It will be 488 metres long (the length of more than four

The Prelude FLNG facility, based on actual design data

football pitches) and 74 metres wide. Fully ballasted, it will weigh 600,000 tonnes (around six times as much as the largest aircraft carrier), making it the largest offshore floating facility ever built. It has been designed to withstand Category 5 cyclones. It will be permanently moored over the Prelude gas field for 25 years or longer if other gas fields are tied back to it. LNG carriers will moor alongside the FLNG facility for cargo offtake. During steady state operations, the Prelude FLNG facility is expected to load around one LNG cargo a week, one condensate cargo every two weeks and one LPG cargo every month. Once operational, up to 240 crew will be directly involved in operating the FLNG facility in shifts, with half of those onboard at any time.

2.3 Downstream FLNG infrastructure

The FLNG concept, as it relates to downstream facilities, significantly alters the infrastructure downstream of the LNG ship, as follows:

- upstream of the LNG unloading facilities, the same infrastructure as for a conventional LNG project;
- offshore FLNG facility (known as a floating storage and regasification unit (FSRU)), which provides LNG unloading facilities, and stores and regasifies LNG and sends out natural gas; and
- offshore and onshore natural gas pipelines from the FSRU to end users.

FSRUs have similar advantages to FLNG facilities over conventional land-based infrastructure. In addition, FSRUs can be a precursor to onshore facilities (eg, if LNG supply is needed quickly) or supplement onshore facilities (eg, for peak demand).

A good example of such an FSRU project is the Golar Freeze in Dubai, owned by the Dubai Supply Authority, which became operational in December 2010. The Golar Freeze is a specially converted LNG ship with a design capacity to regasify up to 3 MTPA of LNG. It is permanently moored at a dedicated island jetty within the port

The Golar Freeze FSRU in Dubai

of Jebel Ali in Dubai waters, and is connected to Dubai's high-pressure natural gas pipeline system by a 1.4 kilometre subsea pipeline. Unloading LNG carriers moor alongside the Golar Freeze FSRU and then connect directly to the Golar Freeze and discharge LNG into its tanks for storage, prior to regasification. The primary aim of the project was to secure supply of a cleaner source of energy to supplement existing energy supplies to help Dubai meet its peak summer demand for power.

3. Legal and commercial issues

As can be seen from the above examples, FLNG is still at an early stage, and many of the particular commercial, technical, financing and legal issues associated with this new technology and its application are thus also being refined and developed. Highlighted below are some of the issues which will need to be considered in relation to FLNG projects. However, this could be a rapidly evolving area as and when more FLNG projects reach construction and operational phases.

This section focuses on legal, financing and commercial issues relating primarily to upstream FLNG facilities rather than downstream FSRUs.

3.1 What is an FLNG facility?

A question often asked is: is it a ship, an offshore platform, a port or something else? The most analogous infrastructure to an FLNG facility is probably a floating production, storage and offloading (FPSO) vessel for oil; many of these vessels have been safely constructed and operated around the world since the 1970s. However, the categorisation of an FLNG facility will depend heavily on its design – for example, whether it is converted from a ship (ie, an LNG carrier), is a newly built barge-like structure, is self-propelled or has to be towed into place. In addition, it may be regarded as a ship while on the move, and as a platform while moored and operating over a gas field. Therefore, the categorisation of an FLNG facility will heavily depend on applicable laws and conventions, and may vary between different types of FLNG facility and where they are moored.

3.2 FLNG business model and key documents

Each FLNG provider of an upstream FLNG facility (from large independent oil companies to small contractors) will have its own preferred document structure flowing from its particular business model, but the following is an illustration of the types of key documents which may be adopted.

(a) Licence agreement

This is executed between the FLNG provider and the operator of the relevant joint venture, and covers use of the FLNG technology and related IP rights. It may include the following elements.

Intellectual property: This can cover a multitude of different areas, including patent rights, trademarks, copyright, know-how, trade secrets and confidential information. For FLNG, the key IP areas are trade secrets (eg, design manuals, test results and simulation models), copyrighted materials (eg, front-end engineering designs, cost estimates and schedules), know-how (eg, project management, staff experience and operational experience) and patents covering inventions relating to or useful in FLNG facility design and operation.

With such a new and fast-evolving technological solution, from the FLNG provider's perspective, the proper protection of intellectual property will be critical. Having spent considerable time and money on researching and developing an FLNG solution, the FLNG provider will obviously be looking to recoup its costs and make a reasonable return from licensees, as well as to generate future business from other potential licensees. In addition, it can provide an 'entry ticket' for the FLNG provider into upstream offshore joint ventures which are looking for a way to unlock their stranded gas and are seeking an experienced partner to assist them with that.

Licensees, which will be paying for the right to use the intellectual property, will certainly want to use the intellectual property on their particular project, but may also be seeking the right to use it on other projects. The question of ownership of any intellectual property co-developed during the development, construction, commissioning and operation of the project using the FLNG technology will also be a topic for negotiation, with the FLNG provider probably seeking full ownership and the grant of usage rights to the joint venture and the joint venture probably seeking shared ownership and usage rights of such co-developed intellectual property. The FLNG provider will probably view any co-development (or maybe even sole development) by the licensee as being based on the licensed intellectual property – in other words, the co-development would never have happened but for the FLNG provider's intellectual property. Therefore, the FLNG provider will probably seek to own any such co-developed intellectual property. Of course, if the licensee is seeking the FLNG provider's intellectual property to build up its own FLNG capabilities, the FLNG provider will rightly be resistant to allowing this type of usage, as it would seriously undermine both its intellectual property and its competitive advantage. In fact, some joint ventures or partners may be seeking the opposite if they are developing their own FLNG technology, as they would not want to run the risk of 'contaminating' their intellectual property with that of the FLNG provider. The advantages to licensees of using an established FLNG solution include the acceleration of an opportunity, risk reduction through the use of a reliable and robust technological solution, access to state-of-the-art technology and to improvements to that technology, and knowledge and capability building for the joint venture and its partners.

Therefore, from the outset, there may well be a natural tension between the FLNG provider and the licensees over intellectual property.

Performance warranties: These usually set out minimum or maximum performance levels – for example, actual LNG production at or above a minimum percentage of the design case, or actual carbon dioxide content of the processed gas at or below a maximum level. Where these minimum or maximum levels are not achieved, the FLNG provider and/or the contractors will probably be liable for liquidated damages, with an overall cap on liability. For FLNG, these will be consistent with performance warranties offered for onshore LNG facilities, as similar processing, liquefaction and storage technologies are used, although adapted for an offshore environment in the case of FLNG. However, the novelty of FLNG may mean that joint venture operators, partners, financiers and insurers perceive there to be a higher level of risk than with conventional onshore facilities, and this may result in additional focus on performance warranties. On the flipside, FLNG providers may be looking for performance incentive payments if their design case is exceeded.

Liabilities: Aside from the performance warranties, the usual type of LNG liability and indemnity regime will also apply to FLNG, including:

- a 're-do' obligation, with certain limitations, on the part of the FLNG provider and/or contractor where the work has not been undertaken properly;
- a mutual hold-harmless regime for death and injury to personnel, property damage and pollution;
- caps on liability (eg, set at a percentage of the fees received by the FLNG provider and/or contractor); and
- an indirect/consequential loss exclusion.

As with any project, the key issues will be the negotiation of the caps on liability and any exceptions to the usual liability and indemnity regime.

(b) Technical advisory services agreement

This is executed between the FLNG provider and the operator of the relevant joint venture, and covers the design, construction and operational phases of the FLNG project. It may include the provision of secondees from the FLNG provider.

(c) Engineering, procurement, construction and installation contract

This is executed between the operator of the relevant joint venture and the relevant contractor(s), and covers the construction of the FLNG facility. The FLNG provider may be able to provide a pre-agreed engineering, procurement, construction and installation contract with its preferred contractors (eg, a design specialist and an experienced shipyard), which can then be novated to the operator of the relevant joint venture.

(d) Joint operating agreement

Operatorship of the FLNG facility is another consideration. If the FLNG provider is also a partner in the joint venture, it may be desirable to have the FLNG provider as the operator of the FLNG facility, given its knowledge and expertise. However, this arrangement will depend on the set-up of the upstream and midstream elements of the joint venture, as well as the appetite of the joint venture partners for such an arrangement.

The above document structure assumes that the joint venture will license IP rights and purchase a range of technical and construction services so that it can own and control the FLNG facility both during the initial production phase and during any redeployments. This will require significant upfront capital expenditure from the joint venture partners. As an alternative, it may be possible for the FLNG provider to construct and own the FLNG facility and then lease it to the joint venture for a specified duration, with the FLNG facility reverting to the FLNG provider at the end of that duration.

4. Financing FLNG projects

For project participants which are unable or unwilling to undertake corporate financing of the FLNG project (which is generally regarded as a less expensive and simpler alternative to project financing), the external financing and insuring of FLNG projects will need to be carefully considered. Given the recent financial crisis and the technological frontier nature of FLNG projects, lenders' appetites for these types of LNG project may be affected.

The current view is that to succeed, an FLNG project should focus on delivering long-term gas to strategic long-term buyers, as the debt market is unlikely to accept a combination of short-term volatility in the gas market and technological breakthrough. The primary market of the FLNG project is important, and FLNG projects based in the Asia-Pacific region are particularly well placed given the strong current and expected future demand for LNG in that region from established buyers such as China, Taiwan, Japan and Korea, as well as new buyers in countries in that region which are developing LNG regasification facilities.

Key risks which apply to FLNG projects include:

- the risk that completion is not achieved within the specified timeframe in order to allow the generation of revenues to repay the debt (also know as 'completion risk');
- the risk of unproven technology failures during the operational phase; and
- the willingness of the insurance market to insure such projects.

Key factors which will help to mitigate those risks include:

- the demonstrated ability of the project sponsor to deliver high standards of project management in frontier gas technology;
- low risk in relation to the extraction of sufficient volumes of gas, due to the size of the resource and confidence in the sponsor as an established upstream player (also known as reservoir risk); and
- repetition of the same or similar technologies for different FLNG projects, which will allow for learnings to be incorporated in later projects, economies of scale and a robust contractual framework to develop.

In particular, design and construction risks could be mitigated by, say, a combination of turnkey construction agreements and a creditworthy and technically capable FLNG provider and/or contractors. Technology and operational risks could be mitigated by the FLNG provider and/or contractors having a proven track record in the LNG industry, which would be enhanced once their first FLNG facility is successfully operating in a stable state.

Given the frontier nature of FLNG, it is likely that lenders will require some form of sponsor support during the construction/commissioning phase in respect of a project financing. This will also reduce the need for these lenders to do extensive due diligence on the construction/commissioning arrangements. It is also expected that lenders may request more extensive completion tests compared to onshore projects, and for the completion support/guarantees to extend for a limited period over the initial post-commissioning/operational phase of the project. Generally, financing an integrated project (ie, from reservoir to customer) compared to a non-integrated project will be less challenging, as it reduces inter-project risks. Of course, the strength of the LNG sales agreements will be key in determining the bankability of the project and lenders will be looking for long-term, take-or-pay, oil-indexed agreements with creditworthy offtakers.

Finally, environmental and social factors will play an important role. Especially due to the involvement of export credit agencies and a range of commercial lenders, the Equator Principles (a set of standards for determining, assessing and managing social and environmental risk in project financing) and other standards will be applicable to an FLNG project. Given the high-profile nature of these types of project, there will be a significant focus on the project's environmental (including carbon dioxide emissions) and social (including local content/workforce issues) impact.

5. LNG sales from FLNG projects

A technical challenge for FLNG facilities is that LNG will need to be discharged from one floating vessel (the FLNG facility) to another (the LNG carrier) in open seas. The relative motion of the two vessels may make it imprudent and unsafe to discharge LNG; however, it is interesting to note that the motion of the two vessels may in fact be synchronised if the wave and other conditions are right. This is likely to be much less of an issue with larger FLNG facilities, as they will be more stable and could also be manoeuvred to provide shelter for and assistance to the LNG carrier. In the case of the Prelude FLNG Project, significant research and development has been put into:

- the LNG processing (to take account of the roll and pitch of the FLNG facility in a marine environment);
- the LNG storage tanks (to withstand sloshing forces when only partly full); and
- the LNG offloading arms (to enable loading in up to 3.5 metre swells).

The possibility of such prevention or interruption of LNG discharge should be considered in any LNG sales agreements, through the *force majeure*, delivery programme/schedule (eg, extended delivery slots), liability or other appropriate clauses.

If the sea conditions are so severe that an FLNG facility needs to disconnect and move off the field, there may be significant knock-on effects on its ability to discharge LNG when contractually required to do so. Once again, this is likely to be much less of an issue with larger FLNG facilities, as they should be able to remain producing and available for discharge in more severe conditions. In the case of the Prelude FLNG Project, according to its research, there have been few recorded events in the last few decades in the Prelude location which would have necessitated a production shutdown of the Prelude FLNG Facility – which, as previously mentioned, has been designed to withstand Category 5 cyclones. This more significant potential interruption would also need to be addressed in the *force majeure*, delivery programme/schedule, liability or other appropriate clauses.

In addition, smaller FLNG facilities will have limited LNG storage capacity; therefore, the scheduling of LNG carriers for offtaking purposes will be critical and need careful management. Otherwise, if an LNG vessel is significantly delayed, the LNG storage tanks will become full and production of natural gas may need to be reduced or stopped, which would be undesirable from the supplier's perspective.

Finally, another area which may need both contractual and insurance consideration is a potential collision between the FLNG facility and the LNG carrier, and any ensuing liability for damage to facilities, pollution or death or injury to personnel. The liability regime should not be that dissimilar from a conventional onshore LNG facility, where a collision between the LNG carrier and the loading jetty is also possible.

6. Health, safety and environmental considerations in FLNG projects

The LNG industry has an excellent health, safety and environmental (HSE) record and therefore high HSE standards need to be maintained with the introduction of FLNG. The agreement of HSE standards for all phases of any FLNG project between the FLNG provider, the contractors, the operator and the joint venture partners, and the full and proper implementation of those standards, will thus be crucial considerations.

On the environmental front, the design of FLNG facilities should allow for carbon capture and storage facilities to be incorporated so that carbon dioxide extracted from the feed gas can be sequestrated. The Prelude FLNG Project is expected to produce around 15% less carbon dioxide emissions than a similar onshore LNG facility, as it can source cooling water from deep in the sea, which will improve thermal efficiency, and it does not need to use additional energy to 'push' natural gas down a pipeline to shore.

An interesting development in relation to carbon dioxide has recently occurred in Australia, where the Prelude FLNG Project, and maybe other FLNG projects, will be situated. On July 11 2011 the Australian federal government's Clean Energy Package – a plan to address and arrest climate change – was announced. Although Australian emissions account for only around 1.5% of global emissions (compared to the United States and China, which together are responsible for 35% of emissions), Australia is one of the largest per capita emitters in the world. The main mechanism of the Clean Energy Package involves the issue of carbon permits. Businesses will be required to purchase and submit these permits to government annually. Each permit will correspond to one tonne of greenhouse gas emissions. From July 2012, the initial price per tonne of carbon will be A$23. This fixed price will rise at a set amount above inflation until 2015, at which time the price will be set by the market, within a price collar. The plan is for an eventual move to a cap and trade scheme, whereby the government will set an overall cap on total annual greenhouse gas emissions, issuing a fixed number of permits each year, and businesses will trade those permits among themselves. From an LNG export perspective, an average LNG project will probably receive at least 50% of its permits free for the first three years; some may get more. However, there is still considerable uncertainty about the application of the carbon price and any dispensations for the LNG industry. The Australian proposal is applicable only to the emissions of the LNG facility, and not the lifecycle emissions of the hydrocarbon product.

Most analysts expect that the impact of the carbon price on the profit margins of LNG developments will be relatively modest. However, it will still affect the competitiveness of Australian LNG compared to other producing countries supplying the Asia-Pacific market which do not face a carbon price. Interestingly, the Australian Petroleum Producers and Explorers Association argues that the export of Australian LNG actually helps to reduce global emissions where it replaces coal in high-emitting countries such as China and India. Other countries and regions – such as Canada, the United States/California, the European Union, South Africa and New Zealand – have also implemented or are considering implementing similar schemes, so carbon dioxide emissions from LNG facilities will need to be considered in all future LNG projects.

7. Redeployment, disposal and decommissioning of FLNG facilities

One distinct difference between FLNG and conventional onshore facilities is that once the original gas field and any other gas fields tied back to the FLNG facility have reached the end of their economic life (ie, when the cost of producing gas and turning it into LNG for sale is, in effect, greater than the revenue plus minimum acceptable profit margin from LNG sales), then the joint venture will have the possibility either to redeploy the FLNG facility over another gas field or to sell the FLNG facility to interested parties. These could be commercially attractive options allowing the joint venture to exploit other gas fields owned by all or some of them or gas fields owned by third parties (which would provide them with a tolling revenue) or, if sold, to provide them with sales revenue. If these options are unavailable, then the FLNG facility (and any related subsea infrastructure) will need to be properly decommissioned. The fact that it is a moveable structure that can be put into dry dock should make it easier and less costly to decommission than an onshore LNG facility (and any related subsea infrastructure and long pipelines to shore), with far less environmental disturbance.

8. Conclusion

FLNG is an exciting development in the LNG industry that will open up a whole new range of gas reserves for economic production, thereby contributing significantly to increased energy security and diversity.

However, although FLNG is a neat and innovative technical solution, many of the existing LNG liquefaction plant, LNG shipping and LNG sales contracts and legal considerations will continue to apply, with a few adjustments, to FLNG. In addition, many of the lessons learnt from FPSO developments over the years will also apply to FLNG facilities. This continuity and experience should provide significant reassurance to governments, regulators, operators, joint venture partners, contractors, LNG buyers, financiers and insurers that a comprehensive and tested contractual framework will exist for FLNG, as it has for conventional onshore LNG facilities for many years.

Natural gas price reopeners and English law

Paul Griffin
Allen & Overy LLP

1. Discord and long-term gas contracts

The last decade has seen an increasing trend of reference to arbitration or other formal dispute resolution processes in circumstances of commercial imbalance under long-term agreements for the sale and purchase of natural gas by pipeline, and particularly by tanker vessels, as liquefied natural gas. This chapter concentrates on LNG sale and purchase agreements (LNG SPAs), although many of the matters addressed also apply to long-term pipeline gas contracts.

The mis-alignment of many LNG SPAs with prevailing markets and the disparity of commercial balance under those LNG SPAs suggest that this trend of disputes and their external resolution is unlikely to decline, especially in Europe. It certainly appears that few buyers or sellers that have pursued or been forced into these processes have emerged from them with satisfaction, even if the decision has been in their favour. The identities of those appointed to make such decisions, the procedures followed and the nature and effect of those decisions have all been grounds for dissatisfaction.

This chapter considers these issues in the context of price review provisions under LNG SPAs written under English law. It suggests that the typical form of price review provision is often ill suited to its purpose, and that the customary reference to arbitration in the event that the parties do not agree contractual revisions is inappropriate. This chapter also considers alternative approaches which may better reflect the intentions and purposes typically underlying the price review provisions of an LNG SPA.

The wording used in a price review provision is likely to have been the subject of extended discussion and negotiation by the parties over a long period of time. These provisions can constitute some of the most difficult areas of negotiation and may well be left over to the last sessions, when the mood of the negotiators becomes one where the overall enthusiasm to conclude an agreement militates in favour of drafting which, rather than reflecting agreement on these matters, is sufficiently obscure or ambiguous to enable the parties to put it to one side and sign the overall agreement. When such a provision becomes subject to judicial scrutiny, it is for the judge to interpret that provision. A nice judicial acknowledgement of the difficulty of this task in this context arose when, in reviewing a comfort letter in *Chemco Leasing SPA v Rediffusion* [1987], 1 FTLR 201, Justice Staughton commented:

> *when... business men wish to conclude a bargain but find that on some particular aspect they cannot agree, I believe it is not uncommon for them to adopt language of deliberate*

equivocation, so that the contract may be signed and their main objective achieved. No doubt they console themselves with the thought that all will go well and that the terms in question will never come into operation or encounter scrutiny; but if all does not go well, it will be for the courts or arbitrators to decide what those terms mean. In such a case it is more than somewhat artificial for a judge to go through the process, prescribed by law, of ascertaining the common intention of the parties from the terms of the document and the surrounding circumstances; a common intention was in reality that the terms should mean what a judge or arbitrator should decide that they mean, subject always to the views of any higher tribunal.

The scale and significance of these price review provisions can be shown by the recent case concerning Gas Natural Fenosa and Sonatrach. Gas Natural Fenosa is the largest importer of gas into Spain and Algeria's Sonatrach is a major supplier, under both LNG SPAs and long-term pipeline gas contracts. Pursuant to the terms of a long-term pipeline gas contract, a price review was operable in 2007 and, in the absence of agreement, the parties referred the matter to International Chamber of Commerce (ICC) arbitration in Paris. Although there is no report of the case in the public domain, the parties have made a number of public comments on it. The arbitration award was made in August 2010 and provided for the retrospective payment by Gas Natural Fenosa of a sum approaching $2 billion in relation to gas provided by Sonatrach between 2007 and 2009. Following an application in November 2010 by Gas Natural Fenosa to the Swiss Federal Court for the suspension of enforcement of the award, the parties entered into negotiations in relation to the arbitral award. During this period, the share value and earnings of Gas Natural Fenosa fell as the scale of this award and the effect on the company were recognised. In June 2011 the parties announced that they had signed a settlement agreement under which Gas Natural Fenosa was to pay to Sonatrach a sum of $1.897 billion to reflect an amended price during the period from January 2007 to May 2011, and which provided for Sonatrach to take a minority equity stake in Gas Natural Fenosa.

2. Changing circumstances – the 'golden age' of natural gas

The difficulties of containing and transporting natural gas meant for many years a limitation of its markets to those closest to the points of production. But the development of ever-longer pipelines, and particularly the growth of LNG, have enabled natural gas to reach ever more distant markets. While pipeline distances are inevitably limited, this is not the case with LNG, which over the last decade has seen its traditionally regional and separated markets move towards a single, more homogenous global market. There has also been a growing separation of the previously connected markets for oil and natural gas, with natural gas prices in the United States and Europe coming to reflect the value of natural gas in its own right. Perhaps more significantly, the growing development of unconventional gas (in the form of shale gas and coal seam gas and often close to major consumer markets) has also led to a concentration on natural gas in its own right.

The natural gas reserves of the former Soviet Union and the Middle East have long constituted the world's major reserves, but remain comparatively distant from the traditional consumer markets of the United States, Europe, Japan and Korea and the developing markets of China and India. But this balance has been changed by the recent and rapid development of the US shale gas business. While a number of developments are at early stages, it seems that these unconventional gas reserves are very considerable and have the potential to move the projected supply and demand balance in the United States from one of recent shortage to one of abundance for decades to come. Also, Australia is seeing a considerable enhancement of its LNG production capability, with an increase in its LNG production from the traditional natural gas reserves of Western Australia as well as the development of new LNG production facilities in Queensland. These represent the commercialisation of coal seam gas reserves with intended sales towards, primarily, North Asian markets. These effects on supply and demand in the world's LNG sector are combining to create a more global market in natural gas, as well as a connection in the prices of LNG among the previously segregated markets of the Americas, Europe and North Asia.

Although the techniques and operations of North America's shale gas business and Australia's coal seam gas business are much influenced by the local legal and political regimes, the developments which are taking place there are likely to have application more broadly. Already, other jurisdictions – particularly those in Europe – are looking to move towards shale gas production, and similar moves are seen to hold great promise for China and India, both of which have traditionally been import dependent. There are legal and regulatory as well as operational and technical challenges in these onshore developments, but these seem less onerous than those in offshore environments, where the challenges of deep-water exploration and production are growing. The accidents at the Macondo field in the Gulf of Mexico and the Montara field offshore Western Australia have led to a re-evaluation by many states of the terms on which drilling (and particularly deep-water drilling) is permitted.

The oil and gas business has long had a close interrelationship with politics and geopolitics, and times of change have meant uncertainty for the oil and gas business. This has been the case over recent years with increasing state participation in the petroleum sector, whether at the time investments are made or over time, with moves towards resource nationalism in many jurisdictions. In some cases, these steps have taken the form of direct interference or, in the case of a number of Organisation for Economic Cooperation and Development states such as the United Kingdom, Australia and Israel, the retrospective changing of tax rates and fiscal terms. These steps may also lead to distortions in markets, with a notable example being seen in the Middle East, where a combination of domestic subsidised energy prices and a prioritising of export resources is contributing to a need to import LNG into a region of plenty. For different reasons, both Indonesia and Malaysia will be exporters and importers of LNG in the near future.

Oil and gas are traditional fuels for the purposes of power generation and natural gas has been taking a leading role in this sector over recent years. To some extent, this may be a necessary response to an inability to commercialise or use associated natural gas otherwise, but it is also the case that natural gas is an environmentally cleaner fuel and, with pressures in all jurisdictions towards the reduction of emissions and the management of carbon production, natural gas has been making a role for itself as a preferred fuel for power generation. This move towards increasing gas-fired power generation had been matched by a move towards more nuclear power generation, in states with existing nuclear industries as well as newcomers. But this trend is in the throes of reassessment. The recent reduction in nuclear capacity in Japan has led to increased local supplies of LNG in the short term and a wider mood of questioning nuclear developments pending greater certainty over matters of safety and sustainability in the longer term. Even if there is little reason to question the place of nuclear generation in the world's energy supply over time, it does seem that developments are likely to slow in the near term, with natural gas being the most likely (and generally abundant) replacement fuel.

These rapid and profound changes in the markets (and the supply and demand balance) of global natural gas are presenting volatile circumstances for international oil companies, national oil companies, governments and consumers around the world. They are also putting unprecedented strains on the long-term contracts – typically made under English law (or sometimes other common laws) – which underpin the oil and gas sector and which presume to anticipate changes of circumstances over periods of very many years.

3. Effects on existing long-term contracts

In the context of pricing provisions under LNG SPAs, this chapter contemplates how readily long-term common law contract terms can respond flexibly to changing circumstances over time. Can contracts which reflected the parties' intentions at the outset adjust to the changed circumstances, and will that contractual relationship evolve over time or will the changes be such as to bring the contractual relationship to an end? Expressed in more legal terms, can the changes of circumstances be reflected in changed contractual terms while maintaining sufficient certainty to ensure enforceable rights and obligations in accordance with the original intention of the parties as expressed in the words of their bargain?

The petroleum sector is inevitably an area of political and regulatory concentration, whether locally or on a regional scale. Politicians and regulators are not noted for their recognition of the terms of long-term contracts and the traumas of the 'take-or-pay' wars in the United States in the 1980s and the restructuring of long-term contracts in the United Kingdom in the 1990s bear witness to the effects of regulatory change on long-term gas contracts. More recently, the European Union has taken assertive steps against long-term gas contracts perceived as having an adverse effect on trade between member states. The application of the EU competition rules has resulted in forced changes to long-term gas contract terms concerning destination clauses, joint selling and marketing, and pricing. The delivery of natural gas into European markets has been facilitated over many years by long-term contractual arrangements which have become subject to change by reason of EU initiatives. In many cases these forced changes are seen to reduce the value of the contracts for the (usually) state-owned or state-influenced producers. As the Algerian minister of energy has said: "Those that have an impact on the market, that is the European institutions, should be aware of our issues. When they passed their legislation, they never consulted us. They have never thought of talking to the gas-exporting countries before passing their laws."

4. Sanctity of contract and changing circumstances

This consideration of price review clauses contrasts two established principles of law. On the one side is the sanctity of contract (*pacta sunt servanda*) – the principle that contracts are to be performed in the context of the rule of law. On the other is the principle of contracts adjusting to changing circumstances (*rebus sic stantibus*). Both principles are well recognised in civil law systems and in international law, while common law regimes tend to have a lesser regard for the principle of *rebus sic stantibus*.

The principle of *pacta sunt servanda* reflects the parties' freedom to enter into contractual relations on such terms as they may agree, but that having done so, they are then bound by the terms of their contract. A party is in a position to recognise the risks and liabilities inherent in entering into and fulfilling its contractual obligations over time and also to provide for these in the terms of its agreements. The principle is embedded in the customary civil law rule that the contract is the law of the parties.

It is for the parties to a contract to provide within that contract for the regulation of their bargain over time and to maintain their financial or economic equilibrium, despite changing circumstances during the life of their contract. Long-term contracts such as LNG SPAs are particularly prone to the risk of economic, regulatory or other change. To the extent that this change is foreseeable and not exceptional, the parties may be expected to take account of this internally and within the terms of their agreement. But if such changes are unforeseeable and exceptional, and arise from external circumstances surrounding the contract, then how can the breach or non-performance on the part of one of the parties be addressed and potentially relieved? The principle of *rebus sic stantibus* seeks to recognise that a contract remains binding on the parties provided that things remain as they are. According to *Treitel and Guanthar,* this principle contemplates that "contractual obligations may be discharged by supervening events, where these events have brought about a change of circumstances so significant as to destroy a basic assumption which the parties had made when they entered into the contract".

The potential inconsistency of those two principles is exacerbated in long-term and international arrangements such as LNG SPAs, where social, political, legislative, economic and cultural changes will often result in changes to the contractual equilibrium on the basis of which the parties contracted.

Even in the context of civil law regimes and international law, these two principles do not have equal standing. The prevailing norm is *pacta sunt servanda,* with *rebus sic stantibus* having potential effect to relieve adverse economic effects or excuse breach or non-performance in the particular circumstances. The doctrine of *rebus sic stantibus* is seen to be of narrow and strict interpretation as an exception to the principle of sanctity of contracts. The principle is seen to be an excuse for non-performance and its application will ordinarily be limited to cases where compelling reasons justify it.

5. LNG pricing

Over time, the LNG sector has seen a number of approaches to setting an appropriate price for LNG deliveries, including fixed pricing, linking to inflation and, more usually, linking to competing fuels in the buyer's market. The buyer's markets have usually been national or, at most, regional and the most usually chosen competing fuel has been crude oil. In most markets beyond the United States, during the early stages of market development, LNG imports have been the subject of long-term purchase agreements made by national or regional monopoly gas buyers with (often distant) producers. Those buyers have tended not to face competition from other gas sellers in their own markets, but have demand-based concerns – particularly in relation to competition from alternative fuels, principally oil products. Those producers have tended to be concerned with supply-based matters and particularly with costs of production over time.

The principles of pricing LNG have developed differently in different markets. For many years the United States has had a comparatively liberal market for natural gas, where importers and suppliers compete and transparent prices are established through trading exchanges. Accordingly, a reliable 'market price' has emerged. The last decade or so has seen the development of some regional, transparent trading markets in Europe, particularly in the United Kingdom. Elsewhere, Europe has seen local gas supplies, as well as imports by pipeline and as LNG, and with prices being established by reference to the prices of competing fuels (particularly crude oil). The North Asian market (primarily Japan, Korea and China) has not yet seen the development of transparent market prices; instead, LNG has tended to be priced by reference to movements of prices of crude oil, as imported into Japan.

6. Price adjustment and price review

English law is among those legal regimes which are reluctant to modify contract terms as a result of changed circumstances. Parties to long-term contracts written under English law tend to create contractual structures within the terms of those contracts which regulate the relationship of the parties and are intended to maintain the financial equilibrium of the bargain forming the basis of that relationship in the face of changing circumstances over time. LNG SPAs are particularly prone to the consequences of economic change and contractual protection mechanisms tend to fall into two categories.

The first concerns 'price adjustment' clauses which reflect the consequences of foreseeable economic or other change over the (usually very long) contract life. These clauses will ordinarily be detailed and exhaustive, providing for all foreseeable circumstances. Typically, the two components which are addressed within a price adjustment clause within an LNG SPA are:

- the effect of future inflation or deflation on the capital and operating costs of the required production and delivery facilities over the contract life; and
- the relative position of the natural gas which is the subject of the contract with other alternative or competing fuels in the buyer's market over the period of the LNG SPA.

While necessarily different in their terms, many LNG SPAs will include a price adjustment clause reflecting these broad principles.

In certain cases the buyer and the seller will agree that the price adjustment clause is to be supplemented by a further provision, a 'price review' clause. There are several different types of price review clause, but all are different in substance from price adjustment clauses. The price review clause is not intended to address foreseeable economic change, but to have application in circumstances of extraordinary and unforeseeable change. Also, in contrast to a price adjustment clause, the price review clause will apply in the event of changes in the circumstances surrounding the contract and where the resulting economic disruption goes beyond the scope of the price adjustment mechanism. Given the nature of a price review clause, it will tend to be drafted in generality and provide little detailed guidance for what are the (necessarily) unforeseen circumstances of its application. Whereas a price adjustment clause will have regular and frequent application (perhaps quarterly or annually), a price review clause will have a necessarily less frequent and regular application and often by reference to certain events and at the instigation of one of the parties. The nature and operation of those provisions and their potential relationship is perhaps best shown by the review (in Section 13) of *Esso v ESB (Esso Exploration & Production UK Ltd v Electricity Supply Board* (2004) EWHC 723 (Comm), the *Hewett* hardship case (*Superior Overseas Development Corporation v British Gas Corporation* (1982) 1 Lloyd's Rep 262) and the *Atlantic LNG* case (*Gas Natural Aprovisionamientos v Atlantic LNG Company of Trinidad and Tobago* (2008) WL 4344525 (SDNY)).

While not the subject of this chapter, it is also worth touching on a provision which is not unknown in long-term gas contracts and which can be seen to sit between price adjustment and price review: the most favoured nation clause. These provisions are a product of comparatively fixed contractual arrangements and closed markets. They may apply to the benefit of the seller or the buyer. They represent a concern on the part of the seller that it is selling at a price no lower than that paid by the buyer to competitors of the seller and on the part of the buyer that it is buying at a price no higher than that paid to the seller by competitors of the buyer. In principle, the buyer is seeking to avoid the economic disadvantage that would arise if the seller were to supply its competitors in its markets at lower prices. The most favoured nation clause would provide that, in these circumstances, the seller would be obliged to offer that lower price to the buyer. The seller will have an equal and opposite concern and the most favoured nation clause will be expressed to apply in the same way, but to the benefit of the seller.

The practical application of these principles can cause considerable difficulty. In addition to the likely lack of information in relation to the terms of other agreements, the intended creation of comparable terms in diverse circumstances means that matters such as quantities, delivery points, quality and pricing provisions cannot easily be reconciled. It is also important to bear in mind that the operation of price adjustment provisions and price review provisions may have unintended consequences in the case of most favoured nation clauses.

While the main concentration of this article is price review provisions, it is nevertheless instructive to spend a few moments looking at price adjustment provisions, both in their own right and, particularly, to put in context our consideration of price review provisions.

6.1 Price adjustment

In circumstances where a fixed price is as inappropriate as a market price, the buyer and seller will ordinarily look to establish a method of pricing under an LNG SPA which will be effective to address the supply-based concerns of the seller in maintaining over time a price which enables the seller to cover its costs and make an acceptable return on its investment, and to address the demand-based concerns of the buyer to be in a position over time to market the gas economically in competition with competing sources of energy in its markets.

These respective aims are likely to be both complementary and conflicting over time, and the buyer and seller will often seek to manage potential issues from the outset by setting a base price (acceptable to both the buyer and the seller at the outset), and then providing for adjustment of that base price during the contract life by reference to movements in the values of agreed indicators. Likely indicators will include competing fuels such as crude oil, gas oil and fuel oil, with perhaps references to electricity prices or coal prices. Also, the agreed pricing formula may provide for fixed elements (not subject to adjustment) or references to cost-related matters such as general inflation or labour inflation. The parties will agree the frequency of price review (perhaps quarterly or annual), and these reviews will be expected to operate throughout the contract period. This operation will be comparatively automatic, although the LNG SPA will usually provide for revision of these provisions in circumstances of unavailability, change of basis or discontinuance of publication of the agreed values of indicators or more generally, on the occurrence of circumstances where the purpose of the adjustment mechanism ceases to be fulfilled over time.

Most cases of dispute or disagreement in relation to these matters will be seen as matters of expertise rather than judicial consideration, and reference to expert determination (in accordance with the agreed terms of the LNG SPA) will be provided for.

Price adjustment provisions will vary considerably in their terms from agreement to agreement, but the principle of price adjustment will be very common. In essence, these price adjustment provisions will reflect the parties' views of foreseeable changes of economic and other circumstances over the (long) term of their agreement. The seller and the buyer will negotiate complex and lengthy provisions which are seen to reflect their present views of the future events or circumstances which are to be taken into account in the periodic adjustment of the contract price of LNG over the contract life. In this way, the parties are creating an agreed basis of internal regulation of contract pricing within the terms of their agreement. Their expectation is that they will have foreseen and provided for all events (whether internal to their agreement or external) which may be relevant from time to time to the circumstances surrounding their contract and its economic balance over time. They will have acknowledged that no matter what circumstances may lie ahead over the contract's life, their agreement will be read and construed in accordance with the terms they have negotiated and written in their contract, and that (save for the very limited circumstances we will discuss shortly), there will be no prospect of any variation or modification of their bargain, save as may be set out in the specific terms of their agreement or as may be agreed by them in the circumstances prevailing from time to time.

Under the terms of many LNG SPAs, the base price (or 'P0') for the purposes of price adjustment is a number agreed by the parties to represent an agreed price as at an agreed point of time. That number may be expressed to relate to a particular market and purpose, as was the case in *Esso v ESB*. Equally, that price may not be referable to any particular market or purpose, but will instead be a negotiated number at a date and subject to an adjustment mechanism. The negotiation of this number and the adjustment mechanism will reflect the comparative bargaining power of the parties and is likely to take account of the seller's concerns (cost plus) and those of the buyer (competing fuels and alternative markets). The parties' intention is often to maintain the initial price as competitive by reference to inflation measures and market movements, and whether those market movements arise through changes in regulation or otherwise. While it might be said that only local or regional markets were foreseeable at the time that some LNG SPAs were concluded, that ceased to be the case some time ago. The *Atlantic LNG* case is instructive here. It is almost inevitable that the contract price will diverge over time from alternative sources of supply for the buyer and alternative markets for supply for the seller. The parties may console themselves that over a (long) period of time the contract price will tend to fluctuate and be above the market price or below the market price, and that on the basis that these fluctuations will broadly even themselves out, the contract will be in the money for some period and out of the money for another. However, the example of the 'take-or-pay' wars in the United States in the 1980s and the restructuring of the long-term contracts in the United Kingdom in the 1990s show that this can be otherwise.

In relation to some LNG SPAs, the parties are comfortable that such a price adjustment provision is sufficiently appropriate and flexible to address the changes of circumstances that may come to apply to their bargain over the contract life. The period of negotiation of an LNG SPA is itself often lengthy and the parties may take the view that they have considered all foreseeable events or circumstances that may lie ahead, and that their modelling and assessment of the nature and expected operation of their price adjustment provisions provide a basis for confidence that their commercial aims will be fulfilled over the contract life.

On the other hand, the seller and the buyer may recognise that, notwithstanding their best efforts, they lack the clairvoyance necessary to foresee the next 25 years or so. In these circumstances, those lengthy negotiations will become lengthier still as they turn their minds towards the main focus of this chapter – price reopener provisions under LNG SPAs made under the terms of English law.

6.2 Price review

In these cases the parties will look to incorporate in their agreement additional contractual provisions which provide a basis for reviewing (and potentially changing) the price provisions of their LNG SPA in the (unknown or unforeseeable) circumstances and extraordinary events which may come to prevail. In contrast to the price adjustment provisions set out in the LNG SPA, such a price review clause will often operate only on the happening of unexpected, unforeseen or extraordinary events and typically at the request of one of the parties. These contractual clauses have tended to be the product of markets and circumstances which are opaque or certainly lacking in transparency, and will be cloaked in confidentiality provisions. The price review clause will be necessarily general and lacking in precision of expression. It seeks to address matters which are external to the parties' contractual relationship and which occur in relation to the circumstances surrounding that contractual relationship. And while the parties' agreement in relation to uncertain contractual modification may well be legally enforceable, they may not readily recognise that their price review provisions will have the potential to modify their agreement beyond (and potentially in spite of) the wishes of one or both of them. The parties' ability to invoke these price review provisions is likely to be contractually limited in time, in circumstance and in effect, but these limitations are rarely sufficient to manage this risk. For many lawyers trained and practising in the laws of common law jurisdictions, these are provisions with few parallels.

This chapter draws on those contracts which have come into the public domain, typically by reason of public disputes in relation to their interpretation and effect, and those which can safely be anonymised.

While there is no model form of price review provision, periodic application is a common characteristic, with the parties being entitled to serve notice calling for a review every so many years with provision for 'wildcard' reviews enabling a party to call for a review outside the usual cycle and in extraordinary circumstances. Ordinarily, a 'trigger' will need to be fulfilled for the application of the price review mechanism. This may include a change of law, a change of markets or the occurrence of economic disadvantage for one of the parties.

This trigger is likely to be of a quantitative or qualitative type.

In the case of a quantitative trigger, the clause's application will depend on a measurable or mechanistic change such as a divergence of the contract price from the market price by more than a set percentage. *Esso v ESB* is an example here.

More usually, the trigger is of a qualitative type and will have reference to matters such as changes of law or markets or economic circumstances, or economic disadvantage of one of the parties. This provision will be qualified by imprecise adjectives such as material or substantial or significant. The *Atlantic LNG* case is an example of this type of clause.

The relevant trigger will often qualify the events or circumstances of its application by reference to their having arisen beyond the control of either party, their having a fundamental effect on the terms or nature of the parties' bargain and their unforeseeable or extraordinary nature.

Following the service of notice and the fulfilment of the appropriate trigger, the provision will usually have the consequence of requiring that the parties meet and discuss and seek to agree on the appropriate revision of the LNG SPA in the context of the criteria set out in the LNG SPA. In some circumstances, the provision may provide for nothing more than meeting and discussing, with no prescribed effective remedy if agreement cannot be reached. In those circumstances, the agreement will carry on unamended. Conversely, the provision may provide for termination in the event that the parties cannot agree. More usually, though, the absence of agreement will result in the right for either party to have the matter referred to a third party (often an arbitral panel, an expert or a panel of experts) for decision and consequential rewriting of the price provisions of the agreement. In this case, questions of prospective or retrospective application will need to be addressed.

While each price review clause and the relevant criteria will differ in their terms, there tend to be two philosophical approaches: the evolutionary and the revolutionary. As can be seen from the *Hewett* hardship case, a price review clause may contain a mixture of both.

An evolutionary approach will attach considerable weight to the parties' original bargain and recognise that the parties are to be returned to the economic balance as at the beginning of the contract; this is to be done by amending the terms of the contract in order to achieve that original economic balance, but in the new, changed circumstances. Such an evolutionary approach can be seen as maintaining over time the original economic bargain, taking account of changing circumstances. It concentrates on the nature, effect and economic value of the change that has occurred.

Under the revolutionary approach, the agreement will provide for the contract to be revised with little or no regard to the parties' original bargain. Instead, the contract is to be revised in order to make it consistent with the new circumstances. Again, the usual intention is that the provisions of the LNG SPA are to be changed and the parties will remain in contractual relations. But the terms of that contractual relationship are not those which were agreed by them at the outset, but rather those which are subsequently agreed by them or, in the absence of agreement, determined by a third party. This approach refers not to the change, but to the nature and effect of the prevailing market and circumstances, probably regardless of the original market and contractual balance.

Whichever formulation is preferred, the adoption of a price review clause provides the potential for an LNG SPA changing substantially and repeatedly over its long term. And those contractual revisions will be referable not only to the time of their making, but also to the (probably) many years to follow under the LNG SPA.

Within the broad catchment of price review clauses, there are probably two main families: hardship clauses and price reopeners. While these provisions have many common elements, they also have sufficient differences to justify their segregation for the purposes of this chapter.

The typical hardship provision can seem to have similarities with the French law principle of *imprévision* and will be expressed to apply in the event that unforeseen circumstances lead to one of the parties suffering disadvantage or hardship when compared with the economic balance envisaged at the time the contract was made. The trigger for the application of a hardship clause will ordinarily be expressed as a qualitative or quantitative test, referable to a measureable hardship expressed in terms of, say, a required rate of return or to a measure such as 'substantial economic hardship'. Turning, then, to the substantive application of this clause, this will again tend to be expressed in evolutionary or revolutionary terms. Necessarily, the hardship clause will be irregular in its application and it would not be surprising if, despite the long term of a typical gas supply agreement, a hardship clause never came to apply. The *Hewett* hardship case is a good example of the terms of a hardship clause and particularly its nature and effect.

A price reopener will provide for periodic application (perhaps by reference to specific events) and typically will call for a review of the contract pricing provisions in the context of the prevailing market or economic conditions. This review will not be cast in the context of hardship, although one of the parties is most likely to be in a position of economic benefit and the other in a position of economic disadvantage. Similar issues will arise in relation to the nature of the trigger and the substantive operation of this clause. The *Atlantic LNG* case is a good example of the nature and operation of a price reopener.

The following examples of hardship clauses provide some detail on the typical wording of these provisions:

- *Should the occurrence of events not contemplated by the parties fundamentally alter the equilibrium of the present contract, thereby placing an excessive burden on one of the parties in the performance of its contractual obligations, that party may make a request for revision within a reasonable time from the moment it becomes aware of the event and of its effect on the economy of the contract.*
- *If during the term of this agreement, without default of the party concerned, those in the occurrence of an intervening event or change of circumstance beyond the said party's control when acting as a reasonable and prudent operator such that the consequences and effects of which are fundamentally different from what was contemplated by the parties at the time of entering into this agreement which consequences and effects place that party in the situation that then and for the foreseeable future all annual costs associated with or related to the gas which is the subject of this agreement exceed the annual proceeds derived from the sale of such gas.*

Turning to price reopeners, the following examples show some of their typical characteristics:

- *If either party reasonably considers that:*
  - *(i) economic circumstances in the energy market in the [relevant state] generally have substantially changed when compared with those which are reflected in the prevailing price provisions of this Agreement; and*
  - *(ii) such change will continue to have effect for the foreseeable future; and*
  - *(iii) such change has occurred for reasons beyond the control of the affected party acting and having acted as a prudent and efficient gas company;*

  then the parties shall enter into good faith negotiations to seek agreement on a fair and equitable revision of the prevailing price provisions of this Agreement.
- *Either party may require a review of the pricing provisions if it has a good faith basis for believing that, for reasons beyond its control, the method of calculating the contract price does not reflect the market and market practice for the pricing of natural gas for delivery into the Buyer's market at the time such review is requested.*
- *Either party may request a review of the method of determining the Contract Price if it has a good faith basis for believing that, for reasons outside its control, such*

method does not reflect the value of regasified LNG at import points into the European system at the time of such request and such value shall take into account published market prices for natural gas in liquid markets, published access conditions in line with industry-standard costs and industry-standard downstream marketing margins.

These examples suggest a number of common themes, such as changed circumstances, unforeseeability, happenings beyond the control of the affected party, good faith and sometimes specific guidance on matters to be taken into account. These may include matters such as new taxes being deemed to constitute a substantial change, guidance on the nature of comparable contracts, the definition of relevant markets and also greater definition of what is a prudent and efficient gas company and the steps that the buyer will be expected to take in the relevant circumstances.

While this chapter seeks to identify the typical characteristics of price review provisions, and to categorise these and contrast them with price adjustment provisions, reality is rarely so ordered. The pragmatism and necessary compromise of negotiations are likely to lead to a blurring of these principles. An imbalance of negotiating positions and a disparity of market circumstances are sometimes sufficient to mean that the resulting provisions of individual LNG SPAs will pay little regard to this analysis.

In the context of a common law regime such as English law, provisions of this type (and the incorporation of notions such as good faith and principles such as agreeing to agree or agreeing to negotiate) will test the parties' abilities to agree and may have the effect of encouraging the parties towards disputes.

(a) Principles of price review

Before looking at some of the legal considerations and challenges in relation to these provisions, it is timely to consider the nature and scale of the commercial significance of a typical price reopener clause. An LNG SPA will be written to apply for a long period (perhaps something in the order of 25 years), and the value of the contract over its life will often be in the order of several billion dollars. Prior to entering into contractual commitments of this magnitude, the parties will have carried out considerable research and analysis and will have committed considerable time to the detailed writing and expression of their obligations, liabilities and rights from the effective date of the LNG SPA, and then over the contract life. The parties will have paid particular attention to the price to be payable for the specified quantities of LNG and the adjustment mechanism by virtue of which the contract price will come to be revised over time. The management of the respective companies will have recognised the extent of the governance duties bearing on them in obtaining the optimum and most appropriate pricing provisions over time.

But will they have recognised the significance of the presence of a price reopener provision within the LNG SPA, and that it may well mean that, within a short period of time, the carefully refined and expressed provisions may be replaced with alternative wording to a different effect and imposed by third parties, unknown and unidentified at the time of making the LNG SPA? In addition, that third party will not have had a commercial interest in the making of the LNG SPA; nor will it have a commercial interest in the continuing operation over what may be the decades to come. The *Atlantic LNG* case and the *Hewett* hardship case serve to show that these comments are not farfetched.

Once the price review clause has come to have effect (and typically triggered by one of the parties), both parties may find themselves trapped in undesirable price provisions until the end of the contract term or certainly until the next date on which the price review can be implemented. And despite the best intentions of the parties, *Esso v ESB* shows how their intentions can be defeated, and within a short period after entering into a long-term agreement. That case also shows the importance of assessing and reflecting the circumstances of the day and the risks of writing agreements for yesterday's circumstances. Similarly, the anonymised case reports released from time to time by the ICC show examples where parties were aware of the risk of liberalisation at the time of writing those agreements, but nevertheless sought to rely on the effects of liberalisation as a ground for implementing a price reopener.

If the parties wish to maintain relations in the context of unfavourable circumstances, the typical common law requirement of certainty and enforcement of the bargain made will have to be addressed. Several reasons militate in favour of the parties' referral to external (and uncontrollable) indicators or people. While price adjustment is a matter of expertise, price reopeners will ordinarily result in a resetting of the contract terms. This will be a composite matter of judicial consideration and the consequential rewriting of contractual agreements, and often contractual agreements of some complexity. This is not a task which the courts will readily perform. Ordinarily, this is a matter for a panel of arbitrators or an expert appointed in the role of an intervener. While it might be expected that one of the attributes of those third parties would be an expertise in the negotiation and writing of LNG SPAs, a review of the examples to date shows that this rarely seems to be the case. Whereas these third parties will often be in a position to call on expert advice in this respect, that may not necessarily be the ideal balance where this is the core activity for which the third parties are appointed.

For many years, the existence of price reopener clauses led to periodic, agreed resetting of contractual terms in the circumstances of comparatively benign market movements. The economic benefit would be reapportioned from time to time by agreement against the background of potential reference to a third party. But regulatory and market changes over recent years have led to a greater divergence of interest among the buyer and the seller under LNG SPAs and an inability to agree on this resetting. A wish to escape from onerous contractual commitments has led to the parties more readily implementing references to a third party. Where these price review provisions once served a purpose *in terrorem* ("we know this may not be a good deal, but at least it is our own deal"), they have come to be seen to represent a risk worth taking.

The extent of the challenge of capturing a representative test is readily apparent. The relevant contracts will reflect the circumstances at the time they are made, including their long-term nature and the comparatively opaque nature of the relevant market at that time. In circumstances where markets develop and liberalised structures come to have effect, there will be a move to a reporting of prices – perhaps as journalism at the outset, but increasingly with transparency of market prices. In this case, as is discussed later in this chapter, long-term agreements can outlive their circumstances, but continue to apply.

A price review clause is likely to be construed as imposing an obligation on the seller and the buyer to negotiate or renegotiate the price provisions under their LNG SPA. Although this type of provision may be seen to constitute an agreement to negotiate, it will not, under English law, be seen as a duty on either party to reach agreement.

7. Relief and contract modification under English law

The traditional balance of contractual risk under an LNG SPA has often been described in the simple terms that it is for the seller to take and manage the quantity risk and for the buyer to take and manage the price risk. But the nature and value of most LNG SPAs mean that the reality is a good deal more complex than this. Recent changes of circumstances have had an effect on the regulation, pricing and markets in respect of natural gas in many jurisdictions, and LNG SPAs made subject to English law (and with varying terms of FOB, cost, insurance and freight and ex-ship deliveries) have come under increasing stress. Often these tensions have arisen from one party being bound to the terms, and particularly the prices, of an LNG SPA at times when the corresponding terms in the prevailing markets would be more advantageous to it. In these circumstances, the disadvantaged party will typically seek to modify, suspend or terminate those contractual provisions in order to gain relief from its (almost inevitably temporary) disadvantage.

7.1 Governing law

In some case, the governing law of the contract may itself provide such relief by virtue of statute or inherent jurisdiction, even in the absence of express contractual terms. This is the case, for example, in Germany and Spain. While French law is not seen to be a member of this family of jurisdictions, it is principles of French law which have had probably the greatest influence on the nature and intended effect of price reopener provisions under LNG SPAs. French law recognises the principle of *pacta sunt servanda* and resists the judicial adaption of contractual obligations, but does acknowledge two principles which have come to have quite an influence on the parties' negotiations of their LNG SPA and particularly on any price reopener provision:

- *Force majeure* is seen to be an irresistible and unforeseeable event which renders the performance of a contract impossible and provides for suspension of certain obligations or the relief from liabilities that would otherwise arise from a breach of those obligations, but not the modification of the obligations themselves.
- *Imprévison* (or hardship) describes unforeseeable circumstances which affect the economic conditions of the contract and create exceptional changes to

> the contractual balance of the parties which may result in the modification of the obligations, rights and liabilities set out in the contract, to the effect of alleviating the disadvantage of one of the parties or reinstating the original economic balance, but in the changed circumstances.

Another example is New York law. This is more relevant for our purposes, as although it is now rare for new LNG SPAs to be written under the laws of New York, a good number of earlier LNG SPAs written under New York law remain in effect. Again, there is no inherent judicial power to adapt contractual obligations and the rule of *pacta sunt servanda* applies, but this is subject to the potential application of the principle of commercial impracticability. This principle is derived from codified commercial law and is an excuse of performance where the performance of an obligation has become unreasonably difficult or expensive. This doctrine applies in circumstances where performance is still physically possible (and therefore beyond the principle of frustration or impossibility), but is seen as unreasonably burdensome for the party obliged to perform. Impossibility is seen to be an objective measure, whereas impracticability is a subjective measure and one for the court to decide. The doctrine is available to a party in cases where an event has occurred and its non-occurrence is considered a basic assumption of the parties under their agreement, that occurrence renders performance unreasonably expensive or burdensome, and that expense or burden was not anticipated by the parties at the time of making their contract.

The inherent power to adapt contracts in the event of uncontemplated and fundamental change is sometimes seen as a consequence of the duty in many civil law regimes for parties to perform (or in some cases negotiate) their contractual obligations in good faith. In circumstances where it is incumbent on a party to perform its obligations in good faith, it would be inequitable if it did not have a corresponding duty to renegotiate the terms of the contract in circumstances where the other party is suffering from economic hardship for unforeseeable reasons beyond its contract and in extraordinary circumstances. An insistence on strict performance in these circumstances would be contrary to a duty of good faith.

(a) English law

The governing law of most LNG SPAs is now English law; so what is the approach under English law in these circumstances?

Under English law, there is no inherent power for the courts to adapt the terms of a contract to provide for fundamental changes not contemplated by the parties. However, in limited circumstances the courts will be prepared to imply provisions which the parties have not expressed in their agreement. In *Reigate v Union Manufacturing Co (Ramsbottom) Limited* (1918) 1KB592 it was said: "Prima facie that which in any contract is left to be implied and need not be expressed is something so obvious that it goes without saying; so that, if while the parties were making their bargain, an officious bystander were to suggest some express provision for it in the agreement, they would testily suppress him with a common, 'oh, of course'."

A term may be implied where a contract is recognised to be incomplete, but the threshold is a high one and concerns necessity for the contract to function and not mere reasonableness, improvement or convenience. The term will also need to be capable of expression in clear and precise words, although in *Shell UK Ltd v Lostock Garage Ltd* 1976 1 WLR 1187 it was recognised that: "It is no novelty in the common law to find that a criterion on which some important question of liability is to depend can only be defined in imprecise terms which leave a difficult question for decision as to how the criterion applies to the facts of a particular case."

In any event, it is clear that a term will not be capable of implication if it would be inconsistent with the express words of the contract.

In essence, English law will follow the principle of *pacta sunt servanda* and will look to enforce the bargain made by the parties in accordance with the terms of their contract. This chapter subsequently looks at how the terms of their contract might provide for relief from the consequences of breach of a party's obligations or, in certain circumstances, modification of those obligations; but first it considers those principles that are capable of having effect by operation of law and regardless of corresponding contractual provisions.

English law has no written code of general application and does not recognise general principles providing for relief or modification by reason of matters such as economic hardship, *force majeure* or commercial impracticability, unless provided for expressly. While English law recognises the principle of good faith, it sees no place for this principle in the making, performing or enforcement of contracts. English law will look to hold the parties to the terms of their contract in accordance with the principle of *pacta sunt servanda*. This has long been the case. In *Paradine v Jane* 82 Eng Rep 897 (KB 1647), a tenant was held to pay the required rent under a lease in respect of a long period during which he was prevented by military force from occupying the leased estate. It was said: "When the party by his own contract creates a duty upon himself, he is bound to make it good, if he may, notwithstanding any accident by inevitable necessity, because he might have provided against it by his contract."

It was 1863 before Justice Blackburn, in *Taylor v Caldwell* 122 Eng Rep 309 (KB 1863), differentiated the performance of a contract becoming more difficult or impossible and said: "Where there is a positive contract to do a thing not in itself unlawful, the contractor must do it or pay damages for not doing it, although in consequence of unforeseen accidents the performance of his contract has become unexpectedly burdensome or even impossible."

7.2 Frustration

In early steps towards a doctrine of frustration, the court went on to identify a principle that in contracts where performance depends on the continued existence of current circumstances, a condition is implied that the impossibility of performance arising from a change of those circumstances will excuse non-performance. This principle was revisited in the early 1900s as a result of the so-called '*Coronation* cases'. In these cases, apartments were rented in advance for the purpose of watching the coronation procession, although there was no provision to that effect in the contracts. Although the apartments were available and suitable for occupation, the king's ill health resulted in the cancellation of the procession. While this cancellation did not render the intended occupancy impossible, it did frustrate the purpose of that occupancy. Nevertheless, the tenants were bound to make the required rental payments and were not relieved from their obligations.

By 1956, Lord Radcliffe in *Davis Contractors Limited v Fareham* UDC [1956] AC 696 came to express the definition of 'frustration' in the following terms: "Frustration occurs when the law recognises that without default of either party a contractual obligation has become incapable of being performed because the circumstances in which the performance is called for would render it a thing radically different from that which was undertaken by the contract... It was not this that I promised to do."

In 1962, the so-called '*Suez Canal* cases' provided examples where ships which had been chartered in the expectation of following a route through the Suez Canal were, by reason of military action, required to follow a much longer route around the Cape of Good Hope. Despite this consequence and the resulting increased expense and time, the court ruled against frustration, save in one case where the contract expressly provided for a route through the Suez Canal. That a contract had lost its economic balance in such circumstances was not seen as a reason for the grant of relief from the agreed obligations as set out in the contract.

While it is the case that English law does recognise a narrow concept of frustration, a finding of frustration does not result in an amendment or adjustment of the existing contract in accordance with those new circumstances, but provides for a discharge of remaining obligations and the contract coming to an immediate end.

(a) More onerous or impossible

The limited nature of the doctrine of frustration and its application arises from the reluctance of English law to intervene in the sanctity of binding contracts (and to be seen to be relieving the consequences of poor commercial bargains), and the presumption that the parties to commercial contracts will make express provision for changes of circumstances and supervening events. More recent cases have reinforced the limited nature of frustration under English law. In *National Carriers Limited v Panalpina (Northern) Limited* [1981] AC675, Lord Simon said:

> *Frustration of a contract takes place when there supervenes an event (without default of either party and for which the contract makes no sufficient provision) which so significantly changes the nature (not merely the expense or onerousness) of the outstanding contractual rights and/or obligations from what the parties could reasonably have contemplated at the time of the execution that it would be unjust to hold them to the literal sense of its stipulations in the new circumstances; in such case the law declares both parties to be discharged from further performance.*

Material loss and hardship are not enough to invoke frustration and something extraordinary must have occurred which, in the words of *Chitty on Contract*, "renders it physically or commercially impossible to fulfil the contract or transforms the obligation to perform into a radically different obligation from that undertaken at the moment of entry into the contract".

In *Thames Valley Power Limited v Total Gas & Power Limited* [2005] EWHC 2208 (Comm), Total had committed to a long-term supply of natural gas to Thames Valley

Power Limited at prices which came to be considerably less than the market price at which Total could acquire that natural gas for supply. In finding against Total's application for relief on the grounds that it had become uneconomic for it to continue to perform its obligations, Justice Clarke said:

> *This conclusion is consistent with a line of cases to the effect that the fact that a contract has become more expensive to perform, even dramatically more expensive, is not a ground to relieve a party on the grounds of force majeure or frustration. I take as an example Tennants Lancashire Limited v Wilson CS & Co. Ltd. (1917) AC 495 where Lord Lorburn observed:*
>
> > *The argument that a man can be excused from performance of his contract when it becomes "commercially impossible" seems to me to be a dangerous contention which ought not to be admitted unless the parties plainly contracted to that effect".*

In these circumstances where English law provides little (if any) relief or potential for contractual modifications, it is for the parties to provide for these matters in the words of their agreement.

8. Interpretation

Under English law, it is for the parties' contract to regulate their relationship and to provide for those events and circumstances which may give rise to a modification or adjustment of those terms, and the basis on which such modification or adjustment is to be carried out. The words used by the parties for this purpose need to be sufficiently certain and clear to result in enforceable rights and obligations in accordance with their terms and the rules for the interpretation of contracts under English law. What, then, are the rules of construction or interpretation which will apply in relation to these provisions under LNG SPAs?

8.1 General rule

The English law rule of interpretation of contracts can be summarised in the five key principles set out by Lord Hoffmann in *Investors Compensation Scheme v West Bromwich Building Society* [1998] 1 WLR 896:

> *I do not think that the fundamental change which has overtaken this branch of the law, ... is always sufficiently appreciated. The result has been, subject to one important exception, to assimilate the way in which such documents are interpreted by judges to the common sense principles by which any serious utterance would be interpreted in ordinary life. Almost all the old intellectual baggage of 'legal' interpretation has been discarded. The principles may be summarised as follows:*
>
> *(a) Interpretation is the ascertainment of the meaning which the document would convey to a reasonable person having all the background knowledge which would reasonably have been available to the parties in the situation in which they were at the time of the contract.*
>
> *(b) The background was famously referred to .... as the 'matrix of fact', but this phrase is, if anything, an understated description of what the background may include. Subject to the requirement that it should have been reasonably available to the parties and to the exception to be mentioned next, it includes absolutely anything which would have affected the way in which the language of the document would*

have been understood by a reasonable man.

(c) *The law excludes from the admissible background the previous negotiations of the parties and their declarations of subjective intent. They are admissible only in an action for rectification. The law makes this distinction for reasons of practical policy and, in this respect only, legal interpretation differs from the way we would interpret utterances in ordinary life. The boundaries of this exception are in some respects unclear. ...*

(d) *The meaning which a document (or any other utterance) would convey to a reasonable man is not the same thing as the meaning of its words. The meaning of words is a matter of dictionaries and grammars; the meaning of the document is what the parties using those words against the relevant background would reasonably have been understood to mean. The background may not merely enable the reasonable man to choose between the possible meanings of words which are ambiguous but even (as occasionally happens in ordinary life) to conclude that the parties must, for whatever reason, have used the wrong words or syntax. ....*

(e) *The 'rule' that words should be given their 'natural and ordinary meaning' reflects the commonsense proposition that we do not easily accept that people have made linguistic mistakes, particularly in formal documents. On the other hand, if one would nevertheless conclude from the background that something must have gone wrong with the language, the law does not require judges to attribute to the parties an intention which they plainly could have had. ...*

More recently, these principles were applied in *Chartbrook Ltd. v Persimmon Homes Ltd* (2009) 1 AC 1101, which restated the principle that if the interpretation of the wording of the contract is in dispute, the law does not require the contract "to attribute to the parties an intention which a reasonable person would not have understood them to have had".

8.2 Objective test

The essence of the rule is that interpretation involves the search not for the actual intentions of the parties, but for an objective meaning. Essentially, the question asked is: what would the document convey to a reasonable man? In *BCCI v Ali* [2001] 2 ULR 735 the dispute concerned whether payment "in full and final settlement of all or any claims of whatsoever nature that exist or may exist against the bank" covered claims that were not in existence or contemplation of the parties at the time. Here the House of Lords held that: "On a fair construction of this document I cannot conclude that the parties intended to provide for the release of rights and the surrender of claims which they could never had had in contemplation at all. If the parties had sought to achieve so extravagant a result they should in my opinion have used language which left no room for doubt".

8.3 Context

A clause is not to be read in isolation, but must be considered in the context of the document as a whole. This includes admissible background which could reasonably have been available to the parties at the time of the contract. This is a wide means of interpretation, as it includes "absolutely anything" that a reasonable man would regard as relevant. Nevertheless, as was stated by Lord Chadwick in *Megaro v Di Popolo Hotels Ltd* [2007] EWCA Civ 309, it still remains the case that: "The court does not, through the guise of interpretation, make for the parties a bargain which they did not themselves choose to make. It is not for the court, through the guise of interpretation to substitute for the bargain which the parties did make a different bargain which in its view they would have made if they had been better advised or had had better regard for their own interests."

Although words should be given their natural and ordinary meaning, those words may take on a different meaning in the light of circumstances prevailing at the time of the contract.

In *Cosmos Holidays Plc v Dhanjal Investments Lt* [2009] EWCA Civ 316, Sir Anthony Clarke said: "the primary source for understanding what the parties meant is their language interpreted in accordance with conventional usage .... Of course, the particular provision must be construed in the context of the clause as a whole, and the clause itself must be construed in accordance with the contract as a whole, which must in turn be considered in its factual matrix or against the circumstances surrounding it."

8.4 Implied terms

If there is no express contractual provision or an intended express provision is considered insufficiently certain, then in the context of the applicable wording and circumstances, is it possible that appropriate provisions may be implied into the agreement? In short, the courts will be reluctant to imply terms into a written agreement to the effect of providing any form of relief under English law. In order to fulfil the requirements for an implied term under English law, it is necessary to show that such an implied term is both necessary for the business efficacy of the contract and so obviously intended by the parties that it goes without saying. The test was subject to a restatement in 2009 where Lord Hoffman (in *Attorney General of Belize v Belize Telecom Limited* 2009 UKPC 10) said that terms would be implied into written agreements only where such provisions would set out in express words what the instrument, read against the relevant background, would reasonably be understood to mean; and that the usual inference is that if the parties had intended something to happen, then the instrument would have said so.

9. The making of an agreement

Most LNG SPAs will be written subject to the laws of a compromise common law jurisdiction – typically nowadays the laws of England and Wales. As a rule, common law jurisdictions will seek to enforce the bargain made in accordance with its terms. Among other things, this choice of law is generally perceived to avoid the risk of a court or other third party rewriting the agreement, unless that is what the parties provide for in their agreement.

The certainty required at common law to achieve contractual relations will ordinarily militate against relational flexibility over time. Where a contract does provide for the parties to revisit their bargain or renegotiate the agreement made, this will often be done in the context of wording requiring the parties to meet, discuss in good faith and seek to agree appropriate contractual revisions. Where a change of circumstances leads to a fundamentally different situation from that envisaged at the time the LNG SPA was made, it is likely that the effect of this change will be commercially advantageous to one party and commercially disadvantageous to the other. This is likely to strain the relationship of the parties and, in the absence of contractual guidance, make it comparatively difficult at that time to agree new arrangements. In light of the recent upheavals in the natural gas markets in different regions and around the globe, the scale of these commercial differences is forcing parties to LNG SPAs to exercise such contractual and extra-contractual rights of challenge as may be available to them in the circumstances. A price reopener is likely to be an early port of call for a party seeking to reset the economic balance of an LNG SPA. What, then, is likely to be made of those clauses which typically provide for meeting and seeking to agree or negotiate revised terms in the context of unforeseen and extraordinary circumstances or events, beneath an over-arching commitment to good faith with references to other agreements and companies, and with ultimate reference to unidentified third parties vested with authority to resolve disputes and, not unusually, to rewrite the parties' bargain?

9.1 Vague or incomplete

Under English law, an agreement will not be a binding contract in circumstances where it is so vague or incomplete as to lack the necessary certainty for binding contractual relations. A typical price reopener provision is seeking to provide for unforeseeable events in extraordinary circumstances and its terms will often come to be tested in cases where one of the parties considers itself to be suffering financial or economic hardship. It is unsurprising that the interests of the parties are likely to be polarised at the time of the provision's application. In those circumstances, what approach will be taken under English law?

In *Total Gas Marketing Ltd v Arco British Ltd* (1998, *Times*, June 8) Total sought to terminate a long-term gas purchase contract on the grounds that a related allocation agreement was entered into a few days later than provided for. In finding in favour of Total, Lord Slynn said: "I confess that I have reached this conclusion with regret. It is no secret that the reason why the buyer wishes to terminate the agreement is that the market has now turned in its favour. It can obtain gas elsewhere more cheaply than it would have been required to take the gas from the Trent reservoir under the agreement. The buyer is not to be criticised if the wording of the agreement permits this course. But the court should be slow to lend its assistance. Commercial contracts should so far as possible be upheld."

But the requirements of certainty in relation to the expression of the common intention of the parties do not require that the price be specified at the outset for the contract period. In *Mamidoil-Jetoil Greek Petroleum Co SA v Okta Crude Oil Refinery AD (No 1)*, [2001], EWCA Civ 406 [2001] 2 All ER 193, it was said: "Particularly in the case of contracts for future performance over a period, where the parties may desire or need to leave matters to be adjusted in the wording of their contract, the court will assist the parties to do so, so as to preserve rather than destroy bargains, on the basis that what can be made certain is itself certain."

It may be that the parties can reach agreement on the primary matters of principle, but leave other significant matters unagreed so that their bargain fails as incomplete. But an agreement may be incomplete only because it requires further agreement on identified matters.

Venture North Sea Gas Limited v Nuon Exploration & Production UK Limited [2010] EWHC 204 (Comm) provides an example of circumstances where parties making a contract recognised that they were not in a position to agree all terms of their arrangement at the outset, but wished to go ahead on the basis of resolving those remaining matters subsequently. Nuon agreed to purchase several North Sea oil and gas interests from Venture at a price in the order of €100 million; one of those interests was a partial sale of the interest held by Venture in joint venture with two other companies, Ithaca and Volantis.

The agreed sale would necessitate Nuon's becoming a party to those joint venture arrangements. Venture and Nuon were not party to the agreement and the circumstances of the transaction militated against any disclosure to Ithaca and Volantis. The agreement was made subject to a condition that, before a defined 'backstop date', Venture would have entered into a joint venture arrangement with Ithaca and Volantis in substantially the form of a draft joint venture arrangement annexed to the agreement, and which Nuon would be obliged to execute at the agreed completion date. Before agreeing to sign the new joint venture arrangement, Ithaca and Volantis required a number of amendments to be made to the annexed draft. The day before the backstop date, Venture, Ithaca and Volantis executed a joint venture agreement in the form annexed to the agreement, but subject to such changes as had resulted from the negotiations with Ithaca and Volantis. However, Nuon refused to complete on the grounds that although a joint venture arrangement had been executed among Venture, Ithaca and Volantis, it was not "in substantially the form" of the draft annexed to the agreement, as a result of the changes demanded by Ithaca and Volantis.

On a review of the evidence, the court found that by reason of the nature and extent of the amendments, the executed joint venture arrangement was not "in substantially the form" of that annexed to the agreement and therefore the condition in the agreement had not been fulfilled. Nuon could not be compelled to complete the purchase of any of the interests which were the subject of the agreement as it was common ground that the interests under the agreement stood or fell as a "package".

In delivering judgement, Justice Goss said:

that there would be some differences between the executed and draft JOAs was only to be expected ... the parties (i.e., Venture and Nuon) must have contemplated a negotiation between Venture, on the one hand and Ithaca and Volantis on the other. That much is apparent from the requirement that Venture should procure ... JOAs agreed by Ithaca and Volantis and that the executed JOAs should be "in substantially the form" of the draft JOA. In a negotiation of that nature, it would have been remarkable if there were no changes to the draft. Moreover, the requirement was not identity between the executed and draft JOAs but an executed JOA 'in substantially the form' of the draft JOA.

He then went on to say:

> *On the evidence, the market had moved between conclusion of the SPA in June 2009 and December 2009, so that the deal was no longer attractive to Nuon. Against this background, Nuon's resistance to completion necessarily required (and received) close scrutiny. But such considerations can only go so far; either Nuon does or does not have a good defence to the claim; if it does not, it is bound to complete; if it does, it is not bound to complete – and it is nothing to the point that walking away from the deal is now commercially appealing to Nuon. Put the other way, Venture is not the first and will not be the last party to have failed to fulfil conditions precedent to completion in market conditions where it would undoubtedly be to its commercial advantage to have done so. Such things happen. In all this, I have not overlooked the evidence that Venture negotiated with Ithaca and Volantis mindful of the danger of agreeing amendments to the draft JOA which would jeopardise completion; but, at the same time, Venture also needed to accommodate Ithaca's and Volantis's requirements.*

An agreement may fail to specify central matters such as price or quality of goods, but may nevertheless provide for the ascertaining of those matters by reference to identified criteria or a process or machinery for effecting the required certainty, or by reference to a third party or intervener that is invested with the standing to ascertain such matters by reason of the parties' contractual provisions to that effect.

9.2 Agreements to agree

The price reopener provisions in an LNG SPA will reflect a number of areas of uncertainty or vagueness (some intentional) in light of the parties' reluctance to commit themselves to rigid contractual terms in fluid markets and operational conditions. English law is familiar with a number of contractual devices which may be used in these circumstances, such as commitments to meet and discuss, to seek to agree, to use reasonable endeavours to negotiate and to use good faith towards these ends. In many cases the English courts have dealt quite briefly with such "agreements to agree".

Under English law, it is often said that an agreement to agree has no legal effect. In *May & Butcher v The King* [1934] 2KB 17, Lord Buckmaster said: "It has long been a well recognised principle of contract law that an agreement between the parties to enter into an agreement in which some critical part of the contract is left undetermined is no contract at all."

Again, in *Little v Courage* (1995) *Times*, January 6) it was said that: "An undertaking to use one's best endeavours to agree is no different from an undertaking to agree, to try to agree or to negotiate with a view to reaching an agreement; all are equally uncertain and incapable of giving rise to an enforceable obligation."

Also, there are often references in the typical price reopener to good faith. While good faith is a recognised concept in English law, it has no place in the negotiation of agreements. In *Walford v Miles* (1992) 2 AC 128 it was said that an English court: "cannot compel a party to negotiate in good faith... The concept of a duty to carry on negotiations in good faith is inherently repugnant to the adversarial position of the parties when involved in negotiations... While negotiations are in existence either party is entitled to withdraw from these negotiations at any time and for any reason. Accordingly, a bare agreement to negotiate has no legal effect."

This position contrasts with that under a number of civil law systems, where contractual obligations will ordinarily be required to be complied with in good faith and in some cases negotiations will be required to be carried out in good faith. As such, there will be no enforceable obligation on the parties under English law to revise the contractual arrangements, although there may be enforceable obligations to meet and discuss and seek to do so.

9.3 Contractual intention

The courts will be slow to strike down an agreement for uncertainty in circumstances where the parties show an intention that the agreement should have effect, notwithstanding that certain provisions remain to be agreed.

Foley v Classique Coaches Ltd (1934) 2 KB1 concerned the sale of a petrol-filling station and land on terms that the purchaser agreed to purchase petrol for its business exclusively from the seller over time. The purchaser broke that agreement and argued that the continuing commitment to purchase petrol exclusively from the seller was not binding, as it provided that the petrol was to be purchased "at a price agreed by the parties from time to time" and was therefore too uncertain to constitute a binding obligation. The court rejected this argument and found that a reasonable price was to be paid. In coming to this view, the court distinguished a line of earlier cases and was influenced by a number of considerations, including the following:

- Both parties had acted on the agreement for a number of years and believed it to be binding;
- The particular form of arbitration clause was considered to apply to any failure to agree as to price; and
- The agreement formed part of a larger bargain under which the price of the petrol-filling station and land was based on the exclusive purchase of future petrol from the seller.

An agreement may not necessarily be seen as incomplete even if it calls for further agreement among the parties. The courts have shown a willingness to consider the views and intentions of the parties, part performance on the part of the parties and the context of the agreement within wider arrangements. Also, it may be possible to resolve the uncertainty by applying a standard of reasonableness, or in the context where the uncertain matter is of minor importance and will not be seen to undermine the intention to be bound to the overall agreement. Of perhaps greater relevance to a consideration of LNG SPAs are the circumstances where the contract itself provides expressly for the establishing of those matters which are not resolved at the outset.

English law recognises that the presence of a criterion or procedure for ascertaining the unresolved matters may well militate against a finding of unenforceability. Such provisions may entitle one of the parties to resolve the matter (probably under the discipline of an implied duty of reasonableness) or provide for reference of the matter for determination by an expert, arbitrator or other third party. But one of the lessons from a consideration of price reopeners under LNG SPAs is to establish whether the relevant provisions constitute such a criterion or procedure which is itself sufficiently certain to be enforced, or whether they are no more than an agreement to agree.

9.4 Making of agreement

The making of a binding and enforceable agreement under English law depends on the parties reaching agreement through the making and accepting of an offer in terms which are certain and final. An agreement is not binding if it lacks certainty through being too vague or incomplete. Whereas a typical LNG SPA will be a document of some length and complexity, certain parts of that document may not be expressed in detail, despite the inevitably long period of negotiation. Among the reasons for this may be that these provisions look to address circumstances which are themselves uncertain or unexpected, or cover matters which are sufficiently difficult or subordinate to the main elements of the contract that the parties are comfortable to leave them in comparatively uncertain terms in the interests of closing all other elements and moving to sign the overall agreement. In such cases the parties may comfort themselves with the hope that these things 'may never happen'. The express provisions of an LNG SPA dealing with changes of circumstances, hardship and price revision will often fall within this category.

An LNG SPA is a contract for the sale of goods under English law and the primary elements of such a contract are recognised to be description, quantity, delivery point and price. A number of these elements may fall within the uncertainty often present within a typical reopener provision. A typical price reopener will tend to be expressed in terms of reasonable opinions, the assessment of economic circumstances, foreseeability, fairness, equity and meeting, discussing and seeking to reach agreement. Each of these matters in itself could be seen to be at risk to uncertainty, and their combination does little to reduce this risk. So how will English law approach these provisions and their interpretation and potential enforceability?

In essence, there will be no enforceable arrangement among the parties in circumstances where the parties have not reached substantial agreement. But once such substantial agreement has been made, that agreement will not be affected by the fact that certain matters remain to be resolved over time. English law has recently shown signs that it may be coming to differentiate between cases where agreement is required on matters (eg, price) in relation to the making of a contract as contrasted with the continued effect of a contract. The cases also show that while the courts are most reluctant to enforce an agreement to agree in circumstances where no contract has yet come into existence, there is less reluctance to do so where the agreement to agree applies within an existing contract and may be said to form part of the inducement for the original bargain of the parties.

The *Mamidoil-Jetoil* case concerned a provision in the nature of an agreement to agree on pricing, which become operative partway through the contract period. In delivering judgment, the court set out the following principles, identified from a review of the authorities:

- Each case is to be decided on its own facts and the construction of its own agreement.

- Where no contract exists, the use of an expression such as 'to be agreed' in relation to an essential term is likely to prevent any contract from coming into existence, on the ground of uncertainty; and the absence of agreement on essential terms of the contract may prevent any contract from coming into existence, again on the ground of uncertainty.
- In commercial dealings between parties that are familiar with the trade in question, and where the parties have acted in the belief that they had a binding contract, the courts may be willing to imply terms where that is possible to enable the contract to be carried out.
- Where a contract has once come into existence, even the expression 'to be agreed' in relation to future executory obligations is not necessarily fatal to its continued existence.
- Particularly in the case of contracts for future performance over a period, where the parties may desire or need to leave matters to be adjusted in the working out of their contract, the courts will assist the parties to do so, so as to preserve rather than destroy bargains on the basis that what can be made certain is itself certain. This is particularly the case where one party has already had the advantage of some performance which reflects the parties' agreement on a long-term relationship or has had to make an investment based on that agreement.
- An express stipulation for a reasonable or fair measure or price will be a sufficient criterion for the courts to act on, but even in the absence of express language, the courts may be prepared to imply an obligation in terms of what is reasonable.
- The presence of an arbitration clause may assist the courts to hold a contract to be sufficiently certain or to be capable of being rendered so, presumably as indicating a commercial and contractual mechanism which can be operated with the assistance of experts in the field by which the parties, in the absence of agreement, may resolve their dispute.

In 2005 the Court of Appeal considered *Petromec Inc v Petroleo Brasileiro SA Petrobras* [2005] EWCA Civ 891, and particularly an express commitment to negotiate in good faith within the context of a wider contractual relationship which was legally binding. The court identified the three accepted objections to enforcing an obligation to negotiate in good faith as follows:

- The obligation is an agreement to agree and therefore unenforceable by reason of uncertainty;
- It is difficult (if not impossible) to say that a termination of negotiations has arisen in circumstances of good or bad faith.
- As it cannot be known whether negotiations would have reached agreement and on what terms, there can be no assessment of the loss caused by the breach of the obligation.

Where the nature and effect of the contemplated agreement were specifically identified and the potential loss quantifiable, the first and third guards were seen to have little weight. In relation to the second objection, Lord Justice Longmore said:

> *But the difficulty of a problem should not be an excuse for the Court to withhold relevant assistance from the parties by declaring a blanket unenforceability of the obligation…The authority chiefly relied on in support of blanket unenforceability was the decision of the House of Lords in Walford v Miles, which (of course) binds us for what it decides. The main distinction was between that case and this was that in that case there was no concluded agreement at all since everything was "subject to contract", there was, moreover, no express agreement to negotiate in good faith. There were negotiations for the sale of a business in the course of which the defendant prospective vendor agreed not to negotiate with any third party and to negotiate only with the claimant prospective purchaser. All the negotiations were subject to contract and the House of Lords held that the "lock-out agreement" was unenforceable because there was no provision saying how long it was to last. The claimants sought to resolve this difficulty by asserting that it was an implied term of the agreement that, while the defendant wanted to sell the business, they would negotiate in good faith with the claimants. The House of Lords held that it was impossible to imply such a term since it was unworkable in practice and inherently inconsistent with the position of the party negotiating "subject to contract". The lock-out agreement was therefore too uncertain to be enforceable. As Lord Ackner (with whom the rest of their Lordships agreed) said:*
>
> > *"…. While negotiations are in existence either party is entitled to withdraw from those negotiations, at any time and for any reason. There can be thus no obligation to continue to negotiate until there is a 'proper reason' to withdraw. Accordingly, a bare agreement to negotiate has no legal content."*
>
> *That shows the difference from the present case…It is not irrelevant that it is an express obligation which is part of a complex agreement drafted by City of London solicitors… It would be a strong thing to declare unenforceable a clause into which the parties have deliberately and expressly entered. I have already observed that it is of comparatively narrow scope. To decide that it has "no legal content" to use Lord Ackner's phrase would be for the law deliberately to defeat the reasonable expectations of honest men. I do not consider that Walford v Miles binds us to hold that the express obligation to negotiate is completely without legal substance.*

Before moving on to consider some specific examples of price adjustment and price review clauses, it is worth pausing for a moment to reflect on the reality that, in the context of a number of LNG SPAs, English law may well have been agreed upon (and administered in the arbitral process) by those who are not qualified in English law or without recourse to English law advice.

10. Specific contractual provisions

In a contract for the sale of goods under English law, the price to be paid for those goods is a fundamental provision of the agreement. Also, it is clear that gas constitutes goods under English law. Where there is a contractual requirement to renegotiate the parties' bargain, such clauses will typically have a number of components. The first will be the description of those (typically exceptional) circumstances which trigger the commitment to meet and discuss. The second element will be an identification of the effects of the changes on the contract. There will then typically be a statement of the objective of the renegotiation and a setting out of the procedure to be followed in relation to that renegotiation. Lastly (and importantly if there is to be an effective commitment to revise the contract terms under English law) will be a provision setting out the consequences in the event that the parties fail to reach agreement. This may provide for termination of the contract in the absence of agreement or, conversely, that the contract will remain in effect in accordance with its terms. Alternatively, the contract may provide for a reference to a court or an arbitrator or another third party to decide and effect the appropriate changes to the contractual arrangement in the context of the terms of the renegotiation clause. Under English law, the courts have powers to interpret and enforce, but not to make or modify contracts. Powers to rewrite the bargain of the parties will need to be included in the terms of the contract and, in the absence of agreement among the parties, that power may often be reserved to arbitrators or experts. Having reviewed the approach of English law to price review and contractual modification generally, it is timely to turn to some specific examples of contractual modification in the context of price adjustment, hardship and price reopeners.

10.1 Price adjustment: *Esso v ESB*

Esso v ESB provides a good example of a price adjustment provision. Under its terms, the price for gas (described as the 'energy charge') was to be adjusted every six months by reference to four indicators in set proportions:

- the price of gas oil (30%);
- the price of low sulphur fuel oil (30%);
- the price of natural gas (30%) and
- the rate of inflation in Ireland as reflected in the industrial wholesale price index (10%).

In each case the relevant marker was the average over the 12-month period ending three months prior to the review date and the price to be taken for each of the three commodities was in effect the spot price for delivery to Northwest Europe. The buyer and the seller agreed the initial value of the price and represented it as P0. The parties also recorded that this initial value of P0 was set by them: "at a level that reasonably reflected the market price at the date of the agreement obtainable for the sale of reasonably similar quantities of gas over reasonably similar period on reasonably similar terms and conditions between parties of reasonably similar commercial and financial standing for use in a reasonably similar type of power station in the UK or Ireland."

It appears that the price adjustment provisions were intended to maintain the energy charge at a value over time equal to that of natural gas being supplied under other contracts among other parties, but on the agreed basis. However, if that energy charge came to diverge from market price (assessed on that same basis, but by reference to other similar supplies to other similar power stations), then the agreement provided a separate basis for a party to require price review at certain intervals during the contract life. In these circumstances, a price review notice could not be given by the seller unless "it is reasonably satisfied in good faith that the

Energy Charge... is at the time of giving such Price Review Notice eighty five per cent. (85%) or less than the Comparator....and the Comparator is defined as being the...market price...at the date of the relevant Price Review Notice for natural gas being supplied on the basis described above".

The reference to the 'basis described above' is to the market price at the date of the agreement obtainable for the sale of reasonably similar quantities of gas over a reasonably similar period on reasonably similar terms and conditions between parties on reasonably similar commercial and financial standing for use in a reasonably similar type of power station in the United Kingdom or Ireland.

Accordingly, the structure of the clause provides for price adjustment on a six-monthly basis by reference to a set of indicators which appear to have been chosen by the parties as being appropriate to maintain the energy charge consistent with the market price for gas supplied on the agreed basis. However, in the event that that aim was not fulfilled and the energy charge came to diverge from that market price by more than 15%, the seller was in a position to serve a review notice which seems intended to have the effect of re-setting the P0 value to that divergent market price as at the date of the price review notice.

The gas sales agreement was made on November 27 1997 and delivery of gas under the gas sales agreement began on October 1 1999. The relevant price review notice was given by Esso on November 1 2002 – that is, less than five years from the making of the agreement and scarcely more than three years after the beginning of deliveries.

In construing the provisions clause 12.6(6), and particularly the provision that a price review notice could be given by the seller unless "it is reasonably satisfied in good faith that the Energy Charge... is at the time of giving such Price Review Notice eighty five per cent. (85%) or less than the Comparator", the judge considered that it was not enough that the information available to the seller had given rise to a reasonable level of satisfaction on the part of the seller that the energy charge had fallen to that level. Instead, he found that the expression 'reasonably satisfied' meant "satisfied on reasonable grounds".

In construing the meaning of the comparator, the judge considered whether this was intended to be an "actual" market price or a "notional" market price. This was a significant point within the context dispute as, by the date of service of the price review notice, there were no supplies of long-term natural gas being made to power stations in the United Kingdom or Ireland.

In considering whether the words chosen by the parties to express their intention (and with due regard to the commercial and contractual context in which they are found) led to a conclusion that the test was one of a notional price or an actual price, the judge said:

I think it is helpful to begin by considering the fundamental purpose of the price review provisions in so far as that can be discerned from the contract. The price at which gas or any other commodity is sold under a long-term contract will reflect many different factors. These would usually include, among other things, the current price for prompt delivery, any price being quoted for delivery at a future date, the parties' perception of the likely movement of prices in the longer term and the particular terms of the contract.

In the present case the parties took account of future movement in the price of fuel used for power generation by providing for regular adjustment of the price by reference to the average over the previous twelve months of the prices of three different fuels for immediate delivery. It is true that the indexation formula was not linked to the price of gas alone and to that extent could not be expected directly to reflect changes in the price of gas, but that was the parties' choice. It seems to me therefore, that in so far as the parties intended to adjust the price by reference to changes in short term fuel prices, clause 12.2 is the means they chose. What clause 12.2 by its very nature could not achieve, however, was an adjustment to reflect the other factors that have a bearing on the price payable under a long term contract, such as the value to be attributed to non interuptability of supply, to take for one example.

He went on to say: "The effect of a price review under clause 12.6 is to alter the value of the Energy Charge which forms the basis for the subsequent periodic adjustment under clause 12.2. Accordingly, I think its purpose must have been to reflect changes in the value being attributed to the other factors that influence the price payable under long term contract, not simply to reflect changes on the spot or short-term future price for gas."

Having come to the view that the proper construction of the language used by the parties led to a conclusion that the reference to market price was to an actual price and not to a notional price, the judge went on to say:

That in my view is the natural meaning of the words used. But I can see at least two practical reasons why the parties should have looked at similar long term supply contracts to provide the Comparator. First, because setting a long-term price taking into account all the relevant factors involves a substantial element of commercial judgement; it also depends on the parties' negotiating strength. Prices set by negotiation in the market place are likely to reflect on average all factors that have a bearing on the outcome more reliably than the assessment of any arbitrators. Secondly, because if there really is no market in the long-term sale of gas to power stations, there is no need to rebase the price in order to keep it in line with current market prices under contract of that kind. If the parties had intended that the new base price should be calculated in any event using whatever information was available, I think that clause 12.6 would have been worded in a way which made that clear.

One lesson from this case may be derived from the very short period of time from the signing of the agreement to the intended but ineffective operation of the price review mechanism. In a period of less than five years, the basis on which the parties had set their initial price and on which they had expected to reset that price over time had become completely unavailable. Had the parties really carried out a review of the market at the time they entered into the agreement or had they followed precedent more than the activity of the market?

10.2 Hardship: the *Hewett* hardship case

The notion of hardship has its roots in civil law and English law does not recognise the term as having meaning beyond that provided for in the parties' agreement. In English law, such an expression will mean what it is defined to mean in accordance with the terms of the agreement and without reference to any external guidance such as a civil code. The typical hardship clause has effect in circumstances of unusual disruption or unforeseen events or circumstances causing economic hardship for the buyer or the seller.

Perhaps the best example of a hardship clause is the one which was the subject of the *Hewett* hardship case. It is one of the few reported decisions where a hardship clause has been judicially considered. The relevant provision was set out in Article X of the Hewett gas sales agreement and read as follows:

(a) *If at any time or from time to time during the contract period there has been any substantial change in the economic circumstances relating to this Agreement and (notwithstanding the effect of the other relieving or adjusting provisions of this Agreement) either party feels that such change is causing it to suffer substantial economic hardship then the parties shall (at the request of either of them) meet together to consider what (if any) adjustment in the prices then in force under this Agreement or in the price revision mechanism contained in Clauses 4, 5 and 6 of this Article are justified in the circumstances in fairness to the parties to offset or alleviate the said hardship caused by such change.*

(b) *If the parties shall not within ninety (90) days after any such request have reached agreement on the adjustments (if any) in the said prices or price revision mechanism which are to be made then the matter may forthwith be referred by either party for determination by experts to be appointed in the manner set out in Article xviii hereof save that the appointment of the third expert referred to in Clause 1(c) of that Article shall in any event be made by the Minister of Power in consultation with the Lord Chancellor.*

(c) *The experts shall determine what (if any) adjustments in the said prices or in the said price revision mechanism shall be made for the purposes aforesaid and any revised prices or any change in the price revision mechanism so determined by such experts shall take effect six (6) months after the date on which the request for the review was first made.*

The Hewett gas sales agreement was entered into in March 1968 between the Phillips Group and the Arpet Group as sellers and the British Gas Corporation as the buyer and concerned the supply of natural gas by the sellers to British Gas over a period of some 25 years. The Hewett field was developed and deliveries began to the effect that gas was sold and purchased for some years under the terms of the Hewett gas sales agreement. By 1976, the sellers considered that changed economic circumstances had created substantial hardship for them and they made a claim pursuant to the provisions of Article X. British Gas resisted this claim and the matter was accordingly referred for expert determination in accordance with the terms of the Hewett gas sales agreement. The Hewett gas sales agreement provided for determination by a panel of three experts who appear to have arrived at a single decision on a majority basis. In making their determination, the panel of experts decided by a majority that the sellers would suffer economic hardship if the achieved rate of return on capital (ARR) fell below a minimum acceptable rate of return (MARR) and that the sellers would suffer substantial, as contrasted with, insubstantial, hardship if ARR fell below MARR by 2% or more. The determination did not satisfy either party, and the sellers and British Gas entered into moratorium arrangements until October 1980. Shortly after the end of that moratorium period, the sellers again claimed hardship. The differences among the sellers and British Gas included:

- whether substantial economic hardship was a matter of capital costs, operating costs and revenues under the Hewett gas sales agreement or whether opportunity costs could be taken into account in circumstances where the market price of gas was higher than the contract price;
- the 'quality' of hardship – whether all hardship (including the substantial element of the hardship) should be alleviated or only the substantial hardship; and
- the 'quantity' of hardship – whether the date from which the remedy should have effect should be the date on which the hardship began, the date on which the hardship notice was served or the later effective date of that notice.

Following an expert determination, several issues were referred to the High Court. The judgments at first instance and at the Court of Appeal are both reported. Interesting for the purposes of this paper is the characterisation of Article X by Lord Justice Donaldson in the Court of Appeal:

In my judgement, [the hardship provision] is an ultimate safety net. To adopt an analogy which is perhaps appropriate...the parties contemplated that in most foreseeable economic conditions the course of the joint venture would be dictated by the automatic price revision mechanisms...(the agreed price autopilot). But the parties realised that over a period of 25 years economic storms could arise of such severity that the price autopilot would not be able to keep the venture on course. [The hardship provision] provides for a manual override if this occurs and the venture goes so far off course as to cause one of the parties to suffer substantial economic hardship. The experts then take over, correct the course and, if appropriate, revise the settings on the price autopilot.

In characterising the nature of the Hewett hardship clause, Donaldson found that both parties contemplated the extraction and delivery of natural gas at rates designed to exhaust the field in about 25 years, and that while both parties were in a position to discuss the price to be paid for the whole content of the field, they could do so only on the basis of known facts as to capital investment which had already taken place and estimates of future capital expenditure and operating and maintenance costs over that 25-year period. He said: "There were known to be three jokers in the pack. The first was the effect of future inflation or deflation on the capital and operating costs of producing and delivering the gas over the 25 year period. The second was the relative competitive position of gas and other major alternative fuels over the period. Third, and this was more of a 'wild card' than a joker, there was the possibility that the parties had failed to foresee another factor which would falsify some or one or all of their assumptions."

He recognised that the first two of his 'jokers' were addressed by the provisions of the Hewett gas sales agreement providing for price adjustment on a regular periodic basis by reference to future inflation or deflation in the first place and the movements over time of the prices of major competing fuels in the second case. He then went on to note the different approach of the parties to the so-called 'wildcard' where the contractual machinery for this price review clause used wide words which he saw as appropriate when trying to provide for the unforeseen and possibly unforeseeable.

He then went on to identify the following operative words from Clause X: "If at any time or from time to time during the Contract Period there has been any substantial change in the economic circumstances relating to this Agreement and (not withstanding the effect of the other relieving or adjusting provision of this Agreement) either party feels that such change is causing it to suffer substantial economic hardship".

He continued:

This test may be criticised as constituting an unholy mixture of the objective and subjective, but it really does not matter. It is merely a trigger mechanism. If the party is mistaken in thinking that there has been such change in economic circumstances or that such a change is causing it to suffer substantial economic hardship, the mechanism will produce no change at all. Given an alleged substantial change in the economic circumstances relating to the agreement and the feeling by either party that the change is causing it to suffer substantial economic hardship then – the parties shall (at the request of either of them) meet to consider what (if any) adjustment in the prices then in force under this Agreement or in the price revision mechanism contained in clauses 4, 5 and 6 of this Article are justified in the circumstances in fairness to the parties to offset or alleviate the said hardship caused by such change.

In the *Hewett* case, the hardship clause required the expert panel to make such adjustments as were justified in the circumstances, in fairness to the parties, to offset or alleviate the hardship caused by the change in economic circumstances. This was seen to provide for a subjective (and wide) discretion on the part of the panel experts by reference to broad words such as "substantial" and "fairness", and with references back to the original bargain made by the parties (and the economic balance inherent in that bargain) by reason of offsetting or alleviating the hardship caused by the change. While similar provisions may often be subject to contractual qualifications or limitations (eg, floor prices to limit downward price movement or ceiling prices to limit upward price movement, with these provisions having effect at the time of implementation or operating to constrain the decision of the chosen third party), there were no such provisions in the Hewett gas sales agreement.

The Court of Appeal found, on the question of the 'quality' of hardship, that all hardship (and not only the substantial element) should be alleviated (thereby reversing the decision in this respect at first instance); and on the question of the 'quantity' of the hardship, that all such hardship from the date of its first occurrence was to be offset.

10.3 Price reopeners: the *Atlantic LNG* case

A rare opportunity to see the public resolution of a dispute in relation to a price reopener clause arose in 2008 in the *Atlantic LNG* case. The case concerned an LNG SPA entered into in 1995. The LNG SPA was to apply for a term of 20 years from 1999 and provided for the supply to Gas Natural of LNG produced by Atlantic at its Train 1 facility in Trinidad and Tobago. At the time that the LNG SPA was concluded, it was the expectation of the parties that the LNG would be delivered to Spain for consumption in the Spanish market. However, Gas Natural also had the right to take deliveries of LNG at its receiving facilities in New England in the United States as an alternative from time to time. The parties' expectation of delivery and consumption in Spain was reflected in the price adjustment provisions, which consisted of an agreed base price which was to be subject to the quarterly application of a multiplier index related to European prices for certain substitute petroleum products. In addition, the LNG SPA included a price reopener provision which was expressed to have effect subject to the application of particular circumstances which operated as pre-conditions.

The price reopener clause was set out in Article 8.5(a) of the LNG SPA and provided as follows:

> *If at any time either Party considers that economic circumstances in Spain beyond the control of the Parties, while exercising due diligence, have substantially changed as compared to what it reasonably expected when entering into this Contract or, after the first Contract Price revision under this Article 8.5, at the time of the latest Contract Price revision under this Article 8.5, and the Contract Price resulting from application of the formula set forth in Article 8.1 does not reflect the value of Natural Gas in the Buyer's end user market, then such Party may, by notifying the other Party in writing and giving with such notice information supporting its belief, request that the Parties should forthwith enter into negotiations to determine whether or not such changed circumstances exist and justify a revision of the Contract Price provisions and, if so, to seek agreement on a fair and equitable revision of the above-mentioned Contract Price provisions in accordance with the remaining provisions of this Article 8.5.*

The LNG SPA also provided (in Article 8.5(f)) that, in default of agreement, either party could refer the matter for arbitration in New York City and in accordance with the United Nations Commission on International Trade Law arbitration rules. The case report does not state the governing law of the LNG SPA, which is presumed to be the law of New York.

Following the signing of the LNG SPA, the natural gas market in Spain was substantially liberalised to the effect that Spanish natural gas prices decreased. At the same time, natural gas prices in the New England market became comparatively attractive and Gas Natural took most of its cargoes for delivery into the New England receiving facilities. It entered into a long-term sales agreement with GDF Suez to sell its full Atlantic LNG offtake into the US market by means of the New England reception facilities. In April 2005 Atlantic LNG notified Gas Natural that it was seeking a revision of the contract price pursuant to Article 8.5(a). The parties were unable to agree a fair and equitable revision of the LNG SPA (in accordance with the terms of Article 8.5(a)), and in October 2005 Atlantic referred the matter to arbitration and requested an increase in the contract price which would reflect the value of natural gas in the New England market.

The value of the requested increase over the remainder of the period of the LNG SPA was estimated to be in the order of $1 billion. The tribunal was formed in New York City and, having determined that the precondition had been fulfilled, it determined that the pricing provisions agreed by the parties in the LNG SPA should be replaced. The tribunal created new pricing provisions (having retrospective effect back to the date when the price reopener had first been triggered) which constituted a two-part pricing scheme:

- The original pricing formula was retained as the first part of the pricing scheme, but the base price component was revised in order to set a new price to reflect the value of natural gas sold into the Spanish market as it had changed by virtue of market liberalisation.
- The tribunal added a new, second element to the price formula which the tribunal considered to reflect the value of natural gas sold into the New England market.

The tribunal then provided that the price payable for natural gas delivered under the LNG SPA would not be calculated by reference to the quantities delivered to each of the Spanish market and the New England market, but rather that all natural gas deliveries would be paid for at a single price (regardless of the actual place of delivery). This would be the New England price in the event that a certain threshold number of cargoes were delivered to the New England market; otherwise, it would be the Spanish price. It seems that the commercial effect of the determination was to require Atlantic LNG (the party which had initiated the reference to arbitration) to make an immediate, retrospective payment to Gas Natural of a sum in the order of $70 million. Commentators have suggested that over the remaining life of the LNG SPA, the tribunal's decision will have cost Atlantic LNG an additional, aggregate sum in the order of $750 million. This determination by the tribunal to put in place a dual pricing formula was an answer which neither party had requested and both appear to have argued against.

Among the grounds for the challenge by Atlantic LNG of the tribunal determination was that the tribunal had exceeded its authority. In particular, Atlantic LNG argued that: "the tribunal exceeded its powers by imposing a pricing scheme that – in Atlantic's view – skewed the original bargain between the parties and effectively re-wrote their contract. Specifically, the new 'dual price structure' which provides one price when more than the percentage of the Train 1 LNG is delivered into the New England Receiving Facilities, and another price otherwise, gives Gas Natural the unbargained-for ability to determine which of two quarterly prices will apply to all shipments merely by shifting its Train 1 LNG deliveries."

The court found that:

It is undisputed that the Tribunal was specifically charged with the duty to revise the pricing scheme once it determined that the contractual pre-conditions were met. The Tribunal having made that determination, Article 8.5(a) required it to "reach a fair and equitable revision" of the contract price. Neither this standard nor any other contractual provision set a structural limitation on permissible price revision. Indeed, Atlantic's submission for the Tribunal acknowledged the Tribunal's broad authority in this regard. In them Atlantic opined that the relevant contractual terms "do not appear to express limit this Tribunal's award to the imposition of a single price formula". Atlantic's argument concerning the dual pricing formula is better understood as a challenge to the merits of the Tribunal's decision. Such a challenge is unavailing since the court does not review arbitration awards for legal or factual errors.

The *Atlantic LNG* case shows some of the challenges of presuming to write a contract today which will be sufficient to address the changes and uncertainties of a period of 20 years. In 1995 the LNG SPA reflected the preventing markets which were closed and with an absence of gas-to-gas competition or the potential for diversion of cargoes to other markets. Many European gas contracts include price reopener clauses which provide for periodic meetings of the parties for the purpose of re-apportioning the value under the contract by means of an agreed revision of its terms for a future period. But factors such as the liberalisation of gas markets and rising oil prices have led to increasing commercial difference and an incentive to refer disputes for arbitration and the intended re-setting of a contract for a future period. The *Atlantic LNG* case showed that the words used by the parties provided flexibility for the arbitrators to go far beyond the traditional benign resetting of commercial terms and to rewrite the agreement of the parties in the absence of clear limitation or terms of reference.

11. Relational arrangements

For some, a 'long-term contract' is a sociological rather than a legal category, and it is not the duration of the contract which is the definitive characteristic, but the nature and effect of the long-term relationship among the parties. Such a relational contract may not sit comfortably with a common law approach, but it may be more appropriate for the requirements of the production and disposal of petroleum products. Broadly, a relational arrangement would identify the relationship of the parties, but without precision (at the time of contract formation) of the duration, the subject matter or the identification of quantity, price or substance. In this way, it is said to lack 'presentiation', or the making of a present decision about all future elements of a contractual relationship. Also, a relational arrangement would be seen as a consistent cooperation, as opposed to the optimisation of the separate positions of each party. Such an approach is said to provide the ability to adjust the relationship over time and enable the parties to phrase their commitments in aspirational terms, with a corresponding requirement that courts or arbitrators should be slow to strike down such a long-term contract which is expressed in general rather than specific terms. For as long as the parties are willing to see their own interests as subordinate to the overall aspirations of their relationship, their agreement is capable of evolving and developing according to the prevailing circumstances. However, recent events in the gas market have shown that where the parties' financial and economic interests diverge substantially, the parties will be inclined to favour their own interests and enhance their own position. At this point, the parties are likely to look to the specific terms of their agreement, and the interpretation of the meaning of the words used in their agreement will attain primary importance. As we have seen, in the absence of specific wording, the law will be slow to provide relief for a party which considers itself to be suffering hardship or commercial imbalance. The law will be equally slow to find that the contract has terminated in those circumstances – unless that is what the parties have specifically provided in their agreement. It is likely to be the case that, under English law, the parties are necessarily in a relational arrangement from the outset of their LNG SPA, whether or not they recognise this.

Before moving on to conclude, it is also worth touching on another relationship which may well be relevant to LNG SPAs and which also gives rise to potential contract revision and the sorts of issues already aired in relation to price review. Many LNG SPAs are made with state entities and, even if not directly a party, the role of the producer state is likely to have an effect on the commercial balance of an LNG SPA from time to time over its long life. The principle of a state's permanent sovereignty over its hydrocarbon resources is a universal one and, particularly in recent times, has been exercised in a number of jurisdictions regardless of the terms of contractual arrangements, and on occasion in apparent breach of contractual commitments to the contrary.

A producer state will rarely have a contractual right to change agreed terms unilaterally and, over the course of a long-term contract, the principle of sanctity of contract may well clash with the principle of permanent sovereignty. There have been attempts to manage these risks by the introduction of stabilisation provisions into the contractual arrangements. Early examples of stabilisation clauses provided for contractual stabilisation which was to the benefit of the investor only and sought to 'freeze' the contractual relationship and constrain the state from taking contrary steps. In essence, the contractual stabilisation clause reinforced the principle of *pacta sunt servanda* in terms which apparently constrained the state from exercising its rights of permanent sovereignty. Such contractual stabilisation clauses have been questioned on a number of grounds, including that they constrain sovereignty and also that they are contrary to human rights, particularly in relation to matters such as health and safety and the environment. Many states have now moved against any form of host government support, and to the extent that stabilisation provisions came to be agreed, it is now more usual to find an economic stabilisation clause. This is a 'two-way' provision which enables both the state and the investor to seek to rebalance the economic benefits of the contractual relationship to the extent that their respective benefits deviate over time from those expected at the time of the making of the contract. While those economic stabilisation provisions have their roots in public international law, their terms and their effects are closely related to those of price review clauses.

12. Lessons learned – an alternative approach?

Price review clauses will often provide for reference to arbitration in the case of dispute or absence of agreement among the parties. Also, they will invariably become applicable in circumstances where one party is in a relatively better and one in a relatively worse contractual or commercial position than envisaged at the time their bargain was made and expressed in the detailed terms of their LNG SPA. The relevant provisions of their LNG SPA will then be read in the context of an adversarial relationship and a procedure of dispute resolution. This procedure is likely to encourage the parties towards polarised and self-interested positions: the very antithesis of the consensual circumstances in which their original bargain was made. At that time, the price review clause was known to contemplate unknown events or circumstances and to provide a generalised contractual process for the parties to maintain (and not threaten or terminate) their relationship. Among other things, they would meet and seek to recalibrate that contractual relationship in the changed circumstances – the changed circumstances which were necessarily beyond the control of the parties and necessarily unforeseeable at the time their bargain was made.

The revisiting of the price provisions in those unforeseen circumstances will be a matter in the nature of those left over for agreement at a later date by reference to a contractual machinery or reference to a third party. On the face of it, this will be a matter for agreement, not dispute; but the partisan positions will necessarily mean that agreement will be difficult to reach, not least because one of the parties will be approaching these discussions from an 'entitled' position. In these circumstances, where the parties are unable to reach agreement among themselves, the appropriate reference is not so much to dispute resolution, but more to a third-party intervener who does not fulfil a judicial role beyond the relationship of the parties, but derives his standing and his powers from the words of the parties themselves. The arbitrations which have been reported (including those anonymised reports released by the ICC (final award in Case 9812 and final award in Case 13504)) suggest that professional arbitrators (almost invariably disputes lawyers) and retired judges rarely have the appropriate skills to rewrite or provide for the rewriting of an LNG SPA for many years to come in the manner contemplated under a typical price review provision. In the context of a long-term and relational arrangement, theirs is a spot intervention with typically little knowledge of or affinity with the contract's past and no continuing interest in the contract's (often long) future.

Third-party intervention can be invoked by a party only if it is provided for in the agreement. And as the standing and powers of the intervener are derived from the words of the parties as set out in the LNG SPA, it is appropriate for the parties to create and limit those powers as they may see fit and agree at the time of making their LNG SPA. Perhaps it is asking too much of a third-party intervener to decide the principles of the relevant price review and to rewrite the parties' contract accordingly. An alternative, which would be more appropriate for a consensual process, could be for the intervener to set out the principles to be applicable to the price clause and then encourage the parties to negotiate the detailed text, with a right to return to the intervener in the absence of agreement. No LNG SPA has been made without extensive negotiation and all pricing previsions will be complex, having many different elements and component parts. Such an approach (of having the intervener establish only the principles initially) might also have the benefit of enabling the parties to trade or enhance the component parts of the price 'package' in a way which would optimise the commercial balance of the agreement and provide the basis for a more enduring relationship than would be likely with the 'spot' approach of a traditional arbitration. This is, after all, the process and discipline which characterises the initial negotiation process and which leads to the finding of the optimal position for both parties in the context of the original making of their agreement.

And the simple re-expression of price review provisions may be helpful. The traditional process starts from the probably flawed premise that the price adjustment provisions will be appropriate to reflect the parties' bargain for a period of some decades, save in extraordinary and unforeseeable circumstances. It then falls to one of two parties (the one which, at that particular time, sees itself as being 'out of the money' or suffering hardship) to prove the application of the clause and argue for the appropriate contractual revision. That it is for one party to prove its case – in the context of imprecise notions such as extraordinariness and unforeseeability, and in the context of unfriendly market circumstances – can serve only to create adversarial relations and a concentration on short-term and partisan interests. Would it not be more appropriate to a 25-year relationship to recognise at the outset that the parties' clairvoyance is limited and to provide for periodic recalibration of their agreed pricing provisions? This would not be a matter of proof among adversaries from comparatively advantaged and disadvantaged contractual positions, but a resetting of the parties' bargain at that time in the circumstances of the day and in the context of potential reference to an intervener in the event that negotiation was unsuccessful. The words of the price review clause would set out the objectives of the parties in broad but representative principles, and would provide specifically for the intervener to recognise the nature of its role in the context of a relational contract. It would also set out the nature of the relationship of the parties in the context of their LNG SPA and their respective commitments as part of this periodic recalibration. While these duties would be unlikely to be cast in terms of good faith, they would include meaningful commitments towards the adaptation and maintenance of the agreement in the unforeseen circumstances.

Whereas the courts and arbitrators applying English law are unlikely to have power to adapt contract terms to take account of uncontemplated fundamental changes (even if this is what the parties request), that is not the case with a third-party intervener who may be appointed by the parties to facilitate the implementation of their contract. Such an appointment would be akin to those well-recognised mechanisms to 'fill in the gaps'. For example, the decision in the *Atlantic LNG* case (which seems to have been sought by neither party) might have been different had the arbitral panel been constrained to choose the new provision proposed by either the seller or the buyer, and not permitted to act as a judicial decision maker and prescribe a new price provision requested by neither the seller nor the buyer. There are reasons which argue for and against this form of ('pendulum' or 'baseball') resolution, but the overall principle may well be more acceptable to the parties at the outset than would be the contemplation that, after a comparatively short period of years, an unknown group of third parties will be rewriting their agreed price provisions for, potentially, the majority of the life of their LNG SPA. While it is possible to criticise this form of pendulum arbitration for its lack of sophistication and its forcing of the arbitrators to forgo seeking an appropriate resolution and instead simply to choose the 'less wrong' of the two alternatives, the disciplines that such a clause would have imposed on the parties to the *Atlantic LNG* case could well have produced an enduring answer preferable to that decided by the arbitrators: that is, for the buyer to choose which of two prices applies by reference to the majority of cargoes delivered (at the buyer's election) to one of two points over a period of time.

The available cases suggest that the individuals appointed as arbitrators under LNG SPAs are often lawyers skilled in the interpretation and application of contracts in the context of resolving disputes. They are rarely individuals (lawyers or otherwise) skilled in the making or negotiation of LNG SPAs or, less still, their writing. In Europe, there remain many LNG SPAs (and even more long-term pipeline gas contracts) of enduring effect which deny the patent existence of gas-to-gas competition and market prices in their relevant markets. Many of these contracts are coming before arbitral tribunals, and the available evidence suggests that these tribunals tend not to be well versed in these markets and their effects, however eminent they may be in relation to legal matters. And although it will be usual for those tribunals to have the power to seek expert support on these matters, this does not always appear to have remedied these difficulties. As these contracts continue to make their ways towards dispute resolution by means of arbitration, there seems little prospect of a change of these customary procedures. The restructuring of long-term gas contracts in the 'take-or-pay' wars in the United States in the 1980s and in the North Sea in the 1990s was carried out before local courts, with local judges and within a common political and regulatory environment, but the restructuring of LNG SPAs and long-term pipeline gas contracts in Europe is being carried out against a background of private arbitration and before arbitrators who may well have little familiarity with the relevant contracts or markets, and who are likely to be approaching their task from a background of diverse legal, political and regulatory influences.

In a European context, these issues have arisen as much under long-term pipeline gas contracts as they have under LNG SPAs, although the usually international context of each of those types of agreement has meant that dispute resolution has taken place before arbitrators rather than the local courts. To date, Asian markets have not shown the same moves towards liberalisation as have those in Europe; but if those moves do come, then the use of similar structures and provisions in those LNG SPAs suggests that the troubles that now beset LNG SPAs and long-term pipeline gas contracts in Europe may well come to visit those Asian LNG SPAs at that time.

US LNG import terminals: the perfect storm

Donna J Bailey
Chevron Gas and Midstream, Chevron USA Inc

1. Introduction

The current environment facing US LNG import and regasification terminals offers yet another reminder of the inherent unpredictability of energy markets. The US LNG import sector finds itself caught up in the perfect storm created by the unforeseen confluence of several developments:

- the dramatic increase in natural gas supply within the United States from unconventional gas sources;
- the resulting downward pressure on domestic natural gas prices;
- the continuing growth in global and particularly Asian LNG demand, fuelled in part by the Fukushima nuclear incident in Japan; and
- the growing price differential between US and non-US landed LNG prices.

As a result, US LNG terminal owners and capacity holders find themselves with largely stranded assets, with not only little commercial upside but also increasing operational risk. These factors are forcing terminal owners and their capacity holders to consider a variety of commercial, legal and operational measures in an effort to maintain the operational integrity of the US LNG import terminals and enhance the economic value of such facilities. This chapter explores:

- the reasons for the sector's rapid and dramatic change in outlook;
- the operational and commercial issues presented by the current environment;
- the various measures being taken or considered by US LNG import terminal owners and capacity holders to deal with the current situation; and
- some of the legal issues presented by those measures.

2. The change in the US LNG outlook

In the aftermath of the US energy crisis in 2000 and 2001, several factors contributed to the rush to build LNG import and regasification terminals in the United States. Among them were:

- the projected growth in US natural gas demand, particularly to fuel gas-fired power plants;
- the projected decline or stagnation in US natural gas production;
- projected high gas prices in the United States, making US energy markets more attractive to global LNG suppliers;[1] and
- the relaxation of federal regulation over greenfield LNG import terminals.[2]

These factors led owners of the three of the four existing LNG import terminals in the United States to reactivate their facilities[3] and developers to seek authorisation to construct and operate new LNG import terminals. At the height of the US LNG import terminal rush, there were over 35 proposals to construct and operate LNG import terminals in the United States, including several proposed offshore terminals.[4]

Despite an eventual slowdown in the number of proposed LNG import and regasification terminals, by 2011 a number of LNG import terminals had been authorised and 11 terminals had actually been built in the continental United States.[5] The map below shows the location and capacities of existing LNG import terminals in the United States.

By late 2007, however, storm clouds were beginning to develop that would eventually result in a steep decline in the quantity of LNG imported into the United States.[6] One of the most significant factors contributing to the decline of LNG imports has been the dramatic increase in the natural gas supply produced from unconventional gas sources, particularly shale gas.[7] Increasing domestic gas production has in turn led to downward pressure on US domestic gas prices at the

---

1 Randolph McManus, *The Liquefied Natural Gas Industry: A Changing Regulatory and Commercial Landscape*, 49 Rocky Mtn Min L Inst 20-1 (2003).

2 *Id.* The Federal Energy Regulatory Commission (FERC) has jurisdiction under Section 3 of the Natural Gas Act over the siting, construction, expansion and operation of LNG import and export terminals. 15 USC § 717(b) (2006) With the issuance of the *Hackberry* decision in 2002, FERC permitted, for the first time, the construction and operation of an LNG import terminal on a proprietary, individually negotiated basis, not subject to open access or service or rate regulation. *Hackberry LNG Terminal, LLC* 101 FERC ¶ 61,294 (2002) *order issuing certificates and granting reh'g*, 104 FERC ¶ 61,269 (2003). Prior to that time, import terminals in the United States were subject to full Natural Gas Act regulation, including the requirement that service be provided on an open access basis subject to rate and terms of service filed with and approved by FERC. Section 311(e) of the Energy Policy Act of 2005 codified and extended the relaxation of FERC's jurisdiction over LNG import and export terminals by:
- requiring that FERC approve all applications for the construction and operation of new LNG import or export terminals unless it finds the application is not in the public interest;
- prohibiting (at least until 2015) FERC from denying an application for a new import or export terminal or terminal expansion solely on the basis that the facility will be used wholly or in part by an affiliate; and
- prohibiting FERC from conditioning an order approving an application for a new or expanded import or export terminal on any regulation of the rates, charges, terms or conditions of service of the LNG terminal, or a requirement that such rates, charges, terms and conditions be filed with FERC, or a requirement that the terminal offer services to customers other than the applicant or affiliates. 15 USC § 717b(a) and (e)(3)(B).

As a result of this change in regulatory regimes, there are now essentially two categories of import terminal in the United States: pre-*Hackberry* heavily regulated terminals and post-*Hackberry* lightly regulated terminals.

3 Three of the four pre-*Hackberry* terminals had been mothballed for some time. See, for example, *Cove Point LNG LP*, 97 FERC ¶ 61,043, *order granting and denying reh'g in part, granting and denying clarification*, 97 FERC ¶ 61,276 (2001), *order denying reh'g and granting and denying clarification* 98 FERC ¶ 61,270 (2002) (authorising reactivation of the Cove Point terminal); *Southern LNG, Inc*, 89 FERC ¶ 61,314 (1999), *reh'g denied*, 90 FERC ¶ 61,257 (2000) (authorising reactivation of Elba Island terminal); *Trunkline LNG Co*, 49 FERC ¶ 61,199 (1989), *clarified and amended*, 69 FERC ¶ 61,129 (1994) (authorising service resumption).

4 FERC Office of Market Oversight and Investigations, *State of the Market Report* at 96 (2004), available at http://www.ferc.gov/legal/maj-ord-res/land-docs/som-2003.pdf.

5 The Gulf LNG Energy, LLC import terminal in Pascagoula, Mississippi is in the final stages of commissioning. When completed, it will be the 12th operating terminal in the United States. See Monthly Status Report No 46 filed by Gulf LNG Energy, LLC with FERC in Docket No CP06-12-000.

6 See FERC Division of Energy Market Oversight, *Natural Gas Market National Overview* at 8 (August 2011), available at http://www.FERC.gov/market-oversight/mkt-2011/08-2011-ngas-ovr-archive.pdf (hereinafter August 2011 Overview).

7 See John Meagher, "LNG's Unexpected, Unconventional Shift", *Petroleum Economist* (May 2010).

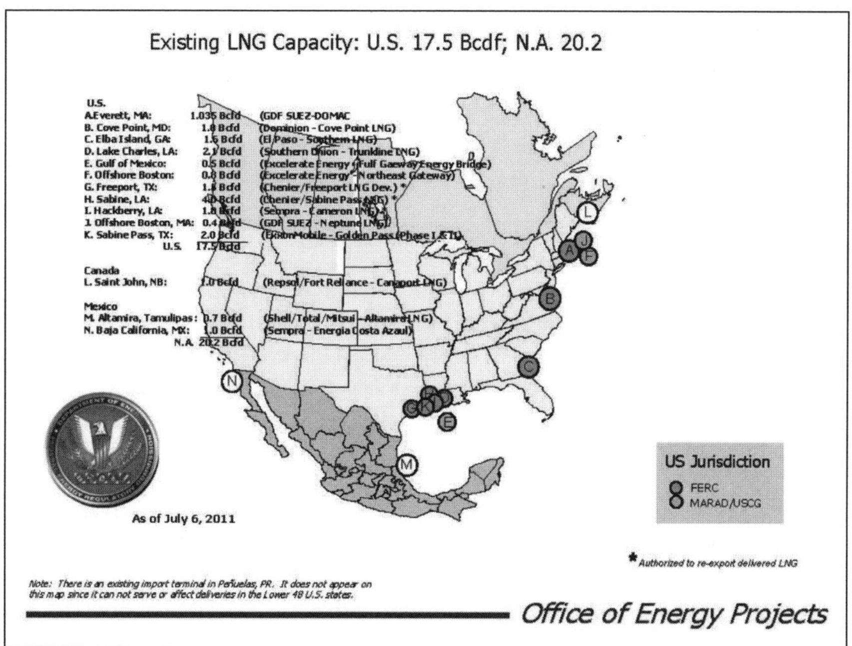

Source: http://www.FERC.gov/industries/gas/indus-act/lng.asp

same time as Asian and European markets were seeing price escalations.[8]

The global disparity in LNG prices received at the terminal ('landed' prices),[9] and the increasing demand for LNG in Asia due to economic growth in the region and the decreased reliance on nuclear generation in Japan following the Fukushima nuclear incident, are all factors contributing to the decline in US LNG imports.[10]

Exacerbating the consequences of these developments is that, as a result of the rush to build LNG import terminals in the early part of the decade, there now exists a significant surplus of import and regasification capacity in the United States relative to current and projected levels of LNG imports. Based on import capacity existing in 2010, it is estimated that LNG imports in that year represented an average utilisation of less than 7%.[11]

8 *Id.*

9 FERC's estimate of LNG landed prices for September 2011 demonstrates the competitive disadvantage LNG imports into North America face relative to other global markets:

Cove Point	$4.39	United Kingdom	$8.39
Lake Charles	$3.89	Belgium	$8.63
Altamira	$4.53	Spain	$9.18
India	$13.73	Korea	$14.38
Japan	$14.47		

See http://www.ferc.gov/market-oversight/mkt-gas/overview/2011/08-2011-ngas-ovr-archive.pdf.

10 See *August 2011 Overview, supra* note 6, at 11.

11 This estimate is based on a comparison of LNG import data made available by FERC and the Department of Energy as compared to the capacities listed in the immediate preceding diagram.

3. Unexpected operational issues

As a result of the factors discussed above, LNG import terminals in the United States are operating in a very different environment from that contemplated at the time they were reactivated or constructed. These factors have forced US LNG import terminals to address several unexpected operational issues, which fall into several general categories.

3.1 Maintenance of cryogenic conditions

It is axiomatic that in order for an LNG import terminal to be able to perform its primary functions – the receipt, storage and regasification of LNG and send-out of natural gas – the facility must be maintained in a cryogenic condition. In other words, a sufficient quantity of LNG must be retained in the LNG storage and piping systems for the terminal to maintain a constant minus 260º Fahrenheit temperature. This is necessary to ensure that the facilities do not experience repetitive thermal expansion (due to warming) and contraction (while in a cryogenic state).[12] The maintenance of a constant cryogenic state within the terminal can become a challenge without an inventory management system for static volumes or a continuous throughput of LNG. As more and more terminals are experiencing a reduction in imported LNG, the need for cryogenic maintenance has become an increased operational issue and burden.

Although the terms and conditions applicable to LNG terminals built after the *Hackberry* decision are not generally publicly available, it does not appear that many LNG import terminals in the United States have the ability to require their capacity holders to import any minimal quantity of LNG into their terminals for the purpose of ensuring that the facility is maintained in a cryogenic state. Nor does it appear that many US LNG import terminals have the ability to require their capacity holders to maintain any minimum LNG inventory at the terminals.[13]

Indeed, one of the key legal issues posed by the current environment is who has the contractual obligation to maintain the LNG import terminal in a cryogenic condition: the terminal owner or capacity holder. In the absence of an express obligation on the part of the capacity holder to import certain quantities of LNG to the terminal or to maintain a minimum inventory level at the terminal, and where the terminal owner has the obligation to provide terminalling services, the burden of maintaining the cryogenic conditions necessary to provide the services would appear to reside with the terminal operator. In any event, as is discussed below, terminal operators are taking a number of steps that address, directly or indirectly, the need for additional LNG imports to maintain cryogenic conditions in their terminals.

---

12 Aside from an import terminal's inability to provide LNG terminalling services if it is not maintained in a cryogenic state, the repetitive warming up and cooling down of import facilities presents significant safety and equipment integrity issues and may adversely impact on equipment warranties.

13 As an example, recent proceedings initiated by Dominion Cove Point LNG, LP (Cove Point) have confirmed that Cove Point does not have the ability to require its firm capacity holders to import LNG to maintain the operational integrity of the import terminal. See *Dominion Cove Point LNG, LP*, 135 FERC ¶ 61,261 (2011). A review of FERC gas tariffs of the four regulated terminals did not reveal a clear right of the part of any of the terminal owners to impose a requirement that firm capacity holders import LNG or maintain LNG inventories for the purpose of maintaining the terminals in a cryogenic state.

(a) Weathering

In order to maintain cryogenic conditions in the face of uncertainty as to the timing of LNG imports, terminals must maintain and circulate a minimum volume of LNG inventory. During those instances of static throughput, terminal operations must contend with the boil-off gas resulting from the continual circulation of LNG for thermal maintenance. The constant vaporisation of the LNG light ends that naturally occurs during this process is commonly referred to as 'weathering'. A consequence of weathering is that the gross heating value (GHV) and Wobbe index of the LNG increase.[14]

At some point, depending in part on the blended GHV of the LNG receipts into the terminal, weathering may cause the GHV or Wobbe index of the LNG inventory to exceed the maximum quality specifications of the downstream pipelines. This could result in the LNG owner having difficulty selling the regasified LNG, incurring pipeline penalties for delivering gas that does not comply with the pipeline's quality specifications, or incurring costs to implement British thermal unit management processes downstream of the LNG import terminal so that the gas can meet such quality specifications. The current low volume of LNG imports and the need to maintain a minimum LNG inventory have combined to exacerbate the magnitude of weathering and its associated impact on LNG quality. As with the issue of who has the burden of maintaining cryogenic conditions at an LNG import terminal, one of the key issues in today's environment is who should bear responsibility for any inventory weathering that causes contamination of incoming LNG or causes the natural gas sent out from the terminal to fail to meet downstream pipeline quality specifications. If multiple parties bear the responsibility, another issue is how the cost of remediating the situation should be allocated among them.

3.2 Boil-off gas

Small quantities of LNG will naturally vaporise or 'boil off' through normal agitation and heat ingress as LNG comes into contact with warmer atmospheric conditions during unloading, while in storage or while circulating through the cryogenic facilities at the import terminal. This boil-off gas must be dealt with in one of several ways. It may be recondensed and combined with larger volumes of LNG being sent to the vaporisers to be regasified and sent out to downstream pipelines, or the boil-off gas alone may be compressed and delivered to downstream pipelines. The boil-off gas may also be re-liquefied and returned to storage in the terminal. Which of these options is selected depends, in part, on what facilities are available at the particular terminal.

Although common to any LNG operation, boil-off volumes can have unexpected implications for LNG terminals operating at low import levels. First, when boil-off gas is sent out of the terminal as natural gas without inventory replacement, the

14 The weathering effect results from the fact that methane, which comprises the majority of the LNG stream and has a lower boiling point than the other components of the LNG stream, will vaporise first, leaving a proportionately greater percentage of other components – typically ethane and propane – in the LNG inventory. Since those components have a higher heating value than methane, as the quantity of methane decreases, the heating value of the residual LNG rises.

daily send-out of the boil-off gas will eventually deplete the LNG inventory, which in turn will threaten cryogenic conditions at the terminal.[15] This depletion of the LNG inventory will be accelerated if the import terminal does not have the capability to compress only the boil-off gas and instead has to use vaporisers to handle the send out. This is because vaporisers typically require a higher level of send-out than boil-off gas compressors.

The other impact of boil-off in terminals operating at reduced inventory levels is its detrimental effect on the quality of LNG in inventory and in send-out. For example, because nitrogen, like methane, vaporises before other components of the LNG stream, there tends to be a much higher percentage of methane and nitrogen in the boil-off gas composition. Depending on the level of nitrogen in the LNG in inventory, the boil-off gas that is sent out from the terminal may exceed the downstream pipeline's nitrogen quality specification. Similarly, because ethane tends not to be one of the components that vaporise early, the concentration of ethane in the remaining LNG inventory may increase, potentially to levels where the ethane content might exceed the quality specifications of foreign LNG buyers, which can become relevant for LNG that is re-exported.

4. Economic impact

The operational issues discussed above are not the only issues arising out of the current LNG market. In fact, the operational issues themselves contribute to the other major issue arising out of the current import market: the decreased ability to extract commercial value from US LNG import terminals. The economic consequences of the current LNG import market in the United States are significant for terminal owners and capacity holders alike. Although the most significant economic burden falls on the terminal capacity holders, because they generally bear all or most of the terminal's costs regardless of the extent to which they use the terminal,[16] the current environment also has an undeniable effect on terminal owners. The current low utilisation level of US LNG import terminals limits the terminal owners' ability to generate incremental revenue associated with higher throughput, additional services or trading activities at their terminals. Thus, in the absence of a change in the status quo, terminal owners in the United States face the prospect of both no real economic growth in the foreseeable future and additional costs to address the operational issues caused by low levels of LNG imports.

One of the largest potential costs faced by terminal owners, unless that cost can be shifted to their capacity holders, is the cost of acquiring LNG to maintain cryogenic conditions. Given the price levels that LNG currently commands in the European Union and Asia, if an LNG terminal owner has to acquire a cargo of LNG for operational purposes, the costs associated with the acquisition of the cargo are significant. FERC recently estimated that an average-sized cargo delivering into the Cove Point LNG terminal could cost the acquirer of the cargo in excess of $30 million.[17]

---

15 This reduction in inventory, as well as the weathering discussed in the preceding section, may be eliminated by the installation of a boil-off gas re-liquefaction unit.

16 See, for example, *Trunkline Gas Co, Trunkline LNG Co*, 108 FERC ¶ 61,251 (2004) at p 17 and p22; *Dominion Cove Point LNG, LP, Dominion Transmission, Inc*, 115 FERC ¶ 61,337 (2006) at p26.

Acquiring LNG to keep LNG import terminals cold is not the only cost that terminal owners face in the current low import environment. In order to preserve LNG inventory for as long as possible, some terminal owners have elected to install facilities to reliquefy the boil-off gas so as to avoid the depletion of inventory associated with send-out of the boil-off gas.[18] Other terminal owners have installed one or more compressors for the purpose of compressing boil-off gas rather than utilising vaporisers.[19] Both installations – compressors and reliquefaction equipment – represent additional capital and operating costs.

Like terminal owners, US terminal capacity holders face significant costs and lost economic opportunities in the current LNG import environment. Since the rates for capacity typically recover all of the terminal owner's costs associated with the construction and operation of the terminal, the capacity fees paid by the firm capacity holders are high.[20] In addition, capacity owners face the same constraints on extracting value through the use of terminal capacity as terminal owners. Thus, in the current low LNG import environment, the economic burden associated with terminal capacity payments often cannot be offset by importing LNG and selling it into US domestic gas markets, since the price of LNG available for purchase so far exceeds the price for which it could be sold into US markets. Capacity holders also face the prospect that their LNG inventory will weather to the point of requiring downstream processing, nitrogen injection and/or incurring pipeline penalties or damages due to failure to meet specifications. Additionally, they face the potential commercial burden of importing sufficient LNG to keep the terminals cold.

5. Commercial solutions

Given the operational challenges faced by LNG import terminals in the United States today, the significant economic costs associated with these terminals and the incentive of terminal owners and capacity holders alike to find ways to make the terminals more operationally and commercially viable, it is of little surprise that in recent years a number of strategies have emerged to address the changed commercial landscape.

5.1 Re-exports

The first strategy to deal with the challenge of keeping terminals in a cryogenic state in an era of unpredictable LNG imports began to emerge in 2008, when two applications were filed by terminal owners with FERC seeking authorisation to amend their Section 3 authorisations so as to permit the terminals to engage in the re-export of foreign sourced LNG.[21] This strategy had the dual benefit of facilitating

17 *Dominion Cove Point LNG, LP*, 135 FERC ¶ 61,261 at p34 (2011).

18 See, for example, *Freeport LNG Dev*, LP, 127 FERC ¶ 61,105 (2009).

19 See, for example, *Dominion Cove Point LNG, LP*, 127 FERC ¶ 61,258 (2009) (installation of additional compression at an estimated cost of over $7 million to better manage boil-off in low flow periods); *Trunkline LNG Co*, 123 FERC ¶ 61,152 (2008) (installing compressor unit to compress boil-off gas when the terminal is not sending out through its vaporisation); *Cameron LNG, LLC*, 132 FERC ¶ 61,070 (2010) (installing a back up boil-off compressor unit to avoid having to flare the gas if its first boil-off compressor failed); *Sabine Pass LNG*, LP, 134 FERC ¶ 61,091 (2011) (installing redundant boil-off compression and related facilities to avoid use of vaporisation facilities).

20 See *supra* note 16.

21 *Freeport LNG Dev, LP*, 127 FERC ¶ 61,105 (2009); *Sabine Pass LNG*, LP, 127 FERC ¶ 61,200 (2009).

the terminal's ability to attract the quantities of LNG required for operational purposes by permitting any LNG in excess of that required for operational purposes to be redirected to higher-value markets outside of the United States, and increasing the overall commercial utilisation of the capacity in the terminals.

By mid-2011, a number of the existing LNG import terminals in the United States had sought, indicated they were planning to seek or received authorisation to re-export LNG imported into the United States.[22] Export authority provides both terminal operators and capacity holders (to the extent that such re-export services are provided to capacity holders at the terminals) with improved tools to address the terminals' operational needs and compete for LNG within the global marketplace.

5.2 Liquefaction

Ironically, the factor that has contributed so significantly to the decreased utilisation of US LNG import terminals – that is, the increase in US downstream gas production from unconventional sources and the resulting decline in gas prices – has given rise to the latest strategy for revitalising US LNG import terminals. Cheniere Energy, Inc, parent of Sabine Pass LNG, LP, led the US market with its innovative proposal to convert the Sabine Pass LNG import terminal to a fully bi-directional import-export LNG facility through the addition of liquefaction equipment for the export of domestically produced natural gas.[23] The liquefaction project is clearly viewed by Sabine as an opportunity to enhance the utilisation and commercial value of the Sabine terminal.[24] One of the issues related to the proposed liquefaction and export project is whether and how it will impact on existing operations and customers, since Sabine's filing contains little explanation as to how the operations of the existing terminal and the proposed bi-directional terminal will be integrated and coordinated.[25]

Following closely behind Sabine in the move to convert existing LNG import facilities to bi-directional import-export facilities was Freeport LNG Development LP. Freeport has announced its plans to convert its import terminal in Quintana Island, Texas to a bi-directional terminal and is currently undergoing FERC's Natural Environmental Policy Act pre-filing review in connection with that proposal.[26] Trunkine's Lake Charles, Louisiana LNG import terminal is also in the process of seeking authorisation to engage in the liquefaction and export of natural gas

---

22 See, for example, *Cameron LNG, LLC*, 134 FERC ¶ 61,049 (2011); see also *Application for Blanket Authorization to Re-Export Liquefied Natural Gas* filed by Dominion Cove Point LNG, LP, with the US Department of Energy in FE Docket No 11-98-LNG on August 8 2011, and related letter filing dated July 8 2011, submitted by Dominion Cove Point to FERC requesting a determination as to the applicability of FERC's mandatory pre-filing National Environmental Policy Act process to its proposed re-export of LNG.

23 See Application of Sabine Pass Liquefaction, LLC and Sabine Pass LNG, LP for Authorisation Under Section 3 of theNatural Gas Act, filed on January 31 2011, in Docket No CP11-72-000.

24 See Letter to Shareholders from Charif Souki, Chairman of the Board of Cheniere in Cheniere's 2010 Annual Report.

25 See, for example, Motion of Chevron USA Inc to Intervene and File Comments in Docket No CP11-72-000 filed on March 4 2011 (seeking a better understanding of how the liquefaction project at Sabine's import terminal would be operated and integrated into existing operations and how existing customers would be impacted).

26 See Letter Order dated January 5 2011 in Docket No PF11-2-000 approving use of FERC's pre-filing review process for Freeport's liquefaction project.

produced in the United States.[27] Similarly, Sempra Energy's chief executive recently suggested that Cameron LNG, LLC, which already has authorisation to engage in exports of previously imported LNG at its terminal, would consider adding liquefaction equipment at its terminal for the export of domestically produced gas if there were adequate customer subscription to the services.[28]

5.3 Trucking

In what appears to be yet another strategy to revitalise a US LNG import terminal and capture additional economic opportunity, Southern LNG Inc, owner of the LNG import terminal at Elba Island, Georgia, has requested authorisation from FERC to reactivate truck loading facilities at its terminal for the purpose of leasing the facilities to, and operating the facilities on behalf of, an affiliated joint venture.[29] Southern LNG states that a newly created joint venture between Southern LNG or one of its affiliates and a third party would purchase LNG imported into the terminal and take delivery of the LNG at the truck loading racks for sale to peak shaving facilities and as an alternative to diesel fuel for heavy duty vehicles.

Although the liquefaction applications proposed by Sabine and Freeport, and the trucking application proposed by Southern LNG, are still pending before FERC, it is likely that once approved, they and the other similar proposals that have been announced will begin to produce a substantial change in operations at US import terminals.

6. The contractual and regulatory implications of the current environment

It is scarcely surprising that at the same time as terminal operators are exploring ways to address current operational and commercial issues affecting US import terminals, they are attempting to pass on the resulting costs and risks to their capacity holders. However, it appears that, for the most part, the current environment is not one that was previously envisioned by many terminal owners or their capacity holders. Hence, there has been little indication – at least for the greenfield LNG projects built over the last decade – that there was any clear contractual assignment of responsibility for the issues arising out of the current import environment.

This has left terminal owners and capacity holders at odds over the many issues arising out of the current situation:

- who should bear the costs of maintaining the terminal in a cryogenic state;
- whether capacity holders can be required to import certain quantities of LNG and, if so, which capacity holders should be required to do so;
- if a terminal owner acquires LNG for thermal maintenance, whether the costs of acquiring the LNG should be allocated to its capacity holders and, if so, how such costs should be allocated among them;

---

27 On July 22 2011 the Department of Energy Office of Fossil Energy granted long-term authorisation to Lake Charles Exports, LLC, a jointly owned subsidiary of BG Group plc and Southern Union Company, to export LNG on its own behalf or as agent for BG LNG Services, LLC. *Lake Charles Exports, LLC*, DOE/FE Order No 2987, FE Docket No 11-59-LNG (July 22 2011).

28 *LNG World News*, August 10 2011.

- whether capacity holders can be required to send out boil-off gas;
- who should bear the costs of gas reliquefaction equipment or boil-off compressors;
- how costs associated with any physical enhancements at LNG terminals should be allocated among a terminal's existing and potential new customers;
- if a terminal's operation is fundamentally altered, what rights existing customers have either to participate in any additional services or to protect themselves from such services adversely impacting on their existing services; and
- who bears the economic consequences of in-tank weathering.

The above list is not all-inclusive, obviously, but is sufficient to highlight the contractual complexity behind the current import situation. Several different approaches are being utilised to bring some of these issues to resolution.

6.1 Negotiated resolution

As with most contractual disputes where both parties have an incentive to reach a commercial resolution, it would appear likely that one resolution of the various issues currently affecting US import terminals has been or will be through individual negotiation among the affected stakeholders. In the United States, where terminals constructed after 2002 are not required by FERC to make their terms and conditions of service public, it is generally difficult to access information about whether and how the parties have dealt with these issues. This suggests that approaches taken in this context to resolve operational and commercial issues are likely to be terminal specific, with limited opportunity for each terminal to learn from and build on the approaches taken by other terminals.

The Sabine terminal does offer one window into how these issues have been dealt with, since its terminal use agreements and related amendments are publicly available.[30] A review of amendments to certain of Sabine's terminal use agreements in the summer of 2010 reveals the parties' agreement to certain basic measures to preserve the operational integrity of the facility. Those measures include:

- Chevron USA Inc's and Total Gas & Power North America, Inc's[31] general agreements to maintain a minimum inventory level;
- Chevron's and Total's agreements to send out their *pro rata* share of boil-off gas; and
- Chevron's and Total's agreements to pay their *pro rata* share of the costs to cure off-spec gas due to weathering associated with boil-off.[32]

---

29 See *Southern LNG Inc*, 131 FERC ¶ 61,155 (2010) and related application in Docket No CP10-477-000.

30 Cheniere is required by US securities laws to file all material agreements with its periodic reports to the Securities and Exchange Commission. As a result, its terminal use agreements and amendments thereto are a matter of public record.

31 Chevron and Total are two of the firm capacity holders at the Sabine terminal.

32 See Exhibits 10.2 and 10.3 to Cheniere's 10-Q Quarterly Report filed on August 6 2010 with the SEC.

6.2 Regulatory resolution

A different approach from resolution of the current operational and commercial issues affecting US LNG import terminals is necessary for terminals that continue to be subject to significant regulatory oversight by FERC:[33] the Dominion Cove Point, Southern LNG Elba Island, Trunkline Lake Charles and Distrigas of Massachusetts Everett terminals. These import terminals, like interstate pipelines in the United States, are generally not free to engage in privately negotiated changes to their terms and conditions of service, but instead are required to seek approval from FERC to implement any changes to their terms and conditions of service.[34]

Recent filings by Dominion Cove Point LNG, LP are illustrative of the process that is required by the owners of such terminals to effect changes in their conditions of service and the substantive approaches that may be proposed by them to address certain of the current operational issues.

On May 27 2011, in Docket No RP11-2136-000, Cove Point filed revised tariff sheets proposing, among other things, that it have the ability to issue an operational flow order[35] requiring firm capacity holders at the terminal to deliver certain quantities of LNG to the terminal when necessary to maintain the operational integrity and performance of the terminal.[36] Cove Point's transmittal letter accompanying its tariff filing stated that the filing was necessitated by the threat that it would be unable to keep the cryogenic portions of its terminal cooled to the necessary temperature due to the lack of LNG imports to the terminal.[37] Specifically, Cove Point proposed that when the issuance of an operational flow order became necessary in order to maintain operational integrity of the terminal, the order would identify the minimum quantity of LNG that must be delivered and the time period by which delivery would be required. The order would be directed to the capacity holder that had not delivered an LNG cargo to the terminal in the longest period of time.[38]

In the event that the capacity holder to which the operational flow order was issued failed to comply with the order, Cove Point proposed that the responsible party be assessed a penalty equal to the product of the minimum quantity stated in the order multiplied by the higher of $25 or the price of natural gas for the day after the date on which the LNG was required to be tendered.[39] To the extent that the penalty failed to cover Cove Point's cost of itself purchasing the quantity of LNG that the responsible party failed to deliver, Cove Point proposed to bill the responsible party for the difference between the costs that it incurred and the penalties collected.[40]

---

33 See, *supra* note 2.

34 See, for example, *Southern LNG Inc*, 89 FERC ¶ 61,314 (1999).

35 An operational flow order is an order issued by an interstate pipeline or terminal operator requiring its customers to take certain action or refrain from taking certain action or restricting service as a result of an operational problem on the pipeline or at the terminal.

36 See the May 27 2011 Tariff Filing of Dominion Cove Point LNG, LP, in Docket No RP11-2136-000 at p2.

37 *Id.*

38 As one would expect, there was significant disagreement as to whether the terminal operator or its customers should bear the costs of any required import and, if the customers are required to do so, how such costs should be allocated among them.

39 See section 12(c) of the General Terms and Conditions of Cove Point's FERC Gas Tariff as proposed in its May 27 filing.

40 *Id.*

In addition, Cove Point proposed to revise its scheduling provisions to require that capacity holders be subject to a penalty if the capacity holder failed to bring a cargo to the terminal as scheduled unless notice of diversion was given at least 15 days before the scheduled delivery date.[41] Cove Point further proposed that it be able to require the party that failed to bring in a cargo as scheduled to pay all of the costs incurred by Cove Point as a result of that failure.

On June 24 2011 FERC issued an order rejecting Cove Point's proposed operational flow order related tariff revisions, without prejudice to Cove Point's ability to make a future filing proposing the ability to recover its operational costs, suspended the remaining tariff sheets for the maximum statutory period to be effective November 26 2011, subject to refund and the outcome of a technical conference in the proceeding.[42]

Although a technical conference has been held in the proceeding, to date little public progress has been made in resolving the disagreements between Cove Point and its firm capacity holders relating to Cove Point's proposed tariff revisions. The parties did, however, agree to an interim settlement that, on a non-precedential basis, addressed Cove Point's need to keep its terminal cold in the near term by arranging for a one-time operational purchase of LNG by Cove Point and a one-time procedure to recover the cost of the LNG.[43] Although not a permanent solution, the interim settlement did buy time for the parties to continue to explore resolution of the issues prompting Cove Point's filing in Docket No RP11-2136-000.

Although this regulatory struggle is not over, it – and the other commercial, contractual and regulatory efforts underway to deal with issues arising out of the current import environment – make it clear that these issues will continue to plague US terminal owners and their capacity holders until some resolution is reached or the US LNG market changes yet again and presents a different set of challenges.

---

41 See section 5.3(g) of Cove Point's FERC Gas Tariff, Rate Schedule LTD-1 and LTD-2.

42 *Dominion Cove Point LNG, LP*, 135 FERC ¶ 61,261 (2011).

43 *Dominion Cove Point LNG, L.P.*, 136 FERC ¶ 61,059 (2011).

Shale gas for LNG

Vivek Bakshi
Jeff Scobie
Ron Stuber
Fraser Milner Casgrain LLP

1. Introduction

This chapter focuses on the part that shale gas will play in the development of the global gas industry and examines how exploration, development and production are being approached in Canada.

When the International Energy Agency declared in mid-2011 that the years leading to 2035 may be a "golden age for gas",[1] the full impact of the Tohoku earthquake and resulting tsunami on the global nuclear power generation industry was yet to be determined. The prospect of natural gas-fired power as a partial replacement for nuclear power will further play into the evolution of the energy supply mix. Against this backdrop, shale gas may well play a significant part in this 'golden age'. Moreover, no chapter on shale gas would be complete without a reference to the game-changing nature of the shale gas sector on global energy markets. Overall, international shale gas resources[2] are estimated at over 22,000 trillion cubic feet[3] (tcf), with as much as 6,622 tcf estimated as being technically recoverable.[4]

The vast majority of shale gas resources production to date has taken place in the United States, with as much as 25% of total US production represented by shale gas in 2010[5] (a twelvefold increase over 10 years). The development of this significant resource has contributed to a fall in natural gas prices in North America and has had a significant effect on its nascent LNG industry. This chapter examines the effect of the exploitation of US and Canadian shale gas resources on what was, in the mid-2000s, a vibrant LNG import market in those countries, and the resulting changes to the LNG value chain.

2. Production of shale gas

2.1 Extent and location of shale gas in Canada

It is estimated that there is potentially 1,000 tcf or more of shale gas in place in Canada.[6] Normally, only 20% or so can be economically recovered (as compared with

1 "Are We Entering a Golden Age of Gas?", International Energy Agency, Special Report World Energy Outlook 2011.

2 "Shale Gas and the Outlook for U.S. Natural Gas Markets and Global Gas Resources", presentation by Richard Newell, US Energy Information Administration, June 21 2011.

3 These figures do not purport to cover all potential regions where shale gas might be found and exclude, most notably, Russia and the Central Asian and Middle Eastern regions.

4 *Ibid.*

5 "Shale Gas and the Outlook for U.S. Natural Gas Markets", *supra* note 3.

90% or so of conventional natural gas), but this percentage could grow with further advancements in recovery technology.[7]

Shale gas is just one of a number of 'unconventional' sources of natural gas in Canada, the others being coal bed methane, tight gas and frontier gas.

There are a number of known significant reserves of shale gas in Canada, including Horton Bluff in the Maritime provinces, Utica in Quebec, the Colorado Group in Alberta and Saskatchewan, and (most significantly for prospective west coast LNG export projects in Canada) the Horn River and Montney areas of northeast British Columbia. There are currently several proposed LNG export projects to be located on Canada's west coast, all of which will (to varying degrees) depend on gas supplied from shale gas reserves. For example, Kitimat LNG will acquire its gas from numerous sources, including its sponsors' shale gas reserves in the Horn River area, and the Petronas/Progress Energy LNG project will utilise Progress Energy's shale gas reserves in the Montney area.

2.2 Differences between shale gas and conventional natural gas

Shale stone is a sedimentary rock that was originally deposited as clay and silt. It is much less permeable than conventional sandstone natural gas formations, so only minor amounts of gas can flow naturally to a wellbore. In order for gas to flow through shale, it must pass through much smaller pore spaces than conventionally sourced gas. Because conventional gas reservoirs result from the migration of methane to such areas where they are trapped, they have a higher concentration of methane molecules than shale sources.

2.3 Ramifications for gas producers

For the reasons set out above, shale gas has historically been regarded as too difficult and uneconomic to produce. However, new technologies – principally horizontal multi-stage hydraulic fracturing (commonly known as 'fracking') and horizontal drilling – have now made production from shale easier and cheaper. With conventional North American natural gas reserves in decline, and shale gas now easier and cheaper to produce than was previously the case, there has been a recent sharp escalation in shale gas production in North America. Shale gas production costs per volume are still higher than those for conventional gas, but will likely decrease as technology improves.

Shale gas development in Canada is still in its early stages, so the industry a whole is still somewhat uncertain with respect to its economics, recoverability rates of gas, quantities of reserves, environmental impact and the functionality and safety of technology.

2.4 Ramifications for LNG export projects

Gas from shale is similar to gas from conventional sources in the sense that its composition, or specifications, can vary widely from area to area. Natural gas is a

---

6 "Understanding Canadian Shale Gas", National Energy Board of Canada Energy Brief, November 2009.

7 *Ibid.*

mixture of hydrocarbons primarily composed of methane, but also containing smaller amounts of heavier hydrocarbons and other gases or impurities. The percentages of each can vary widely from source to source, and shale gas is no different in this respect. Shale gas, however, is typically a dry gas, although some formations do produce wet gas.[8]

Once extracted, the shale gas is processed to remove other gases, water, sand and impurities. Some heavier hydrocarbons – such as ethane, propane and butane (natural gas liquids (NGLs), and condensates – may be removed near the production site and sold separately; such removal is often performed in order for the gas to meet pipeline specifications. Alternatively, if the gas is destined to be delivered to an LNG liquefaction plant, they may be removed close to or at the plant. The specifications of the inlet gas to the plant will need to match the plant's design, and possibly the specification limitations that may be imposed on the LNG seller by the downstream LNG buyer or pipeline specifications at the delivery point. Typically, the liquefaction process will require the prior removal of almost all non-methane components. LNG is usually composed of a very high percentage of methane (well over 90%) as compared to typical natural gas. If LNG is shipped with non-methane components such as NGLs, those components may (depending on the relevant import facility) be removed on receipt.

As is discussed in further detail in section 3.4, the proposed LNG liquefaction plants in western Canada clearly consider certain Asian countries as their primary markets for LNG. Typically, the heat content requirements of LNG import terminals in these countries are higher than for terminals in Europe and the United States. This means that more NGLs may be left in the LNG than would be the case for LNG destined to other countries.

LNG liquefaction projects rely on a large and stable source of natural gas supply as feedstock for their plants, and for significant periods of time (traditionally, 20 to 25 years). Shale gas resources, as dedicated sources of supply, could present some challenges, considering that decline rates for shale resources seem to be far more accelerated than those for conventional resources (largely due to the effects of horizontal drilling). If one were to look at a production curve for a shale gas resource as compared to one for a conventional resource, the curve would be significantly steeper and shorter in duration. This could present difficulties for an LNG producer that is reliant on specific shale gas reserves for long-term supply. A related concern is that, partly because the decline rates for shale gas are more difficult to predict than those for conventional reserves, attempts to estimate total recoverable reserves are also much more difficult.

2.5 Location of shale gas plays relative to LNG facilities

Given the necessary coastal location of LNG production, storage and loading facilities, shale gas reserves that are closer to such coastal locations are obviously preferable from the LNG producer's point of view. The locations of the shale gas

---

8 "An Overview of Modern Shale Gas Development in the United States", J Daniel Arthur, Bruce Langhus and David Alleman. http://www.all-llc.com/publicdownloads/ALLShaleOverviewFINAL.pdf.

reserves and the facilities will dictate the necessary infrastructure required to bring the shale gas to the plant. With respect to the Montney and Horn River shale gas areas of northeast British Columbia (which would be the sources of gas for most of the proposed Canadian west coast liquefaction plants), the Canadian National Energy Board (NEB) has stated that "there is not enough infrastructure in northeast [British Columbia] to handle growth in shale gas production beyond the next few years".[9] However, there have since been a number of pipeline proposals and approvals which will help to alleviate this concern.

2.6 Particular environmental issues faced by shale gas producers

The process of hydraulic fracturing involves the injection of fluid (usually water), proppant (often sand) and gels into the well at high pressure in order to fracture the shale. This creates conduits through which the gas can flow to the well. Horizontal drilling captures the gas from the fractured shale. These combined processes have raised numerous environmental concerns, including the concern that the quality of groundwater may be adversely affected by the fracturing fluid. *The Economist* recently reported that "although widespread pollution of groundwater by hydraulic fracturing seems unlikely (shales that hold gas typically lie far deeper than groundwater supplies), such risks have raised a great deal of environmental concern about the technology…this has led to a moratorium on shale gas exploration in France".[10] Similar complete or partial bans or moratoria on hydraulic fracturing (or related processes) have been imposed in other jurisdictions, including the US states of New York and New Jersey, the Canadian province of Quebec, the Karoo region of the Republic of South Africa and the Australian state of New South Wales. There have even been suggestions that hydraulic fracturing has, in some cases, caused earthquakes.

Other environmental concerns regarding shale gas production include:

- the potentially higher carbon dioxide emissions associated with shale gas production as opposed to conventional gas production;
- the heavy draw on freshwater resources due to the large quantities required for hydraulic fracturing fluid; and
- the proper disposal of water used in hydraulic fracturing operations.

Although far more wells are required for shale gas production as opposed to conventional gas production, the land-use footprint for shale gas is not expected significantly to exceed that for conventional gas, due to advances in horizontal drilling technology that allow for up to 10 or more wells to be drilled and produced from the same well site.

2.7 Regulatory environment facing Canadian shale gas producers

The exploration and production of oil and gas in Canada, including shale gas, are regulated by a number of federal, provincial and in some cases municipal laws and

9 "A Primer for Understanding Canadian Shale Gas", National Energy Board of Canada Energy Briefing Note, November 2009.

10 "Coming to a Terminal Near You", *The Economist*, print edition, August 6 2011.

governmental agencies. Canada's Constitution dictates the split of legislative authority between the federal and provincial governments. Shale gas development is regulated under the same legislation, rules and policies as apply to the development of conventional natural gas.

Provincial rules and regulatory agencies are of primary relevance to Canadian gas producers. For example, most aspects of the oil and gas industry in Alberta are regulated by the Alberta Energy Resources Conservation Board, and in British Columbia by the BC Oil and Gas Commission. These agencies set requirements for drilling and production operations, including strict requirements surrounding well bore casings and their depths to ensure that gas and fluids cannot migrate up a well bore to contaminate groundwater. They also set requirements concerning the safe disposal of fluids that return to the well head as part of the hydraulic fracturing process. Fluids that cannot be treated and recycled must be injected into approved deep disposal wells. Most permits and leases to exploit oil and gas resources are granted by provincial governments, as the owners of those resources, and royalties are payable to such governments.

The NEB is a federal agency, established under federal legislation, which regulates several aspects of Canada's energy industry, including the construction and operation of interprovincial and international oil and gas pipelines, as well as the tolls and tariffs for such pipelines. The NEB also regulates the import and export of gas, LNG and NGLs.

Environmental matters are regulated by a mixture of federal and provincial laws. With Canada the home to many indigenous peoples, the use of land traditionally occupied by such indigenous bands ('First Nations') is governed by federal law, being for the most part the Indian Act, the Indian Oil and Gas Act and the Indian Oil and Gas Regulations – the latter of which sets out the procedure for the acquisition of permits, leases and contracts for the exploration and development of natural gas on First Nation lands.

3. Impact of shale gas on gas markets

While the production of shale gas is subject to numerous challenges – the most significant of which relates to its environmental impact, as discussed – the prospect that shale gas might be an importing nation's solution to addressing energy security and independence is significant. Depending on the source and reliability of those imports, shale gas production that displaces such imports may have material geopolitical consequences.[11]

The prospect of energy independence in Poland through the production of shale gas has been pursued with vigour, and while significant challenges to the development of the polish shale gas industry remain, the geopolitical consequences of being independent of Russian-supplied natural gas are being debated avidly.[12]

11 See "Application of Sabine Pass Liquefaction, LLC and Sabine Pass LNG, LP, for Authorization under Section 3 of the Natural Gas Act", January 31 2011, p31, in which the applicant identifies certain geographical and national security interests that might be advanced by LNG exports from the United States.

12 "Dash for Poland's gas could end Russian stranglehold", *The Times*, April 5 2010.

3.1 Effect of flood of supply into markets supplied by pipeline

In North America,[13] shale gas production has not been viewed in the same way as in Poland, largely due to a lack of reliance on 'foreign'[14] natural gas (but see section 3.2 below). Shale gas production has offset declining production from conventional natural gas resources (due to basin maturity), but coincided with the global financial crisis and falling natural gas demand. Recent figures suggest that natural gas demand was down 2% in North America in 2009, compared to 2008 figures.[15] In the same period, combined North American domestic natural gas production increased by 1%.[16] The development of the shale gas sector in North America has therefore addressed concerns over declining production, but due to declining demand over the same period, this has had a downward impact on price.

Notwithstanding location, natural gas prices were low in both the United States and Canada in 2009 and 2010 (when compared to prices from earlier that decade) – remaining, on the basis of simple 12-month averages, in the $4/million British Thermal Units (mmbtu) to $5/mmbtu range.[17] During the same period, crude oil prices recovered from the immediate aftermath of the global financial crisis to return to pre-recession levels. As the 2009 and 2010 figures for natural gas show, natural gas prices did not recover. This enduring decoupling of natural gas and crude oil prices in North America may accordingly be here to stay, while shale gas production levels remain high; this certainly appears to be the view of both the US and Canadian governments.[18]

The flood of supply of shale gas in the United States has resulted in the US Energy Information Administration revising, in its Annual Energy Outlook 2011, its projected Henry Hub spot price for 2035 to $7/mmbtu, from $9/mmbtu a year earlier, with prices not expected to cross the $5/mmbtu threshold until 2020.

3.2 The LNG market pre-shale and post-shale

The consequences of sustained low natural gas prices in the United States and Canada have had a significant effect on their respective LNG industries. Apart from Alaska's Kenai LNG (which appears to be at or near to the end of its useful life), investors in the United States and Canadian LNG sectors had focused their attentions on the importation and regasification of LNG; as a result, there are presently 11 LNG regasification terminals in the United States (with one new terminal under construction), and one terminal in Canada. However, while a number of these regasification terminals have been in existence for a considerable amount of time,

---

13 In this chapter, 'North America' is taken to mean the United States and Canada.

14 There has historically been a fluid natural gas market between the United States and Canada, such that the two are often viewed as a single market. The availability of cheap natural gas from the United States is now resulting in natural gas being imported into Southern Ontario from the United States, rather than such Canadian demand being satisfied by production from the Western Canada Sedimentary Basin (which has been declining consistently since 2005).

15 "Canadian Crude Oil, Natural Gas and Petroleum Products. Review of 2009 and Outlook to 2030", National Resources Canada, Petroleum Resources Branch, Energy Sector, May 2011, page 13.

16 *Ibid.*

17 "Natural Gas and Sulphur Price Forecast" GIJ Petroleum Consultants Ltd, effective July 1 2011.

18 "Annual Energy Outlook 2011" US Energy Information Administration, p78 and Natural Resources Canada Review, *supra* note 15, p19.

investment decisions were taken to build a significant portion of the newer regasification capacity during periods of high domestic demand, declining conventional supply and high natural gas prices. In fact, in 2004 the US Energy Information Administration was aware of as many as two dozen proposals to build new regasification terminals, and projected that by 2010 the new terminals would be collectively importing into North America 812 billion cubic feet (bcf) of LNG annually.[19]

In Canada, the position was much the same in the mid-2000s. The outlook for LNG import was bullish and as many as nine regasification terminals were proposed at one time (of which Canaport LNG, Rabaska LNG and Cacouna LNG were approved).[20,21]

We now know that part of the game-changing nature of the growth of shale gas production was to stop a number of the proposed North American regasification terminals in their tracks, as production grew, prices fell and demand declined (quite the opposite of the previous situation). Of the liquefaction terminals that are now in commercial operation, there appears to be significant overcapacity. The latest figures available are for 2009 and appear to show US and Canadian LNG import capacity standing at around 13 bcf per day, but with actual imports of less than 2 bcf/day – an over-capacity of around 85%.[22]

While it may be fair to say that shale gas production is directly responsible for the overcapacity being experienced by LNG regasification terminals in North America, a number of other factors are also at play.

The tolling model employed by a number of the regasification terminals may well have also contributed to the diversion of cargoes originally destined for the United States and Canada to those markets where natural gas prices did not suffer quite the same fate as in North America. The evolution of the value chain from its point-to-point origins has given rise to LNG buyers not necessarily being equity owners of regasification terminals (or having 'in-house' demand for the LNG purchased for delivery to such regasification terminals). Rather, the tolling arrangements appear to have been specifically designed to provide access to a lucrative downstream North American market for natural gas. As that downstream market has weakened, further recent modifications to the value chain have contributed to diversions; spot trading, flexible shipping arrangements and flexible regasification capacity. While it is not possible to say whether any one of these factors was the key ingredient to the extent of the overcapacity now seen in North American regasification terminals, the 'new', more flexible supply chain certainly permitted capacity holders/tollers to abandon capacity and seek arbitrage opportunities.

---

19 "The Global Liquefied Natural Gas Market: Status and Outlook", Energy Information Administration www.eia.doc.gov/oiaf/analysispaper/global/us/lng.html.

20 "Canadian LNG Import and Export Projects Update", Natural Resources Canada, www.nrcan.gc.ca/eneene/sources/natnat/imppro-eng.php.

21 Canaport LNG, the Repsol and Irving Oil sponsored regasification terminal, was the only terminal constructed in Canada and presently supplies regasified LNG principally to the northeastern United States.

22 Natural Resources Canada Review, *supra* note 15, at p15.

3.3 The case for LNG export in the United States and Canada

As the North American LNG sector continues to grapple with import overcapacity and shale gas production shows little sign of falling off (despite increasing environmental concerns and public interest in this issue), sponsors have turned their attention to exports. In a quite truncated timeframe, the value chain has been turned on its head, and the United States and Canada are now in the headlines for the number of new projects seeking export licences from their respective regulators. Freeport LNG, Sabine Pass LNG and Cameron LNG have been authorised to re-export delivered LNG and new applications have been made for the construction of liquefaction plants. In Canada, Kitimat LNG, which was originally proposed as an import terminal, has applied to Canada's National Energy Board for an export licence and several other new liquefaction plants have been targeted for Canada's west coast, including Shell Canada's Prince Rupert Island Project, BC LNG Export Co-operative's Douglas Island Project and the Petronas/Progress Energy LNG Project.

The abundance of shale gas resources is the key driver for the new liquefaction terminals proposed in the United States and Canada; in fact, in the export licence applications filed by Kitimat LNG in Canada and Sabine Pass LNG in the United States, the applicants are careful to impress upon the regulators the overall abundance of supply of natural gas.[23] Sabine Pass LNG even argues that the promotion of an LNG export industry will further encourage the development of marginal shale gas resources and increase shale gas production further.[24]

3.4 Destination markets for US and Canadian LNG and project structures

New Canadian liquefaction terminals being located on Canada's west coast (in proximity to the Canadian shale gas basins) are likely to supply Asian markets, due to their proximity and the perceived favourable oil-indexed pricing structures available in those markets.

In the case of the new US liquefaction plants proposed, destination markets would likely include South America, Europe and potentially (subject to the post-Panamax expansion of the Panama Canal) East Asia. The new liquefaction plants proposed in the United States are associated with existing regasification terminals, and accordingly will share existing marine, storage and pipeline infrastructure. This will create the first wave of bi-directional LNG facilities in the world and usher in a further evolution in the value chain, moving it further from its original point-to-point model. In this new world, capacity holders will be able to use these new bi-directional terminals to access the US downstream market when US natural gas prices are attractive, use the terminal as storage for LNG pending re-export or regasification as prices dictate or waive regasification capacity altogether when US natural gas prices are low and instead toll cheap natural gas through the liquefaction trains for export. For such optimisation to be effectively managed, access to flexible transportation arrangements will be key for both pipeline and shipping capacity.

---

23 "Kitimat LNG NEB Licence Application", KM LNG Operating General Partnership, May 16 2011, p10 and Application of Sabine Pass Liquefaction, LLC, *supra* note 11, p 5.

24 Application of Sabine Pass Liquefaction, LLC, *supra* note 11, p5.

This will be the first time that one market will present such significant optionality, being at once a source and destination for LNG.

Meanwhile, the Canadian experience falls squarely within the traditional value chain – at least in the case of Kitimat LNG, which appears to be adopting a point-to-point approach, with the upstream reserve owners owning (through a project vehicle) the liquefaction plant and its capacity, and targeting long-term (Asian) offtakers.[25] This differs from the approach in the United States, where sponsors appear to be adopting a liquefaction tolling model.

3.5 Re-coupling of oil and natural gas pricing?

While the liquefaction of shale gas remains in its infancy in North America, the prospect of LNG export prices affecting domestic pricing is unlikely. If there is significant growth in the volume of LNG exports to markets which tie LNG prices to crude oil, there may be scope for the argument that crude oil prices may once again affect North American natural gas prices. Clearly, there will need to be appropriate flexibility in the value chain to permit diversion of natural gas from domestic consumption to export (and vice versa), to facilitate such price signals to be made. This may be a topic that is ripe for discussion some years from now, once the North American LNG export market has developed further.

4. Financing shale gas LNG projects

4.1 Financing

Financing for proposed shale gas LNG projects may take different forms dictated by the nature of the project sponsor and the proposed project structure, including integrated project financing of various facilities in the value chain or separate financing of individual facilities. Some participants may use corporate equity or debt financing for upstream facilities and access commercial credit for downstream facilities, while others may use commercial credit for all of the facilities.[26]

In the past, access to Canadian project finance debt was somewhat limited as Canadian banks have been reluctant to provide long-term debt. Thus, Canadian projects often received long-term financing led by foreign lenders with some Canadian bank participation, or more often long-term financing from a limited number of Canadian life insurance companies. More recently, the pool of Canadian providers of long-term debt has increased significantly and Canadian projects have been financed by a combination of short-term bank debt and long-term project bonds. While Canadian banks have become more competitive for short-term debt, long-term debt availability remains a challenge without strong bank relationships.

As the Canadian project bond market becomes more competitive, more infrastructure projects are being financed by project bonds. Canadian and foreign institutional investors have provided long-term bond financing for several Canadian infrastructure projects. Financing for any proposed shale gas LNG facilities would

---

25 Kitimat LNG NEB Licence Application, *supra* note 23, pp 5-6.

26 See Salt, Section 3 generally.

likely come from a combination of foreign banks, as well as foreign and Canadian institutional investors, with potential support from Export Development Canada.

4.2 Project risk

Foreign lenders in Canada enjoy the same property rights as are available to Canadian citizens. The risk of nationalisation or government expropriation of assets is considered extremely limited. Any political risk related to a shale gas liquefaction plant is more likely to be related to individuals or groups that exercise property or other rights or seek to shape public opinion in an attempt to halt a project. As liquefaction plants would require different levels of government approval, elected government officials may decide to abandon support for such project in the face of great public opposition.[27]

A unique risk factor for any project in Canada, including in any shale gas liquefaction plant, is the government's duty to consult with, and in certain circumstances accommodate, First Nations on projects that have an effect on lands traditionally used or occupied by such First Nations. The nature and extent of this duty have emerged through a number of decisions of Canadian courts which have recognised certain rights of First Nations peoples in relation to large projects that may affect traditional First Nations lands. This is usually determined during the environmental assessment process and, for a project involving pipelines, is likely to involve more than one First Nation. Thus, an element of risk exists for a project until the consultations with the relevant First Nations and any necessary accommodations have been completed. There is additional risk of delay during construction if there is any allegation that the project sponsor is not meeting the agreed accommodations or opportunities to the First Nations groups involved. Thus, Canadian financing documents for proposed shale gas liquefaction plant projects are likely to include conditions precedent requiring that First Nations consultations and key permits such as environmental permits have been achieved prior to financial close.

Another risk factor for shale gas liquefaction projects in Canada is the various government permits and approvals required at a number of governmental and quasi-governmental levels, including environmental approvals, zoning, development and construction approvals, water allocation, navigable waters approval and energy approvals – all of which adds potential delay and administrative discretion to the development process. Environmental approvals are particularly relevant, since most Canadian environmental statutes prohibit the release of any substance that can cause an "adverse effect" on the environment unless authorised by regulation or permit. As a result, almost all construction sites will require an environmental permit from a government regulator. Due to its potentially harmful effect on the environment, a shale gas liquefaction project will likely trigger the need for an environmental assessment before many other permits can be obtained, having the potential to delay a project.

As the Equator Principles have been adopted by many of the world's leading financial institutions, they will likely be applicable to shale gas LNG project finance

27 See Salt, Section 5.6, pp107-108.

transactions in Canada and lenders must be certain that they have done the due diligence necessary to confirm that the buyer has complied with its commitments under the Equator Principles.[28]

In addition to the foregoing risks, shale gas LNG projects will involve additional project risks. For example, the lenders will require independent confirmation of the shale gas reserves to ensure that projected feedstock volume is sufficient to satisfy the commitments made to the LNG purchasers – as discussed in section 2.4, difficulties in determining decline rates and recoverable reserves will be key project risks for lenders. Given that the level of reserves may change or may need to be proved as the project proceeds, the potential reduction of the estimated reserves for a shale gas LNG project presents additional risk for the lenders.[29] In mitigation of such risk is the fact that, unlike the vast majority of liquefaction plants, plants in the United States and Canada will have access to an extensive pipeline network and should be better able to manage production declines or periodic shortfalls in the supply of feed gas through spot purchases on the market.

Lenders must also address risks related to the projected LNG pricing. As discussed above, shale gas production has caused domestic natural gas pricing to fall, making it important that the LNG pricing for export purposes remains high enough for the project to service its debt. LNG price formulas pegged to domestic North American hub pricing may give lenders some concerns. In the case of Kitimat LNG, destination-market based price formulas appear to be the approach that will be proposed to LNG purchasers.[30]

In financing LNG liquefaction plants, lenders will likely consider the number of firm, long-term commitments from offtakers for minimum quantities of LNG on a take-or-pay basis. However, if there is high demand for gas, lenders might consider financing even on shorter-term contracts.[31] Lenders will also take into account the contract terms that the offtakers have negotiated to determine whether they have the ability to sell spare capacity to other buyers. Additionally, lenders will look at the creditworthiness of the offtakers to ensure that they will be able to meet their obligations in the take-or-pay context.

Other considerations for the lenders will be the level of security available from the project sponsors and the terms of any agreements with federal, provincial/territorial governments. In particular, lenders will take care to ensure that regulatory approvals in respect of the upstream production of shale gas are watertight in the face of environmental concerns over hydraulic fracturing.

4.3 Financing documentation

Financing documentation for proposed shale gas liquefaction plants should be designed to allocate the project risks, while also supporting the leveraged capital structure necessary to finance the project. In addition to credit facility agreements and inter-lender agreements, there will be security agreements, subordination

28 See Salt, Section 5.4, p105.
29 See Salt, Section 5.1(a), p102.
30 Kitimat LNG NEB Licence Application, *supra* note 23, p7.
31 See Salt, Section 6.2 at p110.

agreements, guarantees, collateral agreements, hedging agreements and a host of other agreements designed to ensure the greatest possible security for the lenders. There will also likely be equity contribution agreements with the project sponsors backed by pledge agreements for *pro rata* or back-end equity contributions. For proposed liquefaction plants in Canada, there will likely be separate credit facilities agreements for the Canadian lenders and for the international lenders, but with all parties joining a common terms agreement.[32]

4.4 Security and collateral

Generally, all forms of personal and real property are available as collateral in Canada. The power over the creation, perfection and priority of security interests in personal property rests with the provinces and territories, which have enacted Personal Property Security Acts setting out comprehensive rules respecting security against personal property.

Contracts relating to security and collateral for financing a shale gas LNG project will be negotiated depending on the structure of the project. All or part of the project may be legally and economically self-contained through a private entity whose only business is to develop and operate the project. For example, in a proposed liquefaction project, the lenders will likely take security over any land held by the project entity, over the liquefaction plant, over the pipeline, over bank accounts, over the feedstock gas and LNG, to the extent possible, and over any other property available. Because project lenders anticipate being repaid from revenues generated by a project's operations, security will be tied to the revenue generated by the completed project, not just the value of the secured assets. Security will be taken over all of the project's assets, as well as the project entity's contracts, licences and property rights, to allow the senior lenders to take control of the project assets for operation or resale on default. Lenders may also require limited guarantees from project sponsors as well as direct agreements with the key project participants consenting to an assignment of the project contracts as additional security, so that in the event of a default, the lenders can step in to preserve the project contracts.[33]

32 See Salt, Section 4.2, p98.
33 *Ibid.*

LNG master sale and purchase agreements

Steven Paul Barra
Eni SpA

1. Introduction

Liquefied natural gas (LNG) is expected to play an increasingly important role in the natural gas industry and in global energy markets in the coming years. As the global market becomes more liquid and, accordingly, reliant on short-term transactions, master LNG sale and purchase agreements will become increasingly in demand for the implementation of short-term transactions. As negotiating an LNG transaction can be time and resource intensive, the master agreement has emerged as a tool for short-term transactions, providing a defined framework that permits expeditious transactions and significant flexibility.

A master agreement is, in essence, a framework agreement, in which the majority of terms and conditions that would govern any transaction between the parties are established in advance, while the specific parameters of any single transaction – such as volume, price and delivery point – are set out in individual confirmation notices executed if and when a transaction is agreed.

As the master agreement is designed to capture a short-term opportunity, it typically has a different focus from a long-term sale and purchase agreement. The long-term transaction is typically designed to achieve a balance between, on the one hand, the exigencies of projects (including the need to commercialise satisfactory gas reserves for a long period to justify the investment, bankable terms and a guaranteed net rate of return to sponsors) and, on the other, the buyer's need for a price that permits long-term saleability of its take-or-pay commitment. Conversely, short-term transactions under a master agreement do not typically provide the cornerstone of a significant capital investment for the seller or for a long-term market risk assumption by the buyer. As such, these agreements can provide for significantly greater flexibility for both parties.

Unlike long-term sale and purchase agreements, which generally come into force upon the taking of a final investment decision, the terms and conditions of a master agreement are not binding until the parties enter into a confirmation notice or memorandum for a particular transaction for a single or several cargoes. Most confirmation notices cover a short period of time, even for multiple cargo deliveries. As such, there is no long-term volume risk for the buyer or price risk for the seller, since volume and pricing are defined on a discrete basis, not significantly in advance of the delivery date.

These spot deals nevertheless form part of the LNG chain and the master agreement will need to take into account the interlinking contractual arrangements on either side in order to function effectively.

To be in the best position to capture a short-term deal, it is useful to have in place arrangements (under which either party may buy or sell and deliver on an ex-ship or free on board (FOB) basis) with a range of potential parties to avoid 'running from one bushfire to another'. Like most forms of energy agreement, master agreements should be regularly reviewed to ensure that they are in step with market developments. Templates that have been developed are a useful starting point, although an international standard does not exist, and parties are often wedded to their own 'standard' and time still needs to be invested in reaching an agreement. In the case where both parties can purchase a cargo, it is more likely that they will find agreement on the allocation of risk.

This chapter focuses on some of the more important issues for LNG master agreements within the context of the current LNG market – in particular:

- delivery point;
- failure to deliver or take remedies;
- remedies in the case of off-spec LNG;
- *force majeure*;
- payment, default and credit support terms;
- legal issues;
- transportation issues; and
- form of the confirmation notice.

2. Delivery point

The delivery point is an important cornerstone of the deal. Ex-ship deliveries tend to dominate short-term opportunities. In the case of FOB deliveries the buyer will take delivery at the loading port and provide the shipping capacity, while in the case of ex-ship deliveries the seller has the obligation to deliver to berth at the receiving terminal. In common with gas contracts in general, the seller usually gives a representation that it has title to all such LNG and covenants that such LNG is free from all liens, adverse claims and proprietary rights when the title passes at the delivery point.

For ex-ship cargoes, title and risk traditionally pass from the seller to the buyer as the LNG passes the delivery point – that is, the point at which the flange coupling of the unloading line at the buyer's unloading facilities joins the flange coupling of the LNG discharge manifold onboard the LNG tanker – while for FOB cargoes the delivery point is the loading port at which the flange coupling of the seller's loading line at the seller's facilities joins the flange coupling of the LNG loading manifold onboard the LNG tanker.

Most spot transactions are ex-ship or FOB; however, some LNG producers will prefer a cost and freight (CFR) based agreement. In this case title and risk pass to the buyer in the loading country; however, the seller is responsible for delivery of the cargo to the unloading port. The buyer will need to ensure that any insurance risk is adequately covered during the voyage from the loading point.

One also needs to bear in mind that the International Chamber of Commerce's Incoterms 2010 became effective on January 1 2011. These changes do not represent a dramatic departure from previous versions of the rules, but instead update the rules to accommodate recent developments, and there is usually no need to revise existing master agreements. However, it may be appropriate to consider the new Incoterms 2010 when negotiating new master agreements.

The revised rules replaced the delivered ex ship (DES) term (often quoted in master agreements) with delivered at place (DAP), which provides that delivery occurs when the seller puts the goods at the disposal of the buyer at a named place, on a vehicle ready to be unloaded. The new delivered at terminal (DAT) term replaces the term delivered ex quay (DEQ), and provides that delivery will take place when the seller puts the goods at the disposal of the buyer after unloading at the named terminal.

For FOB and CFR terms, reference to "across the ship's rail" has been replaced with "on board", meaning that the goods have been delivered when they are on board the vessel, so the risk of loading the goods on board is now on the seller's shoulders.

Driven by largely fiscal considerations in the United States and in other select countries, the master agreement or the confirmation notice often provides that title and sometimes risk of loss pass from the seller to the buyer at the last point where the LNG tanker is entirely outside the territorial waters of the buyer's country en route to the unloading port (referred to as the title transfer point). Title (and risk of loss) of any LNG and natural gas vapour remaining on board the tanker after discharge of the LNG cargo at the delivery point returns to the seller at the title transfer point. In these circumstances we also often see a provision that the buyer grants to the seller a licence to use as fuel such quantities of LNG and natural gas in the LNG tanker as are reasonably required to enable the LNG tanker to complete its voyage to the unloading port and outward bound until the tanker passes the title transfer point. With any passage of title comes an insurance undertaking and the associated financial implications will need to be carefully evaluated.

If a cargo is cancelled after the LNG tanker has passed the title transfer point inbound, title (and risk of loss) will revert to the seller as the LNG tanker passes the title transfer point or, if the tanker does not exit the territorial waters prior to discharging its LNG cargo, immediately upon notice from the seller to the buyer.

3. Failure to take and failure to deliver

These provisions are an important facet of a master agreement and each party will want to ensure that it has adequate protection. In the rare case that these provisions are invoked, the potential impact on the parties can be significant. Accordingly, it is important that an adequate credit assessment of the other party be undertaken, and if necessary, appropriate credit support be put in place. Master agreements usually provide that the remedies of the seller or buyer, as the case may be, are the sole and exclusive remedies for the failure to take or deliver the cargo. Although a duty to mitigate forms part of the underlying applicable law, for the avoidance of doubt these provisions restate the duty on the part of the non-defaulting party. In the absence of a provision in the contract, the parties will be required to demonstrate their losses by making a claim under New York or English law. If New York law applies, the New York Uniform Commercial Code should not be excluded.

In the past, master agreements typically provided for compensation to the buyer in the case of a failure to deliver equivalent to a fairly low percentage of the product of the contract price multiplied by the quantity not delivered. This was particularly the case for arrangements with gas producers and was a structure essentially designed to emulate a liquidated damage calculation. As gas markets have become more liquid globally, and both gas and LNG are more readily available to offset the buyer's exposure, the remedies provided in master agreements have evolved over time to more accurately reflect the ability of a buyer to mitigate its losses.

Below are the key elements which are normally addressed in these provisions in the current climate, although they will be subject to negotiation and the peculiar circumstances of the parties.

3.1 Failure to take by the buyer

This provision usually comprises the following elements:

- A buyer that, for technical or operational reasons, is delayed or otherwise unable to take delivery of an LNG cargo within the arrival period (ie, the period during which an LNG tanker is scheduled to arrive at the pilot boarding station outside the unloading facility, as specified in the confirmation notice) is required promptly to notify the seller. The parties are required to use their reasonable endeavours to agree a new arrival period or to have the LNG cargo delivered at an alternative delivery point at no additional cost to the seller. As long as the buyer is available to take delivery within the arrival period, the seller's remedy is to receive demurrage.
- If a buyer fails to receive a cargo within a short period (usually two days or 48 hours after the end of the arrival period), for any reason other than *force majeure*, adverse weather conditions, reasons attributable to the seller or the buyer exercising its right to reject off-spec LNG, the parties are required to use reasonable endeavours to reschedule the relevant LNG cargo, subject to the buyer reimbursing any reasonable actual documented costs incurred by the seller as a result of such rescheduling. As a practical matter, agreeing on a new arrival period or alternate delivery point is challenging, given the scheduling complexities at LNG terminals, physical constraints of storage and the long distances between terminals. A buyer is deemed to have failed to have received a cargo if:
  - it fails to issue a notice to the seller to proceed to berth, in the case of ex-ship deliveries; or
  - it has not tendered its notice of readiness, in the case of FOB deliveries.
- This must be done within a short period, which can be 48 hours after the end of the arrival period, or in any event shortly after the notice of readiness has been issued at the pilot boarding station by the master of the LNG tanker to the buyer, confirming that the LNG tanker is ready to deliver LNG. In any event, a period of 48 hours will be normally unreasonable.
- If the parties are unable to reschedule the relevant LNG cargo, then the buyer is liable to pay for the cargo at the agreed contract price, less any amounts that the seller can recoup by reselling the cargo. This is known as a cover cost methodology.

- The seller is required to use reasonable endeavours to sell the cargo to a third party, exercising a duty to mitigate. The seller credits the buyer with the proceeds of the mitigation sale; however, the seller can keep the 'cream' – that is, the excess over and above the contract price, less any reasonable additional transportation and logistic costs, if any, incurred by the seller to effect the sale of the deficiency quantity, and any reasonable legal costs incurred by the seller, if any, due to such failure.
- The buyer then pays to the seller the positive difference, if any, between the contract price for the cargo, as stipulated in the confirmation notice, and the sum of the resale price and permitted additional costs specified above.
- The payment to the seller is the seller's sole and exclusive remedy with respect to the buyer's failure to take the scheduled LNG cargo.
- The buyer may appoint, at its own expense, a third party to audit the seller's accounts, subject to the auditor executing a confidentiality agreement acceptable to the seller and provided that such request for audit is made within a relatively short period of time, as agreed by the parties in the master.

Important practical issues arise in this context. For example, where the seller has a duty to mitigate, what is the appropriate timeframe during which the mitigation must occur? Determining what this timeframe should be requires striking a balance between:

- the buyer receiving the benefit of the mitigation where appropriate; and
- the need to close out the transaction within a reasonable timeframe (ie, the seller should not have to keep a loaded vessel floating around the ocean indefinitely while waiting for a potential buyer).

3.2 Failure to deliver by the seller

This provision usually comprises the following elements:

- A seller that for technical or operational reasons is delayed or otherwise prevented from delivering an LNG cargo within the arrival period is required promptly to notify the buyer. The parties will use their reasonable endeavours to change the arrival period or to have the LNG cargo delivered at an alternative delivery point at no additional cost to the buyer.
- If the seller is unable or is deemed to have failed to receive a cargo within a short period (usually two days or 48 hours after the end of the arrival period), for any reason other than *force majeure*, adverse weather conditions or reasons attributable to the buyer, the parties are required to use reasonable endeavours to reschedule the relevant LNG cargo, subject to the seller reimbursing any reasonable actual costs incurred by the buyer as a result of such rescheduling. The practical issues mentioned above in relation to failure to take are also relevant here. The trigger is either the seller:
    - not tendering its notice of readiness, in the case of ex-ship deliveries; or
    - failing to provide the buyer with the notice to proceed to berth, in the case of FOB deliveries.

  This must be done within a short period, which can be 48 hours after the end of the arrival period or shortly after the end of the arrival period. In any event, a period of seven days will be normally unreasonable.

- If the parties are unable to reschedule the relevant LNG cargo, the seller is liable for the buyer's reasonable actual documented direct costs with respect to the expected delivery quantity, less any quantity of LNG delivered to the buyer, such costs being limited:
  - to the amount, if any, by which the buyer's cost of acquiring substitute LNG or natural gas required as a result of such failure to deliver exceeded the contract price, multiplied by the quantity of such substitute gas, which in any case shall not exceed the expected delivery quantity, less any quantity of LNG delivered to the buyer;
  - to the extent that the buyer is unable to purchase substitute LNG or natural gas, to the buyer's costs associated with adjusting, reducing or terminating its resale arrangements in respect of the undelivered quantity;
  - alternatively, in highly liquid natural gas markets with mature price indices, to a remedy tied to a published index price, as a proxy for a liquidated damage mechanism; and
  - to all incremental costs, including (if any) additional port charges, cool-down costs, demurrage and boil-off, and the associated transportation costs in order to have the replacement LNG or gas reach that specific location.
- Where both parties are entitled to purchase a cargo under a master agreement, the sum of these amounts is usually limited to the contract price multiplied by the quantity not delivered. LNG producers will try to limit their economic exposure to a significantly lower percentage of the product of the shortfall quantity and the contract price.
- The payment to the buyer is the buyer's sole and exclusive remedy with respect to the seller's failure to deliver the scheduled LNG cargo, with the exception of payments such as demurrage and excess boil-off if the LNG tanker's used laytime is greater than its allowed laytime, as further discussed under the section "Transportation issues".
- The seller may appoint, at its own expense, a third party to audit the buyer's accounts, subject to the auditor executing a confidentiality agreement acceptable to the buyer, and provided that such request for audit is made within a relatively short period of time, as agreed by the parties in the master agreement.

4. Remedies for off-spec LNG

The starting point is that the LNG delivered must comply with the specifications contained in the relevant confirmation notice, either at the loading port for FOB deliveries, or at the unloading facility in the case of ex-ship deliveries. It is important clearly to define the applicable specifications to ensure that they are consistent with the specifications applicable at the receiving facility. In case of inconsistency between the two, the buyer will find itself obligated under the master agreement for the cargo, while being unable to accept delivery at the receiving terminal. As a result, it will pay damages for failure to take delivery of the cargo.

The general principles which should feature in this provision are as follows:

- If either party determines, prior to loading or unloading (depending on whether the cargoes are FOB or ex-ship), that the LNG will be off-spec upon loading or unloading, it is required to notify the other party of the extent of the expected variance as soon as possible and discuss ways that a mitigating solution can be found.
- The buyer is required to use reasonable efforts, including coordinating with the operator of the buyer's facilities, to accept the off-spec LNG. There is no obligation on the buyer to accept delivery if the operator does not agree, or to take delivery if the seller refuses to assume any conditions – financial or physical – imposed by the terminal operator as a condition to acceptance of an off-spec cargo.
- As soon as practicable after becoming aware that the LNG is expected to be off-spec, the buyer is required to inform the seller as to whether it rejects the LNG cargo containing the off-spec LNG or is willing to accept all or any part of the off-spec LNG, in which case the buyer needs to provide the seller with its good-faith estimate of any losses that the buyer will incur as a result of accepting such off-spec LNG.
- The buyer will be reimbursed by the seller for any losses incurred as a result of the receipt and treatment of such off-spec LNG. There will be a discussion as to the extent of the seller's liability – one often sees a cap being a percentage of the expected delivery quantity multiplied by the contract price which will need to be negotiated between the parties. Practically speaking, this limitation does not have a significant value, as the buyer does not and will not accept delivery unless the seller agrees to assume whatever economic consequences the buyer faces. The limitation is thus artificial, because at the end of the day if it does not waive the limitation, the buyer can reject the cargo and the seller faces damages for failure to deliver.
- A party which reasonably believes that off-spec LNG exists after the commencement of loading or unloading should notify the other party as soon as reasonably possible, with either party having the right to suspend delivery of off-spec LNG.
- The seller is required to reimburse the buyer for losses as a result of the buyer accepting the off-spec LNG, including:
    - the loss in value of any other LNG supplies at the buyer's unloading facilities (other than the off-spec LNG unloaded by the buyer), and in the case of FOB deliveries the loss in value of any LNG heel in the tanker, where such loss in value results from blending such off-spec LNG with other LNG supplies at the buyer's unloading facilities;
    - any reasonable actual documented direct costs incurred by the buyer (including payments to third parties such as the operator of the buyer's facilities) in treating or disposing of the off-spec LNG; and
    - all losses relating to damage caused to the buyer's facilities and, in the case of FOB deliveries, the LNG tanker.

Further, a master agreement sometimes provides that if the buyer unknowingly accepts the LNG, the seller is required to indemnify the buyer for all losses suffered, as the buyer has not had the opportunity to agree the costs upfront.

- If the buyer rejects any off-spec LNG, the seller is deemed to have failed to have made available the quantity available and the failure to deliver provisions apply. For the avoidance of doubt, the buyer is not responsible to the seller under the master agreement for any delays associated with delivering such off-spec LNG, provided that the buyer has acted as a reasonable and prudent operator.
- In keeping with the overall principle, any consequential losses are excluded, such as damages suffered by third parties using the same terminal. Regasification capacity arrangements between the terminal operator and the buyer will need to be taken into account.

5. *Force majeure*

Apart from master agreements with gas producers, both parties are often buyer or seller and accordingly will want good coverage for *force majeure* events. A *force majeure* event which is threatening to cancel a cargo can generate a lot of discussion between the parties. Another issue that the parties will need to find agreement upon is where the *force majeure* coverage should stop along the LNG chain. By way of example, if for some reason the seller is unable to provide the cargo due to a *force majeure*-related problem upstream with its supplier, the seller will want to claim *force majeure*, while the buyer will not want to assume an unmanageable risk. Similarly, the buyer will want protection in case there is a problem with a downstream pipeline that prevents evacuation of gas (and may in turn prevent the buyer from accepting a cargo due to lack of available storage). Meanwhile, the supplier will not want to be held ransom to an event which affects the ability to evacuate regasified gas through the interconnecting pipeline. The crafting of *force majeure* provisions should take into account *force majeure* clauses in contractual arrangements immediately upstream and downstream, and a significant negotiation will ensue as to how far upstream and how far downstream the protection should extend. It is generally accepted that the loading and unloading terminals are covered, but everything beyond that is open for negotiation. At a minimum, it should be balanced.

As English and New York jurisdictions do not provide for *force majeure* relief, it is important to review carefully the terms of the *force majeure* clause. Typically, the clause covers the following key areas:

- definition of '*force majeure*';
- examples of *force majeure* events, provided that they fall into the definition;
- specific events which are not usually classified as *force majeure*;
- how parties should react to *force majeure*; and
- the consequences of extended *force majeure*, including the parties' right to terminate.

5.1 Definition of '*force majeure*'

The definition of *force majeure* should be read with care, taking into consideration how the master agreement and confirmation notice will engage with the rest of the LNG chain upstream and downstream. A good definition of *'force majeure'* is as follows: "A party is generally not liable for any failure to perform, or delay in performance of its obligations under a master agreement and a confirmation notice if and to the extent that its performance is prevented, impeded or delayed by any act, event, or circumstance that is beyond the reasonable control of the party claiming *force majeure*; which cannot be avoided by the exercise of due diligence, acting as a reasonable and prudent operator; and is not due to a party's fault or negligence."

It is also important to provide that the event or circumstance shall not be considered to be *force majeure* unless it is beyond the reasonable control of the party and the operator of the relevant seller's or buyer's facilities (the seller's facilities include the transporter in the case of ex-ship deliveries), acting as a reasonable and prudent operator. In other words, a party cannot obtain protection for breaches or other unexcused non-performance by its various subcontractors.

Examples of *force majeure* events include the following:

- acts of God such as fire, flooding, atmospheric disturbances, lightning, storms and hurricanes;
- acts of war, civil disturbances, sanctions on the import or export of goods, acts of privacy, or any criminal acts against an LNG tanker; and
- acts or omissions of a governmental authority related to any governmental approval that were not voluntarily induced or promoted by a party or brought about by the party's breach of its obligations under the master agreement or under any international standards.

Specific examples applicable to LNG master agreements include:

- loss or failure of or serious accidental damage to the seller's facilities (covering the liquefaction facility, berthing facilities at the loading port), or the buyer's unloading facilities;
- loss or failure of, or serious accidental damage to, the LNG tanker specified in the relevant confirmation notice; and
- breakdown or unavailability of marine services.

The definition of the seller's and buyer's facilities will be subject to negotiation. On the seller's side, to the extent that *force majeure* protection extends beyond the loading terminal itself, some parties will prefer to use a definition of a specific gas supply area so that a party is entitled to relief only if the *force majeure* relates to a cargo sourced from the gas supply area forming part of the upstream facilities nominated in the confirmation notice. Whether the buyer is entitled to relief downstream of the regasification terminal will also be debated. The buyer must argue the need to have such relief, and it is unlikely to secure *force majeure* coverage for events affecting the transmission network. Even in long-term sale and purchase agreements, sellers will be reluctant to grant relief to the buyer beyond the interconnecting pipeline that extends from the tailgate of the regasification terminal to the point of interconnection with the downstream primary gas transmission facilities. The seller is likely to request that any such interconnecting pipeline forming part of the buyer's facilities be specified in the confirmation notice.

Events which are not usually *force majeure* include the following:

- failure to pay moneys then due to any party, including unavailability of funds;
- the ability of the buyer or seller to obtain better economic terms for LNG from an alternative supplier or buyer;
- any change in law after the execution of the confirmation notice that does not prevent performance, but merely renders such performance more costly;
- natural depletion of, or the absence of, economically recoverable natural gas; and
- unexcused failure to perform by a contractor of a party.

Producers may push to have reservoir relief omitted from the list of *force majeure* excluded events, or seek to have loss or damage of the natural gas reservoirs or depletion of the proved remaining recoverable reserves included as a *force majeure* event.

5.2 How parties should react to *force majeure*

As soon as a party considers that a *force majeure* event has occurred, the party must give notice to the other party stating:

- the particulars of the act, event or circumstance giving rise to the claim in as much detail as is then reasonably available;
- the obligations the performance of which has been or is reasonably anticipated to be hindered, delayed or prevented as a result; and
- the estimated period of the *force majeure* and the specific actions that the party affected will take to resume performance of its obligations.

Where a cargo is not available due to a *force majeure* event by the seller's supplier, the specific action that the seller can take to remedy the event and the information available in respect of the event are limited. A party affected by *force majeure* is normally required to supplement and update the notices at intervals during the period of *force majeure*. One often sees a requirement to inform the other party at defined intervals by way of supplementary notices, or alternatively to require the party affected to inform the other party as new information comes to light.

The *force majeure* provision often provides that the party affected by an event of *force majeure* has an obligation to provide or procure access at the expense of the non-affected party to the extent possible, in order to examine the scene of the event which gave rise to the *force majeure*. In practice, this provision is unlikely to be invoked, especially where the *force majeure* is related to events or circumstances at the seller's suppliers' facilities.

It is important to use the term 'reasonable endeavours' rather than 'best endeavours' when developing the *force majeure* clause, as a party affected by *force majeure* will not want to be compelled to incur significant additional economic costs as a result. Some master agreements may specify that in order to fulfil the 'reasonable endeavours' obligation, a party may be required to incur a reasonable sum taking into account the value of the transaction. The parties are required to exercise reasonable endeavours and diligence to resume normal performance after the occurrence of an event of *force majeure*. In the case of master agreements with gas producers, one often sees an allocation of supplies provision in the case of *force majeure*, with priority being given to long-term agreements.

Prior to resumption, the parties are required to continue to perform their obligations to the extent not prevented by such event of *force majeure* (eg, the *force majeure* event may affect the delivery of one cargo while other cargoes scheduled under the confirmation notice(s) may still proceed as expected). In the case of a *force majeure* event affecting an LNG ship, reasonable endeavours may include acquiring a substitute LNG ship when available, provided that the non-affected party pays the additional costs.

5.3 Consequences of extended *force majeure*

Master agreements will also provide for termination of the relevant confirmation notice in the case of prolonged *force majeure*. In contrast with long-term sale and purchase agreements, the period is short. There may be a debate about the period (usually seven to 14 days, although sometimes longer), and whether the right to terminate is held by the affected party as well as the non-affected party. More often than not, this right is shared by both parties.

6. Payment, default and credit support terms

While payment, default and credit support provisions are dealt with separately in a master agreement, given the link between the clauses, it is useful to consider them together.

6.1 Credit support

Traditionally, credit support was relevant to the buyer to underpin the buyer's primary obligation to pay for the LNG, while the seller's fundamental obligation was to deliver the LNG to the buyer. Credit support is an example of where master agreements have evolved, as certain buyers have been able to persuade their counterparties that credit support provisions should be reciprocal, but at the same time flexible, given that a buyer can be financially exposed in case that the seller fails to deliver LNG or fails to pay an invoice in the case of a failure to deliver, or for costs incurred in the case of off-spec LNG.

A typical credit support provision in a master agreement will provide that either party may request that the other party provide credit support in the form of a parent company guarantee (provided that the company issuing the guarantee is reasonably acceptable to the other party or has an adequate credit rating in line with the parties' corporate policy), irrevocable standby letter of credit or first demand bank guarantee. The template forms for the types of credit support should ideally be attached to the master agreement so that they do not need to be negotiated when the parties are negotiating the transaction. The amount of the guarantee, while being subject to an evaluation of the credit risk of the counterparty, is indicatively capped at the value of the cargo, although this is linked to the credit assessment discussed above. The type of credit support, along with the timing of the provision of the credit support, can be covered as special provisions in the confirmation notice. Unless provided otherwise, the party requesting the credit support is likely to stipulate that the credit support be furnished simultaneously with the execution of the confirmation notice.

Should the credit rating of a party or its provider of credit support fall below the threshold credit rating during the period of the confirmation notice, the other party may request additional security.

If the credit support is not provided within a short period, the requesting party has the right immediately to suspend its obligations and subsequently terminate.

6.2 Payment

The due date for payment of an invoice is usually 10 business days (when banks are open for business in the jurisdiction where the parties are operating) after receipt of the invoice. The invoice is paid in full without netting or offsetting (eg, the buyer would need to honour an invoice in full for delivery of a cargo without offsetting an invoice due to the seller for costs incurred for off-spec gas). A party disputing an invoice is nevertheless required to honour the invoice in full by the due date pending resolution of the dispute, except in the case of manifest error. Unpaid invoices (or refunds of contested amounts paid, but later determined not to have been due) accrue interest at the agreed interest rate set out in the master agreement. A party has the right to suspend (either immediately or following a cure period) any and all LNG deliveries under all confirmation notices entered into under the master agreement if the other party – usually the seller – has not received payment within five business days of the due date, or at any time that the other party is in breach of its credit support obligations.

6.3 Default

A master agreement will set out the termination events which give the non-defaulting party the right to issue a notice of early termination of a confirmation notice or, as may be selected by the non-defaulting party, all confirmation notices entered into under the master agreement. A failure to take or deliver *per se* does not constitute a default event. In contrast with some long-term sale and purchase agreements which provide for substantial liquidated damages in the case of termination, the non-defaulting party will need to make a claim under the applicable law. The types of termination event typically present are as follows:

- failure by the other party to perform any of its material obligations under any confirmation notice, including payment of any amount when due, where such failure or refusal is not remedied within a period of five days;
- acts of insolvency;
- failure by a party to provide credit support;
- a false representation or warranty given by the other party (these relate mainly to title and corporate standing); and
- a violation of the anti-bribery and anti-corruption provisions.

7. Legal issues

As discussed above, flexibility has an increasing value. For a range of reasons, it can be useful in large companies for transactions to be entered into by affiliates under an existing master agreement, in order to avoid the need to negotiate a new master agreement in a short space of time. This is effectively a temporary assignment of the rights and obligations under the master agreement, and the party which executed the master agreement remains responsible for the obligations of the affiliate which enters into the transaction. The master agreement may have been entered into some time ago and should be checked to see that it represents current practice. The confidentiality clause shall also allow a party to release information to an affiliate without the necessity to obtain the consent of the other party.

In contrast to long-term sale and purchase agreements, destination flexibility and cargo diversion clauses are not usually part of the master agreement, as the delivery point is agreed as part of the spot transaction and confirmed in the confirmation notice. In the relatively rare instance that a profit-sharing mechanism is included in the master agreement, the construction of the mechanism should take into account EU competition law. In brief, profit sharing for diversions between European ports or into or out of the European Economic Area are permitted for ex-ship cargoes on a net basis, taking into account any additional costs incurred in delivering the cargo to the alternative destination. In the case of other types of diversion – such as for FOB cargoes or ex-ship cargoes where the additional costs are not taken into account in the calculation of the profit sharing – the risks of being in violation of EU law can be significantly more elevated.

English law is the accepted governing law in the LNG industry. The majority of master agreements are governed by English law, although some are also governed by New York law. Disputes are governed by international arbitration – either the International Chamber of Commerce or the London Court of International Arbitration. Expert determination can also be useful to resolve technical disputes – for example, in relation to off-spec gas or if there is a dispute in relation to measurement.

Preferably, there should be a provision waiving sovereign immunity for jurisdiction or enforcement of arbitration awards or expert determination, especially where a party is a national gas company.

Consequential loss should be excluded. Increasingly, the parties will seek representations that each party will not be in violation, or cause the other party to be violation, of any applicable law related to the business practices of the parties, including the US Foreign Corrupt Practices Act and EU member country anti-bribery and corruption laws. Some parties may also seek to have an acknowledgment of their business practices compliance programmes.

8. Transportation issues

This section usually covers a range of topics relating to the procedures for delivery, including:

- when arrival notices by the LNG tanker need to be given;
- requirements that the LNG tankers and unloading ports must meet;

- the buyer's and seller's obligations at the unloading port;
- the notice of readiness and berthing priority;
- demurrage and excess boil-off; and
- determination of quantity and quality.

Important themes include when the notice of readiness is effective, berthing priority and demurrage and excess boil-off.

The notice of readiness is effective:

- in the event that the LNG tanker tendered its notice of readiness during the arrival period, at the earlier of six hours (or such other period required for the LNG tanker to arrive at the berth) after receipt of its notice of readiness or when the vessel is safely moored alongside the berth and is all fast;
- in the event that the LNG tanker tendered its notice of readiness before the arrival period, at the earlier of when the vessel is safely moored alongside the berth and is all fast or six hours after the commencement of the relevant arrival period; or
- in the event that the LNG tanker tendered its notice of readiness after the end of the arrival period or the LNG tanker was required to leave the berth and then returned to the berth, when the vessel is safely moored alongside the berth and is all fast.

Subject to the applicable rules of any governmental authority and those set forth in the loading or unloading terminal's rules (as applicable), and decisions taken by the pilots with respect to the berthing of LNG tankers, the berthing priority is as follows:

- first, to LNG tankers arriving prior to or within their respective arrival periods, and as between such LNG tankers, to the LNG tanker whose arrival period is the first to occur; and
- second, to LNG tankers arriving outside their arrival period, and as between such LNG tankers, on a 'first come, first served' basis.

If the operator of the loading facility, in the case of FOB deliveries, does not follow the normal industry practice of first come, first served, the buyer is relieved of any liability for the period of delay in the berthing of the LNG tanker, and the seller is liable to the buyer for demurrage and excess boil-off if the operator allows any other LNG tanker not on schedule to berth prior to the buyer's LNG tanker. Similarly, if the operator of the unloading facility, in the case of ex-ship deliveries, does not follow the normal industry practice of first come, first served, the seller is relieved of any liability for the period of delay in the berthing of the LNG tanker, and the buyer is liable to the buyer for demurrage and excess boil-off if the operator allows any other LNG tanker not on schedule to berth prior to the buyer's LNG tanker.

This raises a larger issue associated with master agreements, given that they are a framework agreement. Each party could be the buyer or the seller, which means that there is no single 'applicable' standard for the loading and receiving terminals. While some general principles are universally true (eg, the concepts of laytime and demurrage), there are significant differences between terminals and that leaves both parties quite exposed from a risk perspective (except producers). There are a range of possible solutions:

- adding more detail to the confirmation notice, with the understanding that this adds a layer of complexity and therefore time to concluding a confirmation;
- defining the terms generically and accepting that they will possibly not match up with the requirements of any individual transaction; or
- defining different sets of standards, in advance, for different terminals.

For ex-ship deliveries, demurrage is calculated if the laytime used for the unloading of the LNG tanker is greater than the allowed laytime specified in the confirmation notice, where used laytime starts to count upon the notice of readiness being effective, and ends usually when the last unloading arm is disconnected and the LNG tanker is cleared to depart the berth. In this case the buyer is required to pay to the seller a daily demurrage fee as set out in the confirmation notice. There is usually a free extension for reasons attributable to the seller or the LNG tanker, *force majeure*, adverse weather conditions and time during which normal operations are prohibited, such as night-time restrictions. Similarly, for FOB deliveries, the seller is obliged to pay to the buyer demurrage if the used laytime in loading the cargo exceeds the allowed laytime. A demurrage claim is usually time barred if submitted with supporting documentation after a period of 90 to 180 days after completion of unloading of the cargo.

9. Form of the confirmation notice

The form of the confirmation notice should be agreed when the master agreement is negotiated. It should be as clear as possible, and contain all commercial and operational elements relevant to the spot deal, covering the following heads:

- number of LNG cargoes to be delivered;
- expected delivery quantity for each LNG cargo plus or minus a percentage for operational purposes;
- price and bank account details;
- description of the LNG tanker, boil-off and demurrage rates;
- arrival period;
- delivery term and transfer of title and risk of loss (if not addressed in the master agreement);
- unloading port and buyer's unloading facilities;
- allowed laytime and allowed berth time;
- LNG quality specifications and off-spec liability cap, if applicable;
- credit support (eg, type of credit support and date of provision, if not provided at execution date);
- marine terminal liability regime confirming the provision of the marine terminal manual to the seller (ex-ship deliveries) or buyer (FOB deliveries), and that the seller or buyer respectively has entered into the conditions of use or marine terminal liability agreement); and
- special provisions.

10. Conclusion

LNG master sale and purchase agreements will continue to be a key element in the global energy market. In this chapter the more important commercial and legal issues in LNG master sale and purchase agreements are discussed, including failure to deliver or take remedies, *force majeure* and payment, default and credit support terms. Given the dynamic nature of the industry, it is recommended to have in place master agreements with a range of counterparties. Innovative contractual arrangements need to keep in step with the emerging needs of the global LNG market, and master LNG sale and purchase agreements are no exception.

Financing LNG projects

James Douglass
Linklaters LLP

---

1. Introduction

As noted elsewhere in this publication, a successful LNG venture typically involves a series of interdependent projects in order to take gas extracted from the wellhead through processing, liquefaction, storage, transportation and regasification up to delivery of gas to the wholesale end customers. Traditionally, the various links in this chain were separately and independently developed, but this no longer holds true. An LNG project can therefore range from a narrowly defined 'within the fence' construction of, say, a regasification facility to the creation of a massive multi-jurisdictional business that covers all or the major part of the energy chain. Recent examples of this latter type of project include Qatar Gas II, III and IV, Phase 2 of the Sakhalin II project in Russia and the PNG LNG Project in Papua New Guinea.

It is therefore not possible in today's world to speak definitively of a single methodology for financing LNG projects, as the underlying dynamics vary so widely and the scope for innovative funding structures increases – particularly in the wake of the liquidity constraints introduced by the global financial crisis and its aftermath. As the dynamics of the world economy have changed with events post-September 2008, a range of new players have entered the arena with consequences for all stakeholders.

While a number of financings remain structured along relatively straightforward and traditional lines (eg, single facility tolling structures or vessel financings on an asset-backed basis), the headline deals are now far more complicated and require a deep understanding of the issues arising throughout the energy chain. For bankers and their advisers, the day of the 'LNG specialist' is therefore with us and has been so for a number of years.

This chapter cannot cover the whole range of financing options for LNG-related projects. Each project will have its own commercial drivers which influence the financing structure deployed in any given situation. There are, however, some fundamentals from a legal and structural perspective that operate as a benchmark for credit appraisal by the lending community, and a number of developments in the industry are driving further change. This chapter seeks to provide the reader with an overview of these aspects using the financing of a liquefaction plant as the reference baseline model.

This updated chapter explores the financing aspects of the developments in the LNG industry in recent years, with a particular focus on the crucial keystone role of export credit agencies (ECAs) and multilateral lending institutions in LNG financing in the wake of the global financial crisis, and familiarises the reader with the identity and roles of the parties and process involved in a typical LNG project financing.

2. Why obtain external funding for LNG projects?

It is worth pausing on what we mean by the financing of LNG projects. While the term 'project financing' is now too narrow to cover the range of funding options available, we are effectively looking at a financing structure where third-party institutions provide debt to the owner of the project with limited or no recourse to its shareholders. The debt capacity available to the project and its ability to service debt will therefore be dictated by the project's success and the economics which it is able to generate. In a default scenario, lenders will have to look to the project's assets in order to recover any debt then owing to them.

There are many examples of LNG projects being funded by the participants themselves, whether through the use of available corporate resources or through raising funds on a corporate or balance sheet-backed basis in the loan or capital markets. This is the way in which major oil companies have traditionally funded the large part of their upstream operations and, with hydrocarbon revenues seemingly at a sustainably high level for the foreseeable future, cash resources should logically be present to adopt this route. Equity funding of this type provides the project developers with control over the flow of funds and autonomy over the manner in which they develop and conduct their operations.

There are, however, a number of reasons why the raising of project-level debt is attractive to the sponsors of that project. These include the following:

- LNG projects are highly capital intensive and require very substantial upfront outlays of capital. This level of self-funding is beyond the reach of many participants, which would otherwise be forced to compromise equity value through sell-down or some form of carry interest arrangements of the type commonly seen in upstream developments. The cash resources available to larger companies, such as the oil majors, also face internal competition from the range of other projects which they are developing and which may have (in the case of upstream developments) potential for higher economic returns and carry a strategic imperative for extending the company's reserve base.
- The policy and economic goals of participating strategic partners may require the obtaining of external funding in the world markets. In addition to credit limitations on a party's ability to raise funds itself, a number of strategically important partners (eg, state-owned gas companies) wish to raise funding at the project level in order to maintain the sovereign balance sheet and foreign currency reserves, and to develop the international standing and credit capacity of the country in which the LNG project is to be established. In a country such as Qatar, this has proved a phenomenal success over the last decade or so, with a sophisticated and substantial debt programme being rolled out to almost all parts of the global debt markets. A state-owned gas company may also face constitutional or other legal impediments to raising corporate funding on its or the state's 'balance sheet'.
- The availability of debt finance and the appetite of lenders and bond

investors remain strong for well-structured LNG projects. The inherent nature of these projects meets a number of threshold credit criteria for debt providers, including an impressive track record of supply and offtake security across the industry, a 'steady state' business model based on long-term offtake commitments, dollar revenues typically paid to offshore accounts in a 'zero risk' jurisdiction, tested technology and the overall backdrop of increasing demand for natural gas derived from LNG in global markets. These factors normally combine to enable lenders to offer an attractive financing package that contains competitive terms and a flexibility to structure the funding in a way which meets the requirements of the particular project.

- One other factor that has always played a large part in obtaining finance for LNG projects is the active involvement of governments of countries involved in the trade. The importance to a host government of attracting the necessary investment for an export project the size of an LNG project will itself be a source of comfort for financiers and, at least in the early days of establishing the industry in that country, will often manifest itself in the express grant of governmental support to both the project and its financiers. Foreign governments are also prominent in supporting these projects, given the potential for winning very substantial export orders (through engineering, procurement and construction contracts for their national contractors) and, increasingly, for obtaining access to long-term supplies of natural gas.
- The funding sources available to prospective LNG project sponsors have changed due to the impact of the global financial crisis on global commercial banks. A new set of players have emerged reflecting:
    - the importance of energy supply security. particularly to countries in north Asia;
    - the growth in non-commercial bank funding sources, particularly ECAs and multilateral agencies (MLAs);
    - the credit expansion from the policy and commercial banks in China;
    - the appetite of the bond markets for well-structured LNG credits following a number of successful Qatari project bond issuances; and
    - the increased use of Islamic financing.

  Each of these sources is examined in more detail in this chapter. There is also greater familiarity of the issues associated with LNG financing among the providers of these sources of financing, and among the LNG financing community of the process and inter-creditor issues associated with tying these sources of funding into a common finance plan.

3. Sources of funding

The sources of funding available for LNG projects vary significantly depending on a range of credit criteria. These include:

- the identity of the sponsors;
- the country risk involved (together with the overall level of credit capacity available for that country within the lending community);
- the strength of the offtake commitments; and

- the status of the project itself.

The principal sources of credit facilities and their respective participations in the provision of debt capital can broadly be summarised as follows.

3.1 Commercial banks

Commercial banks are major contributors to the funding of LNG projects worldwide. They provide term loan funding for construction and expansions of LNG projects, both on an uncovered basis and under the umbrella of protection provided by ECAs and MLAs. The extent and nature of the financing provided turn on the evaluation of the credit criteria mentioned above and can involve the full spectrum of risk assumption – that is, a requirement for fully covered facilities, a mixture of covered and uncovered facilities (a 'sweet and sour' mix) to wholly uncovered debt for established projects in countries with strong credit standing. Commercial banks will also play a critical role as financial advisers to sponsors and project companies, as lead arrangers in arranging financing and driving it towards financial close, as potential project bond lead managers and more recently as advisers to structuring export credt agencies.

A key feature of the financial universe post-global financial crisis has been the rise in importance of commercial banks from north Asia, which were less (if at all) affected by the turmoil. While the Japanese commercial banks (now concentrated in the three mega banks) have a long track record in LNG financing, the commercial banks of China and Korea have recently emerged (typically in conjunction with financing from relevant governmental agencies) as key providers of debt capital – particularly in projects with Chinese or Korean sponsorship, offtake or contracting.

Although a full discussion is beyond the scope of this chapter, the implementation of Basel III may have a future impact on the pricing and tenor of commercial bank debt available to LNG projects.

3.2 Export credit agencies

ECAs are governmental agencies that seek to facilitate the financing of a project in order to further the commercial interests of its nation in line with the policies of the government of that country. These agencies, which in some cases provide insurance or guarantees and in others provide direct loans, have become cornerstone partners in the financing plans for many of the mega LNG financings, and increasingly play a significant role structuring the commercial terms applicable to all finance providers.

While the approach and detailed policy wording for each ECA differ, key terms are to some extent harmonised through the application of Organisation for Economic Cooperation and Development (OECD) guidelines.

Some of the key elements of the OECD guidelines as they relate to tied support provided by ECAs are set out briefly below:

- Participant ECAs must require that the purchaser of goods and services pay at least 15% of the export contract value as a downpayment on or prior to the 'starting date of credit' (a date that varies depending on the type of contract that the ECA is supporting). Participant ECAs must not provide official support in excess of 85% of the export contract value, including third country supply, but excluding local costs.

- ECAs can provide official support for the 'local' costs of a project, provided that this does not exceed 30% of the export contract value and is not provided on terms more favourable than those agreed for the related exports.
- Generally, the principal sum of an export credit must be repaid in equal instalments at least every six months. For project finance transactions, no single principal repayment or series of repayments within a six-month period may exceed 25% of the principal amount. The first repayment instalment must be made no later than two years after the starting date of credit for project finance transactions (in an amount which is not less than 2% of the principal amount), and six months after the starting date of credit for all other financings.
- Interest must be paid at least annually for project finance transactions (and every six months for other financings), with the first interest payment to be made no later than six months after the starting date of credit. Interest must not be capitalised after the starting date of credit. The minimum interest rates available from an ECA are determined by the length of the loan and the profile for principal repayments.
- The maximum repayment term for project financing transactions is 14 years (provided that the 'weighted average life' of the repayment period does not exceed seven and a quarter years). Where the relevant criteria for a project financing are not met, the maximum tenor is between five and 10 years, depending on the relevant categorisation of the country where the project is being developed.

The OECD guidelines also contain specific provisions and financing terms that apply to certain sectors, such as shipping. Recommendations on certain policy areas, such as environmental issues and bribery, have also been published by the OECD and there is a belief that member countries will do their utmost to implement these. There are also additional specific criteria that must be met when ECAs are participating in project financing transactions in 'high income' OECD countries (eg, Australia).

(a) Tied support

All ECAs generally have products to support the export of equipment manufactured in their country (or, in the case of European exporters, countries within the European Union), as well as a certain amount of content in the country of import (usually the country in which the LNG project is situated). The level of the facility made available for this purpose is generally tied to the amount of 'eligible expenditure', being project expenditure on goods and services sourced from the relevant country, and each ECA has its own detailed procedures to verify compliance with this requirement. The facilities provided by the ECAs in these circumstances typically constitute guarantees or insurance policies to commercial banks that provide the actual funding. This guarantee protection can be 'comprehensive', which essentially passes all risk of non-payment to the ECA (subject to a residual uncovered portion of between 10% and 15% of the loan), or be limited to political risks only. These are discussed further below.

(b) Overseas investment

In addition to tied facilities driven by export orders, certain governmental agencies have additional policies which permit credit to be extended to projects which are considered to be in the national interest of the relevant country. The natural resources sector (and particularly access to oil and gas reserves) is probably the largest beneficiary of these policies, which are particularly deployed by North Asian countries with an ever-increasing requirement for stable energy supply sources. Under these policies, direct loans can be made by the ECA in addition to guaranteed protection of private sector funding, and there is no limitation on the sourcing of the expenditures agreed to be funded. Some economic benefit to the relevant country needs to be shown, whether it be LNG supply or equity investment.

The emergence of the ECAs as structuring lenders has been one of the key trends since the global financial crisis. Although ECAs have long been involved in LNG project financing, their traditional involvement was primarily as providers of political risk cover and comprehensive cover based on EPC procurement strategies.

As the policies of national governments have come to emphasise resource security, and in cognisance of the severe commercial bank liquidity constraints post-global financial crisis, the policies of the ECAs have changed to fill the void left by the departure of a number of commercial banks from the provision of long-term project financing.

As a result, ECAs can generally support the funding of LNG projects anywhere. Previously, their role in the resources sector was essentially limited to projects in developing regions.

Sponsors and their advisers now commonly look to ECA funding as the cornerstone of any LNG finance plan. This is further reinforced by the willingness of several 'structuring' ECAs to play a key role in negotiating the detailed terms of the financing which will apply to all participants. Many of the recent mega LNG and other oil and gas sector project financings have adopted this approach.

The ECAs generally have stringent policies on health, safety, environment and social issues, which are explored in more detail in section 7.4.

3.3 Multilateral agencies

MLAs are made up of members from a multiplicity of participating countries and have a constitutional goal of encouraging investment in developing countries in line with certain policy criteria. Institutions involved in the financing of LNG projects include the International Finance Corporation (IFC), the private investment arm of the World Bank, the European Bank of Reconstruction and Development (EBRD), the African Development Bank (AFDB) the Asian Development Bank (ADB) and the Inter-American Development Bank (IADB).

A key aspect to the participation of many MLAs is the presence of 'additionality', which broadly means that they should add value to the financing in an area where the private commercial sector is unable to do so. The additionality requirement has generally not been difficult to satisfy in the years post-global financial crisis.

MLAs can provide direct funding ('A loans') and a B loan structure under which the MLA is the lender of record but the funding is sourced from, and all exposure to default lies with, commercial banks (the 'B loan lenders'). This relationship is documented under a participation agreement entered into between the MLA and the B loan lenders.

As with ECAs, MLA support can be provided only if the relevant nexus between the project and the MLA itself can be found. For the regionally focused MLAs (eg, the ADB, the AFDB, the IADB and the EBRD), the project needs to be in one of their countries of operation (eg, the AFDB was a financier of the Nigeria LNG project).

The IFC has probably the widest remit of all, as it acts on a truly global basis but generally lends only into developing nations.

MLAs have strict requirements on anti-bribery and corruption matters, and can sanction offending contractors for up to three years. Projects involving a sanctioned company will generally not be eligible for funding from any of the major MLAs.

MLAs, particularly the IFC, are also key players in the development of international standards applicable to health, safety and environment issues, which are discussed in more detail in section 7.4.

Another international finance institution which has participated in LNG project financing is the European Investment Bank (EIB). To be eligible for EIB funding, the project must be located in an EU member state or in territories which benefit from an external EU mandate for development assistance. These development mandates are very extensive.

3.4 Capital markets

LNG projects have in recent years been able to access investment from the global capital markets. This has been successfully implemented, for example, in the funding of the RasGas II and 3 expansion projects and the Nakilat LNG vessel fleet financing in Qatar.

The significance of widening the investor base in this way cannot be overestimated, although the availability of these funds for purely greenfield projects with no track record remains something of an open question. The key to obtaining flexibility to raise funds from the capital markets is to secure a credit rating at or above investment grade from one of the internationally recognised rating agencies. These agencies will undertake an in-depth review of the project and allocate a credit rating based on an evaluation of the project's capacity to meet its existing and planned financial commitments – Standard & Poor's, for example, sets a scale from AAA (which means an extremely strong capacity to meet those commitments) to D (which indicates there has been a payment default on financial commitments). There are various interval grades on this scale, with BBB- being recognised as 'investment grade' and representing a key benchmark below which many institutions will not invest. Pricing and availability of funds from the capital markets therefore depend on the existence of a rating and that rating being investment grade or above.

LNG projects have generally been well regarded by the rating agencies – both the Oman and RasGas projects were rated A- by Standard & Poor's (with Moody's and Fitch in the latter case being a notch above this) and Nakilat rated at A+. Evaluation criteria used by the rating agencies are similar to those employed by bank lenders (see above), but perhaps place a little more weight on the macro factors such as world demand and political will. The other significant feature of the rating of LNG projects is that (at least in the context of liquefaction plants) the sovereign rating of the host country does not necessarily operate as a ceiling and it has been possible to obtain a credit rating for a project which is higher than that assigned to the country in which the project is located. Key credit enhancements to which the rating agencies look in this regard include the elimination of foreign exchange risk through offshore dollar payments and control of these payments through offshore bank accounts.

A key issue in evaluating whether to include a capital markets tranche in an LNG financing is a consideration of the different dynamics in decision making between the commercial banks and ECAs/MLAs on one hand and bondholders on the other.

Banks, ECAs and MLAs are used to (and expect to) monitor operational and credit issues on a real-time basis. Bond investors generally do not and will expect to be notified only of fundamental credit issues such as non-payment of interest (coupon) payments under the bonds.

The bond market presents an extremely deep pool of fixed rate debt at tenors potentially significantly longer than the bank market. This market may not be as affected as the loan markets by the regulatory response to the global financial crisis, but it is conceivable that certain segments of the market (eg. the insurance sector through the Solvency II Directive) could face similar issues.

Another potentially game-changing development in recent years has been the very recent development and expansion of the renminbi bond markets. This market has shown exponential growth since the launch of the first renminbi Eurobond in 2010 (by Hopewell Highways Infrastructure). Although the predominant funding currency in the LNG markets continues to be the US dollar, given the general importance of China in world trade and contracting and its recent emergence as an LNG player, we should expect use of renminbi bonds in an LNG finance plan in the not-too-distant future.

3.5 Islamic finance

The participation of Islamic funding sources in energy and infrastructure financing is a natural development as the banking systems in the Middle East and elsewhere increase in strength. A large number of LNG projects are located in Islamic countries and the use of Islamic funding to support these projects is a welcome progression of regional participation. An Islamic financing (a financing structured in accordance with *Shariah* law) does not permit lenders to lend money or recover interest in the manner conventionally seen in the international commercial markets. Islamic financing structures are varied, but broadly involve the financier taking risk beyond the mere provision of capital. Often the risk is associated with the purchase of project assets, with the financier leasing them back to the borrower for a fixed period.

The structuring and implementation of one or more Islamic finance tranches as part of a wider multi-sourced financing arrangement provide unique challenges for both finance parties and sponsors. Many of these challenges arise from the typical expectation of Islamic finance participants that they will be treated on a *pro rata* and *pari passu* basis to all other senior lenders in all circumstances which may arise. This generally involves, for example, the Islamic and conventional tranches being coordinated in a manner whereby all utilisations and pre-payments are addressed *pro rata* across the facilities, and all payment obligations in respect of the facilities rank *pari passu* at all times. Other significant structural considerations include:

- the need to address the possibility of total loss or destruction of an asset which underpins an Islamic finance tranche (and which may lead to the immediate automatic termination of the Islamic finance arrangements under *Shariah* principles);
- the need to structure the security package to take account of the transfer of ownership interests in the assets (and sometimes the relevant construction contracts relating thereto) which underpin the relevant Islamic finance facilities;
- the desire of Islamic finance participants to seek to mitigate their risks relating to third-party or other environmental liabilities arising from the asset which underpins an Islamic finance tranche (including through additional indemnity protection and introduction of special purpose vehicles into Islamic finance structures);
- the need to factor the *Shariah* approval processes (in addition to the separate credit approval processes) of Islamic finance participants into the overall signing timetable for the financing; and
- relevant tax issues (including those arising from the transfer of assets and characterisation of payment flows from Islamic finance tranches for withholding tax purposes etc).

4. Key players in LNG financing

For those who have not participated in a largescale LNG financing of the type described in this chapter, there can appear to be a bewildering number of parties involved. This section attempts to introduce the reader to the key players in an LNG financing and briefly describes their roles and functions.

4.1 The sponsors

The equity ownership configuration of an LNG project will depend on ownership of the reserves, technology provision, marketing destinations and so on. Typically, an LNG project in a developing company will feature a reserve owning national oil company, together with international players including large international oil companies and equity investors representing key destination markets.

The sponsors will remain engaged as key stakeholders for the duration of the project, providing personnel, expertise, technology and credit support (if appropriate) to the special purpose vehicle project company to which the financing will be provided (see section 4.2).

Financiers will generally require a lock-in of the ownership interests of original shareholders – at least until some time after project completion and thereafter at stepped-down levels. The sponsors may retain direct ownership of the upstream interests for government approval or tax reasons.

Sponsors may also participate in financing an LNG venture, as either senior lenders or subordinated lenders.

The relationship between these parties will generally be regulated by a shareholders' agreement (if the project company is incorporated) or joint venture agreement (if the venture is unincorporated – see section 4.3).

4.2 The project company

The project company will be the borrower of the debt financing or the guarantor of a special purpose vehicle issuer in the context of a bond financing. It will also be the project counterparty to all external project contracts (eg, gas supply (if not self owned), engineering, procurement and construction, offtake, operation or service support, vessel charter (if required for delivered ex-ship (DES) sales), finance and security documentation) and the financing documents.

The project company will initially be staffed by secondees from the sponsors, but over time may develop into a substantial company with a substantial balance street and trading history in its own right. An example of such a company is Sakhalin Energy Investment Company Ltd, which is the owner and operator of the Russian Sakhalin 2 LNG project.

4.3 The unincorporated joint venture

In general terms, unincorporated joint ventures involve the contractual association of joint venture participants in a common endeavour (which is not a partnership for legal or tax purposes) to generate a product to be shared among them. It is not uncommon to conduct upstream oil and gas projects through unincorporated joint venture structures. This is because there can be significant tax and other fiscal advantages for sponsors in adopting this structure for upstream operations. In Australia, the unincorporated joint venture is a typical feature of upstream natural resources projects.

The same benefits of using unincorporated joint ventures for upstream operations may not apply to transportation or LNG infrastructure (or may not outweigh other drivers for choosing incorporated structures for these aspects of an LNG project), and for this reason the market has seen a number of LNG projects (particularly in the Asia-Pacific region – for example, PNG LNG) use a hybrid approach, with the upstream financed through an unincorporated joint venture and the midstream financed through a special purpose vehicle.

If the sponsors decide to pursue an LNG project (or part of a project) through an unincorporated joint venture structure, this will raise a number of structural and legal issues for sponsors and lenders, such as the following:

- rights of co-venturers;
- pre-emption rights;
- ownership of assets;
- use of priority deeds and cross-charges to deal with those rights;

- several (as opposed to joint) borrower commitments and pre-payments;
- separate default mechanisms;
- the structuring of any capital markets tranche; and
- the provision of security for lenders over assets.

Depending on how sales and marketing are structured, it may also result in the sponsors being regarded as 'joint sellers' for the purposes of competition legislation, and this will need careful consideration.

While financial institutions in some jurisdictions (eg, Australia) are familiar with the financing of projects involving unincorporated joint ventures, it should be recognised that such structures are not at all common in international project financings and, while far from insurmountable, a degree of explanation to international financial institutions (where involved) and the impact of this on timing and overall acceptability should not be underestimated.

4.4 The financial adviser

It is relatively common for the sponsors and project company to appoint a bank (usually a commercial bank) to provide advisory services in respect of the raising of debt financing. A number of large international banks employ specialist bankers for such assignments. There are also a number of boutique advisory firms offering financial advisory services without lending capacity. The bankers working as financial adviser will usually be kept functionally separate through information barriers from lending teams at that institution.

It is becoming common to combine the expertise of a global advisory bank with a bank with more localised knowledge – for example, of key offtake destination countries or countries expected to provide the bulk of the financing – as joint financial advisers.

4.5 Other project company advisers

The project company will typically retain a number of other specialist advisers, including legal, accounting and insurance. Often technical, reserves, environmental and market advice can be sourced from the sponsors.

4.6 Host government/approval authorities

As explored elsewhere in this publication, LNG projects have significant footprints, and host governments and their agencies and instrumentalities may be key stakeholders as equity owners, gas suppliers, infrastructure providers, regulators and profit sharers (through production sharing contracts, royalties, taxation or dividend payments).

Government affairs strategies are thus always of key focus for sponsors and project companies.

The lending community will view the identity of the host government and its LNG industry track record as key issues in their credit approval processes. Lenders (particularly ECAs and MLAs) may wish to obtain direct comfort or contractual undertakings from the host government in respect of certain issues. Lenders may also need government consents or approvals for the execution and perfection of finance and security documents.

4.7 Project contract counterparties

The project company will enter into a suite of project contracts with key contract counterparties for gas supply (if not owned by the project company), engineering, production and construction, offtake, services provision and operation, vessel charter for DES sales and so on. The creditworthiness and track record of these counterparties, as well as the terms of the contracts themselves, will be key issues in the credit processes of the lenders.

These parties may also become involved in the financing process if the lenders require direct agreements or notices of acknowledgment and assignment from them.

4.8 Lead arrangers

The sponsors and the project company will usually mandate a small group of commercial lenders (known as mandated lead arrangers) to arrange the debt financing. Before the financial crisis, it was common for mandated lead arrangers to agree to arrange and underwrite the commercial bank debt financing and sell down commitments through a syndication process. The more common approach post-financial crisis is to mandate a number of commercial banks on a 'club' basis to 'take and hold' an agreed level of commitment and agree to arrange the remainder of the facility, sometimes on a 'best efforts' basis. The mandated lead arrangers will usually be appointed as bookrunners for any syndication process and may be given certain rights to bid for hedging business at the appropriate time.

4.9 ECAs/MLAs

ECAs and MLAs with potentially large lending commitments may be brought in at an early stage of development of the finance plan as structuring ECAs or MLAs. Many of the more experienced and significant ECAs and MLAs have undertaken this role in major LNG financings, working closely in conjunction with agency advisers and mandated lead arrangers.

4.10 Lender advisers

Lenders will need access to independent advice in a number of areas in order to take a credit decision on whether to lend to an LNG project. In an LNG project, the typical line-up of lender advisers will include the following:

- Legal adviser – common counsel is often appointed to advise all lender groups, but separate counsel may be appointed to advise individual lenders or groups of lenders if lender interests are felt not to be aligned.
- Technical adviser – the technical adviser's role is to advise lenders on technical aspects (eg, interface issues between engineering, production and construction packages, contracting strategy, adequacy of design, performance issues). and to monitor and report on construction progress. The technical adviser may also have a role in certifying payments to contractors against milestone progress and in inspection of construction work on a real-time basis.

- Environmental consultant – the environmental consultant's key initial role is to provide a report benchmarking the health, safety, environment and social strategy of the project company against the Equator Principles and the requirements of ECAs and MLAs. Lenders will also invariably require, as part of the financing terms, a monitoring role for the environmental consultant on an ongoing basis.
- Insurance adviser – the insurance adviser's role is to advise the lenders on the project's insurance programme, confirm implementation of the agreed insurance requirements for the project and monitor ongoing compliance.
- Reserves consultant – the reserves consultant's role is to advise lenders on the level of initial reserves and potentially confirm reserves levels at key milestone points (eg, project completion).
- Market consultant – the market consultant's role is to independently advise lenders on the project's offtake strategy and terms of offtake.
- Model auditor – the model auditor's role is to audit the logic and calculations in the financial model.
- Shipping consultant – the shipping consultant's role is to advise lenders on the adequacy of shipping arrangements for DES sales, if any.

These consultants will work closely with the various role banks described in section 4.11.

4.11 Role banks

In many major LNG financings a number of 'role banks' will be appointed to fulfil various functions pre-signing and then during the life of the financing. Banks holding these roles will undertake a negotiated scope of work and are usually indemnified by all other finance parties in respect of liabilities accruing from the performance of such roles. The role banks responsible for technical, environmental, insurance, modelling and market matters may also take the lead in the process of appointing the relevant independent consultants described in section 4.10 and in negotiating their scopes of work and liability arrangements.

The scope of work and mandate terms of role banks will vary on a case-by-case basis, but the following general observations can be made.

(a) Intercreditor agent

As is described in section 6, most LNG financings will combine a number of financing sources. The intercreditor agent will be appointed by the various sources of financing to act as a focal point for engagement between the project company and the lenders for matters such as satisfaction of conditions precedent, distribution of reports and data and communications on waivers and amendments to the finance and project documents. The powers and duties of the intercreditor agent will be set out in a document signed by all finance parties or their representatives – usually the common terms agreement. The intercreditor agent plays a key role in successful loan administration, although this is primarily administrative.

(b) Facility agents

Each debt tranche will have its own individual facility agreement and a facility agent will be appointed to administer each facility. The facility agent's functions include dealing with payment flows to and from the facility banks and verifying satisfaction of condition precedents at the facility level. The agent banks for ECA-backed facilities are banks with knowledge of the particular requirements of the relevant ECA.

(c) Bond trustee

The bond trustee is appointed to hold the covenant to pay on trust for the bondholders. If the bond issuance is part of a combined bank/bond debt issuance, it is likely that the project security will be held by the security trustee for the secured lenders as a whole, with its bond trustee holding these interests on trust for the Bondholders.

(d) Security trustee or agent

The main function of the security agent or trustee is to hold the agreed security package on behalf of the secured lenders as a whole. The security is held on trust or as an agent, depending on the nature of the legal systems applicable to the security. In common law jurisdictions, the security will be held on trust. In other jurisdictions, the security may be held an agent through a parallel debt structure.

Exceptions to the principle that the security trustee or agent holds all the security may apply:

- in the case of Islamic financing, in respect of any assets owned by Islamic financers;
- in the case of bond issuance, where the bond trustee will hold the security over the bond proceeds account for the bondholders only and not the secured creditors generally; and
- in situations where onshore security is held by a different institution from the institution holding the offshore security.

(e) Technical bank

The technical bank will liaise with the lenders' technical adviser to deal with matters raised in the technical adviser's report and may have an ongoing role in respect of technical assumptions in the financial model.

(f) Environmental bank

The environmental bank will liaise with the environmental consultant on health, safety, environment and social compliance issues. Many financial institutions supporting LNG projects – including Equator Principles banks, ECAs and MLAs – will have in-house teams to deal with such issues.

(g) Modelling bank

The modelling bank will work with the project company and the model auditor on the cash-flow projections in the financial model and will conduct sensitivity analysis on such cash flows.

(h) ***Insurance bank***

The insurance bank will work with the lenders' insurance adviser on the insurance requirements for the project which will be set out in the finance documentation.

(i) ***Account bank***

A bank will be appointed to hold the project accounts on behalf of the project company. This bank will usually be part of the lending group, although this is not always the case. Such accounts will be secured in favour of the senior lenders. The scope of the account bank's duties and authorities will be set out in an account bank agreement. This agreement also commonly deals with the agreed cash-flow waterfall, which sets out the order and peridiocy of payments from the project accounts.

4.12 Hedging counterparties

It is common practice in LNG financings for the project company and its lenders to agree a hedging or risk management strategy setting out the parameters of any interest rate, currency or commodity hedging to be undertaken by the project company. This hedging is often, although not invariably, undertaken with members of the existing lender group (with teams being separated by information barriers). The hedging counterparties will usually be senior secured creditors and accede to the financing documentation.

4.13 Other stakeholders

In addition to the above parties, other stakeholders may from time to time become involved in the LNG financing process. These range from the courts to the media and non-government organisations. Project sponsors and lenders alike need to devise strategies for dealing with these parties.

5. The financing process

As indicated in section 3 above, the sheer size of LNG projects, coupled with ongoing liquidity constraints, means that the financing plan for an LNG project is likely to involve multiple sources, including a combination of some or all of commercial debt, capital markets, ECAs/MLAs, co-lending from sponsors and possibly Islamic finance.

Accordingly, one of the biggest challenges faced by the sponsors of an LNG project is to find successful strategies to integrate the various funding sources and achieve financial close within an acceptable time period and at the lowest cost (in terms of both upfront execution of the financing and ongoing administration). Such integration is complicated further by the fact that even within the categories of funding sources outlined above, each financial institution will have its own specific characteristics and requirements which need to be accommodated (eg, compare Chinese banks with European banks, each individual ECA and MLA, international bond offerings and domestic bond offerings).

It is not easy to bring these deals to a successful financial closing, but there are strategies to achieve this. Based on recent experience, some possible solutions and related issues are set out below.

As indicated elsewhere in this chapter, the role of ECAs on large-scale multi-sourced energy (including LNG) project financings has evolved since the financial crisis to a point where a core group of structuring ECAs will often play a critical role in the initial structuring of the financing, including (among other things) agreeing:

- the scope of due diligence and other initial risk analyses;
- the key terms of the financing, including security, sponsor support and covenant packages; and
- the key assumptions to be inputted into the banking case model and the extent to which these are to be fixed at the outset.

Further, in an environment of constrained liquidity, procurement strategies should be optimised in order to maximise available funding through a process involving a detailed analysis of individual ECA's eligibility requirements (including the ability of each ECA to support goods and services from each of country of origin, country of export and third countries). Indeed, some recent financings have seen the appointment of a separate 'ECA coordinator' (a commercial bank) to assist the borrower and the financial adviser with this process. It is also becoming more common for the allocations of each ECA to be fixed later in the process to further optimise the procurement strategy.

Further issues for sponsors, project companies and their advisers to consider in developing their strategy to manage the financing process include:

- the availability of sponsor completion support and the approach to lender due diligence on EPC arrangements;
- whether to pre-structure for project expansions and re-financings and additional debt raising, potentially through capital markets issuance; and
- the timing and method of approaches to the commercial bank market for debt once key ECA/MLA commitment are secured.

Effective management of the financing process will require specialist input from financial legal and other advisers experienced in previous LNG financings. Strict adherence to the timetable must be maintained in order to position the relevant LNG project to fulfil its offtake commitments and to effectively take advantage of market opportunities.

6. Combining the financing sources

All recent financings of large LNG projects have obtained funding from a combination of the sources referred to above. The lending structure may therefore look something along the following lines from a documentary perspective as reprsented on the next page.

This mix of facilities can create different interests in terms of risk allocation between lenders and policy requirements in relation to financing terms available to the borrower. It also raises challenges for the structuring and efficient administration of the credit facilities in meeting the day-to-day requirements of the project. Harmonisation of principal financing terms and the establishment of working arrangements for the lender group are therefore important factors in developing the overall structure.

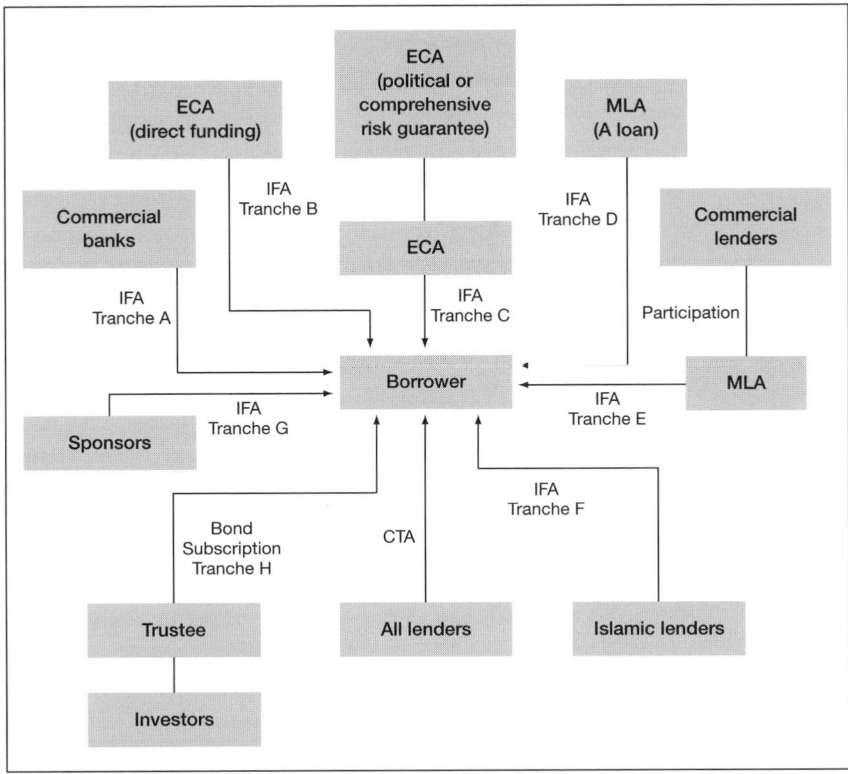

The following represent mechanisms designed to achieve this objective.

6.1 Common terms approach

The primary tool used for harmonising the terms of the financing is a common terms agreement (CTA) to which all finance parties are signatories and which contains financing terms which will be commonly applicable to each of them. The CTA should cover the great majority of the financing terms and essentially define both the commercial parameters of the financing and the 'boilerplate' provisions for all participating financiers.

The CTA will operate in tandem with individual facility agreements (IFAs) between the borrower and the individual lending group, pursuant to which actual loan disbursements will be made. The scope of these IFAs should, however, be limited to dealing with terms specific to that particular loan tranche (eg, margin, interest calculation, fees) and any additional requirements that are bespoke to that financing source (eg, procedures for evidencing the incurrence of 'eligible expenditure' for tied ECA facilities). The following table sets out an indicative list of how key terms are typically addressed in a common terms financing structure.

	CTA	IFAs
Conditions precedent	✔	✔ Limited number of additional conditions to meet ECA/MLA procedural conditions
Drawdown procedures	✔	
Interest calculation and payment	✔ If common base such as LIBOR	Bespoke interest mechanics sometimes offered by ECA/MLAs
Interest margin loan amortisation	✔	
Pre-payments – voluntary and mandatory	✔	
Banking case and budget procedures	✔	
Security requirements	✔	
Taxes, increased costs and market disruption	✔	
Fees	✔ Common fees such as commitment fees	✔ Individual tranche fees and premiums
Agents	✔ Common agents such as intercreditor agent and security agent	✔ Individual facility agents
Transfers	✔	
Representations/ warranties	✔	✔ Limited number of additional representations to meet ECA/MLA procedural condition
Covenants	✔	

	CTA	IFAs
Events of default	✔	✔ Certain bespoke events of default sometimes required by ECAs/MLAs
Environmental requirements	✔ Baseline approach to health, safety, environment and social compliance – representation, covenants and event of default	ECA/MLA specific requirements
Sanctionable practices (anti-bribery and corruption provisions)	✔	Certain specific requirements of ECAs/MLAs

6.2 Intercreditor arrangements

The central objective of a common terms structure is that all of the financiers agree to act collectively through a common agent in making decisions and taking actions in relation to the financing. In general terms, and subject to a limited number of exceptions, no individual lender or lender group is permitted to operate independently in modifying or enforcing its rights under the financing documents. This is generally considered to be beneficial both to the borrower and to the lender group as a whole, as it ensures consistency in approach, provides a single point administrative interface and provides alignment of interests in a stress scenario and avoids the financing being brought down by the 'rogue' actions of a minority.

A separate intercreditor agreement is normally entered into between all lenders to govern these relations and provide the framework within which directions will be given to the common agent. The borrower has a clear interest – albeit indirect – in seeing that these arrangements are workable from an administrative standpoint (eg, timely receipt of any requested waivers), but that it is not exposed to enforcement or other adverse action at the behest of a small minority of the lender group. It can be a point of contention between the borrower and its lenders as to whether this interest is sufficient to justify the borrower's participation in the intercreditor arrangements, whether as a party to the intercreditor agreement itself or as the beneficiary of a collateral obligation from the lenders not to modify its terms without the borrower's approval.

Components of an intercreditor agreement typically include the following.

(a) Voting entitlements

Lenders will generally have a voting entitlement proportional to their exposure in the outstanding debt or, prior to the expiry of the availability period for debt drawdowns, the sum of outstanding debt and outstanding commitments not yet drawn. This is expressed as a percentage and is used in determining whether the necessary voting thresholds have been met. This is, in principle, a straightforward arithmetic exercise, but with the following refinements in the context of a multi-sourced financing of the type described in this chapter:

- Where ECAs or MLAs provide insurance or guarantee protection to commercial lenders, they will generally retain the right either to vote or to direct the voting entitlements of those commercial lenders covered by this protection on the basis that the agencies are carrying the ultimate exposure in a default scenario. This is clear where comprehensive cover is provided, but can result in a split of voting control where a partial risk guarantee is given only in relation to political risks.
- Voting entitlements provided to the trustee of any capital markets tranche are often restricted to exceptional items only. This reflects the lower tolerance of bond investors for involvement in more routine project-related decisions.
- Hedging counterparties typically do not accrue voting entitlements unless and until a termination liability has been crystallised under the hedging arrangements.

(b) Decision-making procedures and thresholds

The intercreditor agreement will establish voting pass marks for decisions to be made on behalf of the lender group. In descending order, these generally fall into the following categories:

- Unanimous consent of all lenders – this normally covers a defined list of matters that go to the fundamentals of the credit approved by each of the individual lenders. A typical list will include proposals to amend payment obligations, tenor, margin or currency of the loan or the release of security ahead of full repayment.
- Super-majority decision – the level of this majority is often a function of the make-up of the lender group: it is driven by a requirement that one or more 'minority' groups vote in favour of the decision. In complex projects, there can be more than one level, depending on the lender composition and the nature of the underlying matter which is being decided. Decisions falling within this category tend to be those relating either to matters of very substantial commercial significance (eg, decision to release sponsor recourse at completion) or to matters which carry particular sensitivity (eg, certain environmental decisions).
- Majority lender decisions – this would be the benchmark level for the taking of decisions which have not been specifically allocated to other categories and would cover the great majority of routine decisions made in the course of administering the loan facilities. This level will again depend on the balance of power created by the voting entitlements of the respective lender groups, but a common starting point customarily used in syndicated loans is a two-thirds majority of all voting entitlements.

- Entrenched decisions – it is not uncommon for the parties to agree a list of decisions of particular sensitivity to ECAs or MLAs which will require their affirmative vote in addition to the requisite majority.

(c) ***Enforcement action***

Given the significance for all stakeholders of a decision to accelerate loans and enforce security against the borrower, it is fairly common to carve out this aspect for specific treatment. While the borrower would clearly wish the level of lender consensus to be as high as possible, lenders themselves also need to strike the correct balance that avoids the overall financing being cratered by the action of a minority group. Against this, a number of lending institutions have a policy requirement that at some point they need to have an independent right to take enforcement action if an event of default has occurred and has not been remedied.

A typical solution to this is to provide for a sliding scale of decreasing percentage voting pass marks required to commence enforcement action on behalf of the lenders. These percentages are a matter for consideration in the context of each project, but may be fixed according to both the length of time for which the event of default has been outstanding and the nature of the default itself.

(d) ***Decision-making procedures***

Given the number and nature of the institutions commonly involved in financing a major LNG project, it is in the interest of all participants to structure decision-making procedures to enable decisions to be made in a timely manner. This is typically a hot issue for borrowers and needs to be balanced with the inherent uncertainty of predicting the complexity of a particular issue and a corresponding time reasonably needed for the lenders to take a properly informed view.

(e) ***Sharing of payments***

The intercreditor agreement will set down the rules governing the allocation among the lenders of moneys received from the borrower. Where the lending groups all have senior creditor status, they will typically rank equally and receive a *pro rata* share of such funds if there is a shortfall against the amount needed to service all debts. If junior debt is involved, the intercreditor agreement will track the subordination provisions agreed between the parties and allocate funds according to the order of priority ranking between lenders.

A specific issue that can arise where an MLA is involved in the financing is the treatment of 'preferred creditor status'. This status is afforded to certain MLA institutions and effectively means that member countries agree to service their debt notwithstanding a general moratorium on payment of foreign indebtedness as a whole. The question therefore arises as to whether the MLA should share the benefit of this privileged status with the other lenders (in practice, they are generally not prepared to do so) and, if this status enables the MLA only to be kept whole during a moratorium, whether the remaining lender group should have a preferential claim on proceeds received after the moratorium to 'catch up' to a position of equal ranking.

(f) Accession of new finance parties

The intercreditor agreement will lay down the procedures for the accession of new finance parties permitted by the CTA. This will generally involve the execution of a deed of accession by the incoming financiers.

The accession of counterparties to swap or hedging agreements introduces a different risk profile into the lender mix, given that these counterparties are not term lenders, but will potentially have a large claim against the borrower in the event of early termination of the underlying swap instrument. This is a highly topical issue, as LNG producers are increasingly considering the use of sophisticated derivative products in the context of LNG sales and pricing arbitrage.

In order to accommodate this flexibility, the agreed parameters for permitted hedging arrangements are often pre-agreed in the finance documentation. This needs to balance market expectations of the hedge providers (which generally includes taking a senior secured position) with safeguarding the project and its lenders from unmanageable exposure to a large claim. This is generally achieved through a combination of establishing acceptable commercial parameters in an agreed risk management strategy document, reducing the grounds for counterparty termination in the standard International Swaps and Derivatives Association documentation and providing for the settlement of any termination payment at an appropriate level in the cash-flow waterfall. Typically, hedge providers would not acquire voting entitlements before the occurrence of a termination event under the relevant hedging agreement.

7. Basic building blocks of LNG financing terms

As mentioned earlier, a broad array of financing products are available to fund projects in the LNG sector and a number of these will carry different market and policy requirements. In addition, a large proportion of the finance documentation follows standard banking principles, with conventions being similar to those prescribed by organisations such as the Loan Market Association or the Asia Pacific Loan Market Association. In this chapter, we do not seek to detail all of these terms, but the following is a subjective attempt to pull out some of the core concepts which customarily appear in LNG financings and are fundamental drivers of the structure.

7.1 The banking case model

The financial model which records and forecasts the economics of the project lies at the heart of the financing and is in large part the preserve of highly skilled analysts. However, as with all forecasts, the key to its veracity is the input assumptions which are made, and it is in this area where care must be taken to ensure that the coverage of these in the documented finance terms accords with the data assumptions being used in the operation of the model itself. This will be an area of key focus for the modelling bank.

These assumptions will cover economic, technical, reserve and market considerations as modified by contractual arrangements in place for the project. The majority of these will be developed in a manner consistent with principles applied across the spectrum of different industry sectors – obvious examples of this are:

- the project's capital and operating costs being taken from budgets approved by the lender's technical consultant;
- operational performance of the facilities, again as approved by the technical consultant; and
- general economic assumptions such as interest rates and inflation either being either pre-set or determined by reference to relevant published indices.

There is, of course, nothing unusual in this and one of the key considerations for agreement between the borrower and the lender group will be the extent to which assumptions are fixed at the outset and the manner in which any discretionary variables are determined in the event of a dispute.

There are, however, a number of points of more particular relevance in the context of an LNG financing, as follows.

(a) Feed stock supply

Lenders will need to be satisfied that the project will have a sufficient and stable supply of feed stock gas in order to support the LNG sales volumes projected in the model and to meet its contractual commitments to offtakers. This will need to be achieved either through a firm supply contract or, in an integrated project, through demonstration of sufficient proven reserves of natural gas certified to the lenders by an independent reserves consultant.

LNG projects have not generally been made on a borrowing base structure (common for upstream oilfield financings), where the amount of debt under the credit facilities varies according to the prevailing level of reserves, which is periodically updated. While the historical approach in LNG projects has been to restrict debt capacity by reference to a reserve certificate provided ahead of financial close, this is not ideal where drilling operations which are expected to 'prove up' probable reserves are being conducted in parallel with the construction of the LNG facilities. If this is the case, sizing the debt capacity against this expected increase in proven reserves should therefore be considered in combination with the provision of a debt buydown obligation (with appropriate credit support), to the extent that this does not materialise by the start-up of operations. This regime would be implemented by providing for a re-test of the reserve base and the issuance of a new reserves certificate by the independent consultant in the period between completion of the drilling operations and start-up of the LNG plant. The reserve assumptions in the financial model would then be adjusted to reflect these results.

(b) Offtake volumes

Unlike crude oil developments, financiers of LNG projects have been accustomed to offtake assumptions being supported by hard contractual commitments from offtakers which extend beyond the scheduled maturity of the debt and are based on a conventional take-or-pay structure. This provides a predictable revenue outlook, with the downside sensitivity of contractual default being softened by the excellent track record to date of the industry. This is accordingly a relatively straightforward process, but with value being placed by buyers on flexibility within the contract terms, there are still areas for discussion that can have a material effect on the level of debt capacity. These include whether, and the extent to which, headline take-or-pay volumes should be adjusted on a predictive basis to recognise any rights of the offtakers to exercise downward flexibility or upward flexibility. In considering this, the presence of 'make good' obligations in the underlying offtake contracts and the ability of the offtaker to sell spare capacity to other buyers and/or destinations will be taken into account. Similar considerations apply where an offtake contract is subject to renewal or termination during the life of the loan, although in these circumstances the potential effect on the project economics is clearly greater.

The much larger question of whether lenders will accept full 'merchant risk' on LNG offtake volumes is discussed at the end of this chapter.

(c) *LNG pricing*

LNG sales in the Asia-Pacific basin have traditionally been structured by reference to a fixed dollar amount (per million British thermal units) adjusted by reference to movements in the Japan Custom Cleared (otherwise known as the Japanese Crude Cocktail) (JCC) benchmark price. This is discussed in some detail elsewhere in this publication, but for the purposes of this chapter, it does mean that the sales revenues available to the project will in practice be closely tied to changes in the price of crude oil during the life of the loan. The manner in which projected oil prices are factored into the model therefore has a significant impact on debt capacity. Notwithstanding the high price of oil at the time of this publication, lenders will naturally take a conservative view on the forward curve, given its significance in underpinning the level of revenues available for debt service. Accordingly, a key area for commercial agreement is whether the financial model should assume a fixed oil price (irrespective of actual price movements) or whether pricing is reviewed periodically by reference to available published data or a price deck or similar approach.

Where projects are selling into markets with a mature gas trading regime (eg, the United States and the United Kingdom), a view needs to be taken on the forward curve for gas prices in that market, as this will likely represent the basis of pricing passed back to the LNG supplier. The issue with which the lending community will have to grapple going forward is whether a particular market is sufficiently developed and liquid to enable reliable forecasts to be made on a gas price index. As is examined elsewhere in this publication, this issue has been of considerable importance in recent years for projects planning to sell LNG into the United States, given movements in Henry Hub and Socal pricing due primarily to the increased production of shale gas in the United States.

The model will, of course, take into account any contractual stabilisation of commodity price risk. This would include specific hedging arrangements with swap counterparties or price management provisions (eg, 'S curves') embedded in the pricing of the offtake agreements.

7.2 Meaning and use of cash-flow ratios

The primary output of the financial model is the production of cash-flow ratios which are designed to test the project's ability to service its debt. In LNG financings, the following two ratios have been commonly used and involve determining for a specific period the cash available for debt service (CAFDS), being net revenues after deduction of all taxes and other expenditures, and measuring this against the debt service requirement for that period:

- Loan life cover ratio – this measures the overall ability of the project to produce enough cash flow to repay its debt over the life of the loan. This is usually achieved by comparing the discounted net present value of projected CAFDS to loan maturity with the aggregate amount of principal outstanding (or available to be drawn) under the debt facilities.
- Debt service cover ratio – this looks at the ability of the project to service its debt out of CAFDS on one or more scheduled repayment dates. While lenders would wish to see in the base case model at financial close that CAFDS is sufficient to meet all scheduled principal and interest payments, the debt service cover ratio is generally used during the term of the facility to operate as a short-term 'health check' on the project's economics in respect of both prior and future periods.

The required level of coverage provided by these ratios will be a matter of negotiation between lenders and borrower in each case and will vary according to their usage. Areas in which the cash-flow ratios play a pivotal role include:

- the sizing of the debt capacity in the initial banking base case model;
- the release of sponsor completion support (see below);
- the ability of the borrower to use insurance proceeds to reinstate the project following a major loss;
- payment of distributions or other subordinated debt by the borrower;
- the introduction of additional senior debt facilities;
- modifications to or replacements of offtake contracts; and
- the occurrence of an event of default for breach of financial covenants.

7.3 Completion support

To date, most major greenfield liquefaction projects have received some form of completion support under which the borrower's shareholders either guarantee the timely completion of the facilities or, more usually, agree to underwrite debt payments until completion has occurred. This therefore sets the expectations of the lending community, but does carry a corresponding benefit to the borrower in providing it with more headroom within which to manage its engineering, procurement and construction operations. Lenders will wish to be satisfied that a sensible contracting strategy is being implemented, but do not generally require the same level of detailed controls which can appear in other sectors (eg, power generation) where the assumption of completion risk within the project itself is more the norm. This approach should also result in cost savings for the project, given the premium charged by contractors to 'wrap' an entire project through liquidated damages and other support.

This approach of offering completion support does pre-suppose that the relevant sponsors' credit standing is sufficient to meet the expectations of the lending community. It will be interesting to see over the next few years how many sponsors of large-scale LNG projects will continue to offer completion support or whether they will seek to match risk-sharing approaches that prevail in the power and other sectors. The latter approach has been more the trend on LNG re-gas terminal financings, where completion risk is easier to pass through to contractors through EPC arrangements, than for large-scale integrated LNG projects.

A number of common issues arise in relation to the structuring of this type of completion support, including the following.

(a) *Equity lock-up*

During the period of completion support, lenders are clearly concerned about the identity and creditworthiness of the shareholder counterparty providing that support. This will generally result in tighter restrictions on sell-down of equity by shareholders during this period and the imposition of credit tests in relation to any incoming shareholder if it is to take over its proportional share of commitments under the sponsor support arrangements.

(b) *Political risk carve-out*

Shareholders will often be prepared to underwrite the commercial risks of completion on the basis that this is where their business expertise lies. They may, however, seek to exclude liability in the event that completion is delayed or otherwise impeded by the occurrence of political risk. This is discussed further below.

(c) *Release conditions*

The conditions to release of the sponsor completion support are viewed as critical by all parties. The central condition will be a technical test demonstrating the operational performance of the facilities and the energy chain to the point of delivery over a period of time and, in relation to the facilities within the project, the borrower will wish to dovetail these requirements with those agreed with its contractors. In addition, lenders will typically wish to see all other major components of the project in place before losing their recourse to the shareholders. Customary additional conditions to release include:

- the project's achievement of cash-flow ratios at a prescribed coverage level;
- all necessary governmental approvals;
- perfection of all required security;
- no continuing default; and
- other requirements specific to the financing, such as funding of any reserve accounts.

From the perspective of the sponsors' corporate exposure, this is a critical document. Certainty of obligation and confidence in the timing of release are therefore likely to be important value drivers in their evaluation of the overall financing structure.

7.4 Health, safety, environment and social considerations

The health, safety, environment and social area is today perhaps the single most important factor to lenders in evaluating their ability to participate in a major project financing. The project's compliance with the highest international standards is now a prerequisite to such participation by the large majority of financial institutions which are involved in this line of business.

While MLAs and ECAs have for many years observed strict environmental guidelines and policies, the significance of this area in terms of legal and reputational impact to the private sector has grown exponentially in recent years and it is no longer sufficient simply to comply with domestic legislation of the country in which the project is located. As evidence of this, over 75 of the world's leading financial institutions have adopted the Equator Principles, which draw substantially from the guidelines issued by the IFC. The principles apply globally to all new project financings at a capital cost of $10 million or more and across all industry sectors. Following the latest revision, which became effective on July 6 2006, the Equator Principles impose the following commitments:

- Each project is to be assessed at either category A or category B risk in relation to its environmental and social impact.
- IFC performance standards must be applied to all projects constructed outside 'high income' OECD countries. In addition to compliance standards in the implementation of operations, these standards set objectives in relation to public consultation, resettlement, impact on indigenous peoples, labour and human rights, as well as biodiversity. These place an increased emphasis on environmental and social management systems, but are not particularly precise so there remains ample room for disagreement on their implementation.
- The borrower must prepare an action plan which draws on the findings and conclusions arising out of the environmental impact assessment and public consultation. Again, this plan must comply substantively with IFC performance standards.
- The borrower is obliged to covenant compliance with local laws, permits and the action plan, and to provide regular information to the lender group.
- The lenders must actively monitor the borrower's activities from a health, safety, environment and social perspective, and have committed to report publicly on the project's implementation of the Equator Principles.

At the timg of writing, the IFC Performance Standards are being reviewed, with new performance standards expected to take effect in 2012. In particular, the standards concerning climate change, biodiversity, human rights and labour and supply chains are expected to be significantly expanded. This will impact on Equator Principles financial institutions, as the Equator Principles incorporate the IFC performance standards by reference.

The imperative to comply with these principles therefore has a profound effect on lender due diligence into the health, safety, environment and social aspects of an LNG project's development. It also drives to a large extent the documentary requirements in relation to such matters, and most recent project financings have included a sophisticated contractual regime in the financing agreements to address all of these issues.

This is, of course, not simply a matter for the financiers, as major developers of energy projects treat health, safety, environment and social matters with the utmost priority and major LNG projects must also comply with the heath, safety, environment and social policies of the sponsors and project company, as well as those of the lenders.

7.5 Security

In a finance structure which does not have recourse to the ultimate owner of the project (at least post-completion), lenders will wish to have the right to enforce against the assets of the project itself should a default occur. The existence of a first-ranking security package in relation to these assets will be an important consideration in credit evaluation.

In an LNG financing, the target security list for lenders is no different from that for any other secured project financing. Conventionally, this package would include first-ranking security over the following assets:

- the share capital of the borrower, together with any subordinated debt provided by shareholders to the borrower;
- land, buildings and physical assets of the project;
- hydrocarbons in transit or storage;
- assignment of borrower's rights in all material project contracts, including the offtake contract and any concession agreement with the host government; and
- security over all bank accounts, including an offshore proceeds account into which LNG receivables are paid directly by the offtakers and any reserve accounts (including, typically, a debt service reserve account).

In many jurisdictions where LNG projects are promoted, the political and legal constraints are such that one or more aspects of this security package are not achievable. The following represent some practical examples of embedded constraints on providing the conventional range of security in the context of LNG financings:

- legal restrictions on foreign ownership or contractual restrictions on any change in control of the borrower which limits the effectiveness of security over the shares in the project company;
- grant of land rights for the LNG facilities which are incapable of mortgage (this is the case for underground pipelines in most parts of the world);
- separation of title between land and buildings which makes it impossible to grant security over a structure until it is complete;
- classification of offshore facilities for the purpose of title and mortgage – again, a common problem in many jurisdictions;
- mandatory requirements for a locally owned company to repatriate all or part of its overseas earnings, which can affect the validity of offshore reserve accounts and subject LNG sales revenues to foreign exchange risks; and

- registration fees, taxes or notarial fees payable as a percentage of loans secured or the property mortgaged, which can add very substantial cost to the project.

Lenders will generally require a borrower to provide as much of the conventional security package as it is practicable and commercially reasonable to do. A number of techniques have been developed in different jurisdictions to address some of the common deficiencies by alternative means. In practice, there is often a need to find a compromise that fits with the legal and political system within which the project is being promoted.

As between the lenders, security will generally be held by a common security agent or trustee, with enforcement being regulated by the intercreditor agreement and the proceeds allocated between the senior lenders on a *pro rata* basis. Where an Islamic financing tranche is included, particular structuring is required to reflect the fact that the Islamic lenders have an enhanced position of asset ownership while the commercial debt is restricted to a security interest. Where trust structures are not recognised in relation to onshore security or other constraints existing on the accession of new lenders, parallel debt structures can also be used to overcome these deficiencies.

7.6 Treatment of political risks

It is stated at the beginning of this chapter that one of the advantages of involving both public and private lending institutions in the financing of a major project is that it provides a 'halo' effect to mitigate the inherent country risk. There is doubtless a range of views on the true value of this, but it is fair to say that it is no longer a factor in a number of the leading LNG producing nations which have established strong international reputations and can attract investment capital without any such fears. Some countries in the Middle East, for example, hold some of the highest sovereign credit ratings in the world.

Nevertheless, political risk does feature as an important consideration in a number of LNG projects and they have historically been fertile ground for the development of structures in which public sector bodies (notably ECAs and MLAs) take the major share of this exposure. This is commonly seen in two principal ways:

- An ECA or MLA provides partial risk guarantees to commercial lenders under which debt service of the commercial lenders will be funded by the relevant ECA/MLA in the event that the borrower defaults due to the occurrence of a specified political risk.
- Where completion support is given by the sponsors as described above, the sponsor is excused liability if the reason for the completion delay or borrower's inability to service debt was attributable to a political risk event.

In a structure that combines both of the above, it is clearly important to harmonise the terms of the political risk protection in both cases in order to avoid commercial lenders being subject to a gap in coverage. It should also be noted that neither of these structures provides any relief to the borrower itself. In a default scenario where the loss is absorbed by a public sector agency, that agency will be entitled to exercise all rights which are available to it and its covered lenders to enforce against the borrower and the project assets. Given the nature of the event which has caused the problem and the likely level of equity invested in the project, it is expected that the borrower, its shareholders and lenders will in practice pursue a more consensual course in seeking to work out the default circumstances.

Most of the institutions that provide this type of political risk coverage have their own prescribed terms which can differ in some important respects. In a multi-sourced financing, it is clearly advantageous to all parties concerned that these terms be made consistent in order to delineate clearly the risk allocation between all stakeholders.

Most partial risk guarantees will generally include the following political risks:

- expropriation of assets of nationalisation of the borrower;
- wars, blockades and embargoes involving the host country;
- political violence covering civil war and politically motivated riots and other civil commotions;
- inability to convert or repatriate foreign currency; and
- revocation or non-renewal of consents.

There are a number of areas which represent important areas of coverage in an LNG project and which may need to be negotiated as extensions into the standard policy. These include terrorism, failure to grant consents and, perhaps most importantly, a failure by the host government to honour its obligations under any concession agreement in existence for the project (so-called 'breach of contract' cover).

As with regular insurance policies, a number of conditions must be complied with in order to make a successful claim under the relevant policy or guarantee. Again, these can vary from institution to institution, but common themes include the following:

- A high standard is applied to the causation of the borrower's default by reason of the relevant political risk event. Standard policy wording can require this to be the 'sole' or 'direct and primary' cause of the default.
- Waiting periods are generally applied which require the political risk event or its consequences to remain in place for a minimum period before a claim can be made.
- Materiality qualifications are imposed on the impact which the political risk has on the project's implementation or operations.
- The beneficiary of the guarantee or policy must not have contributed to the occurrence of the political risk event.

8. Financing treatment of LNG industry developments

This chapter has hopefully demonstrated that the financial sector is continually evolving its products to meet the demands of the industry. It is contended that the financing of LNG projects has been one of the most dynamic and innovative areas of business for major financial institutions in both the public and private sectors. Similarly, LNG developers have played a large part in devising more sophisticated products which are structured to accommodate both the changing patterns of the LNG industry and the particular requirements of individual projects. The sheer scale of funds raised over the last five years stands testimony to the success of this effort.

There are perhaps two particular recent trends in the LNG industry that either have or may lead to a shift in the conventional thinking which has been applied to date. These two areas are both driven by the increase in LNG demand around the world, which has resulted in a large number of expansion projects and a substantial increase in the universe of participants in the LNG industry.

8.1 Project expansions

The economics of a liquefaction project can be exponentially increased by adding further liquefaction trains and storage capacity. The success of expansion projects has been spectacularly demonstrated over recent years in major producing countries such as Nigeria, Oman, Australia and Qatar. Accordingly, it has now become far more common to provide in the original financing terms for the borrower's flexibility to add further capacity and to raise further senior debt to fund this expansion.

In structuring this flexibility, the threshold question is whether the expansion will form part of the original project for financing purposes – that is, whether its assets will be secured in favour of the original lenders and its revenues counted in the financial model. This approach does have certain advantages from a lender's perspective, but the *quid pro quo* should be that external financing raised by the borrower to fund the expansion should then be treated as senior debt on a *pari passu* basis with the existing debt. In these circumstances, lenders will wish to see certain conditions imposed, such as:

- the demonstration of healthy cover ratios for the enlarged project;
- satisfactory environmental reports on the expansion;
- accession of the new lenders to the common terms arrangements;
- the provision of completion support by shareholders in respect of the expansion; and
- debt tenors for the new lenders being at least equal to that of the original financing.

The alternative approach for an expansion is to 'ring fence' it from the initial project and for any expansion financing to be separate from the original financing.

This provides greater flexibility and lessens the involvement of the existing financiers in vetting the expansion's development. In this scenario, the main complication will be to develop satisfactory arrangements for the use of common and interdependent assets which are needed by both the existing project and the expansion. This may also have consequential impacts on the security being provided over those assets which may require an intercreditor arrangement between the existing and new lender groups.

Accordingly, there are complications in pre-legislating for this type of flexibility and there is some argument that this is better worked out at the time of implementing the expansion, given that its likely incremental impact on project economics will be of benefit to all stakeholders. Given the business importance of this flexibility, however, it is an area where project developers increasingly have a preference to set down a framework with largely objective criteria upon which they can place bottom-line contractual reliance. The lending community has generally recognised the commercial importance of this, and that it is not appropriate to impose a single project mindset on a company whose base case business proposition includes this type of expansion. Accordingly, this is an area where finance terms continue to develop.

8.2 The merchant plant

As noted earlier in this chapter, one of the principal attractions of LNG projects to financiers has been the predictable cash flows arising from long-term LNG sales contracts on a take-or-pay basis. However, the LNG market is clearly undergoing change, with an increasing incidence of shorter-term gas supply and LNG offtake contracts, greater diversity of LNG markers and a wider universe of third-party buyers, a ramp-up in construction of LNG terminals and access points to gas markets, and some movement away from traditional crude oil base pricing to domestic gas indices. Coupled with the overall jump in global demand for gas, the LNG trade has become a far more open global market in which the 'volume risk' on unutilised production capacity is lower than it has ever been.

Accordingly, the challenge to the lending community for LNG projects is the extent to which the LNG sales revenues (both volume and price) can be evaluated for financing purposes without the support of conventional offtake contracts in place. This question raises issues of different shades:

- Could an LNG plant be financed without any pre-sold capacity on the basis that its developers judge the best commercial approach is to delay the placement of contracts until a later date?
- If not, is there some mix of sold and unsold capacity at which lenders would be prepared to attribute value to unsold train capacity?
- Where sold capacity is being supplied into new LNG markets with terminals and other infrastructure under construction, how will such risks be valued by lenders for the purposes of financing?
- Can lenders take a 'portfolio' view on a mix of offtakers with varying credit and operational profiles?

At the time of publication, there is no clear answer to these questions and much depends on the particular circumstances of a project. For example, in markets such as the United States and the United Kingdom, the existence of a volume offtake commitment is less important than demonstrating access to that market through committed regas terminal capacity. Similarly, the extent and availability of shipping play a large part in defining the project's ability to service different global markets and even to arbitrage pricing between them.

Ultimately, the physical and commercial arrangements needed to take LNG from its production point to the consumer market are more specialised and in far more limited supply than those applicable to crude oil sales. Given this fact, and the relative lack of liquidity in the short-term market, the assumption of merchant trading risk in its bare form still remains a highly challenging prospect for lenders. However, this is one of the seminal challenges presented by today's industry and it seems likely that, through a combination of financing techniques and contractual mitigations, this challenge will be answered sooner rather than later.

8.3 LNG vessel financing

On a number of large-scale LNG projects with a need to procure shipping capacity for DES sales, separate financings have been undertaken to finance the construction and delivery of an LNG vessel or vessels.

In many projects, the vessels will be owned by a different group of sponsors from those which own the upstream or the LNG plant and will usually include some of the specialist LNG vessels operators. The LNG vessels will be chartered rather than owned by the project company of the LNG plant.

This approach raises intercreditor issues with the LNG plant financiers, particularly to address the situation of a default under the vessel financing (giving rights to vessel lenders to sell the vessel to discharge the loans) which could, without redress, leave the LNG plant project company without dedicated shipping capacity for DES sales. A common solution is to agree quiet enjoyment arrangements which preserve the charters for the benefit of the project company, provided that charter payments are made.

8.4 The challenge of technology

The advance of technologies in areas such as coal seam gas to LNG projects and floating LNG projects will pose further challenges to the LNG financing community. As these technologies have the capability to significantly increase global LNG production and bring it into new territories, the need for a large pool of debt capital for these projects will become even more critical.

As this chapter has hopefully demonstrated, the LNG financing community has continued to meet the technological, political and economic challenges facing the LNG industry at a time of the greatest stress in international financial markets since the 1930s. The emergence of a set of new stakeholders in LNG financing has, and will continue to be, a key success factor in the years ahead.

9. Conclusion

The financing of LNG projects is a business of global standing. It is evolving dynamically and financing terms are being driven by a combination of powerful forces from both within the industry and outside. With an International Energy Association special report of June 6 2011 projecting a 50% increase in the global use of gas (which would account for 25% of global energy demand) by 2035, this evolution is likely to continue with increased vigour in the years to come.

LNG – a minefield for disputes?

James Baily
Paula Hodges
Herbert Smith LLP

1. Introduction

As has been explained in previous chapters, the market for LNG originally developed on the basis of long-term sale and purchase agreements between buyer and seller (typically, 20-year extendable contracts), with the prices under such contracts linked to oil and oil products.

As LNG markets have developed, LNG supplies have increasingly been traded under short or medium-term contracts, or on the rapidly expanding spot market. However, the majority of LNG supplies are still bought and sold under long-term agreements, and longer-term commitments remain important for sellers to secure the project financing necessary in order to bring new production facilities onstream.

During the early development of the global market for LNG, major disputes between buyer and seller were relatively rare. This was a consequence of the small number of buyers and sellers in the market, which tended to be locked together in close contractual (and sometimes shareholder) relationships. However, since the global economic downturn in 2008-2009, there has been a period of significant volatility in world gas markets, which has placed pressure on those relationships and has led to LNG contracts becoming increasingly contentious.

LNG markets have been unsettled in recent years in terms of both demand and prices. The development of domestic shale gas supplies in the United States has resulted in a significant and ongoing contraction in demand for LNG in the United States, proving the former predictions of sustained growth in that market to be incorrect. Conversely, demand in Asian markets recovered strongly following the global downturn. Demand has strengthened further as a result of the Japanese earthquake and tsunami of March 2011 and the effects of that disaster on the Japanese nuclear industry. Increased supplies from new sources of LNG production coming onstream, the impact of shale gas and the growing importance of the spot LNG market globally, as well as an increase in gas-to-gas competition as former regulated markets have liberalised, have all resulted in prices under traditional long-term LNG contracts being placed under considerable pressure. This has particularly been the case where the prices under such contracts are indexed to oil, because both oil demand and oil prices have remained buoyant in the international market during a period when gas demand – at least in some markets – has fallen.

The range of disputes which can occur in relation to LNG contracts is extensive, including technical disputes (eg, those involving measurement inaccuracies), complex legal and factual disputes (eg, those involving *force majeure* issues) and disputes in relation to pricing (eg, price reviews). Against that background, this chapter looks first at the forms of dispute resolution commonly used in the LNG industry and then at certain specific areas of dispute which are particularly prevalent in this sector.

2. Forms of dispute resolution

Given the multinational character of an LNG contract, it is rare for parties to elect to resolve disputes by litigation in the home territory of the buyer or the seller. Parties tend to opt for arbitration as their principal method of dispute resolution, with expert determination being favoured for certain specified types of dispute which are more limited in scope and technical in nature.

2.1 Arbitration

International arbitration allows the parties to select a tribunal to determine their dispute and the place where the proceedings will be held – often a neutral venue. The parties are also afforded the opportunity to agree the procedures to be adopted in resolving any dispute. Unlike litigation, arbitration is a confidential process.

Many states have sought to lay down a framework in which arbitration can take place as a viable (and often preferable) alternative to litigation in national courts, and to provide for the reciprocal enforcement of arbitral awards rendered by tribunals in other countries. This has led to wide acceptance by states of the 1958 New York Convention on the Recognition and Enforcement of Foreign Arbitral Awards. States which are party to the convention (over 140 nations) cannot impose more onerous conditions for the recognition or enforcement of international arbitration awards covered by the convention than are imposed in relation to the recognition or enforcement of domestic awards.

A recent table published by the International Group of Liquefied Natural Gas Importers (GIIGNL) lists LNG importers and LNG exporters in 2010 in percentage terms. This indicates that Japan was the largest importer of LNG in 2010 (31.7%), while Qatar was the largest LNG exporter (25.5%). Only four of the 23 importers and 20 exporters (listed by GIIGNL as at 2010) have not yet ratified the New York Convention: Puerto Rico, Korea, Taiwan (LNG importers) and Libya (LNG exporter).

In the following paragraphs, we outline several key issues for consideration in relation to the arbitration process.

(a) Drafting the arbitration clause

Arbitration clauses are rarely at the top of the negotiation agenda when commercial minds are eager to close the deal and do not want to be held back by considering the possibility of future disputes. It is difficult to legislate for every potential dispute in an arbitration agreement (no less so in the context of LNG contracts), and in some cases attempts to do so can actually increase the possibility of dispute rather than diminish it. However, there are certain minimum requirements that every arbitration agreement should include to ensure that it is enforceable and to facilitate an effective dispute resolution procedure.

As well as clearly delineating the scope of the dispute to be resolved by arbitration, the arbitration clause should make provision for the appointment of arbitrators, the place of arbitration, the language in which the proceedings will be held and the procedural rules to govern the dispute – for example:

> *[All disputes] arising out of, or relating to, this contract, or the breach, termination or validity thereof, shall be settled by arbitration. The arbitration shall be conducted in accordance with [the Arbitration rules of [X institution]]. The [Seat] of the arbitration shall be [London] and the arbitration shall be conducted in the [English] language. The arbitration shall be conducted by three arbitrators.*

Selection of a neutral venue for any arbitration proceedings is usual, but it is also essential to choose a country which has signed the New York Convention in order to take advantage of its beneficial provisions on enforcement. This is, of course, only one half of the equation – the likely place of enforcement of any arbitration award being the other. It is equally important to consider in advance whether the countries where enforcement will potentially be sought are also signatories to the New York Convention.

Additional mechanisms can be employed to allow for the consolidation of disputes along a contractual chain, or where there are a number of contracts between the parties (eg, when several LNG trains are to be utilised over time). This prevents more than one arbitration being commenced in respect of the same issues, which could lead to increased costs and divergent awards.

(b) *The arbitrators*

How an arbitral tribunal is to be constituted should be provided for in the arbitration agreement, which should specify the number of arbitrators and whether they are to be appointed by the parties themselves or by an appointing institution.

A sole arbitrator may be appropriate for relatively straightforward disputes where there is not a significant amount at stake. The appointment of a three-arbitrator tribunal, on the other hand, allows each party to nominate an arbitrator and therefore have equal opportunity to influence the formation of the tribunal. Having been appointed, the two nominee arbitrators will typically seek to agree on a chairman of the tribunal.

There are a number of reasons why the parties' selection of arbitrators is important. An arbitrator's qualifications should be thoroughly reviewed before nomination, particularly in light of the type of dispute which the parties face. Disputes in relation to highly technical issues may lend themselves to arbitrators with scientific or industry-specific knowledge which is relevant to the subject matter of the dispute. Disputes relating to cargo quality and/or quantity might lend themselves to specialists in those areas, while a dispute surrounding the construction of facilities might have aspects that point to an engineer being appointed. A *force majeure* dispute might well give rise to difficult legal issues, but where, for example, it relates to the redirection of a cargo following the closure of a nominated port, the dispute might turn on a point of international shipping custom so that a shipping expert may be better placed than a lawyer to determine those issues.

The constitution of the tribunal is particularly important in the context of international arbitration, where the process may involve people from different cultures and backgrounds. Having someone from a common background on the tribunal may inspire confidence in the parties that their interests and concerns will be adequately represented on the tribunal. Moreover, it is not uncommon to see contracts drafted in dual languages (reflecting the parties' different nationalities), with one language having official status. This might lead to issues of interpretation, with the result that each party may wish to select an arbitrator who is a native speaker of its language.

Arbitrators from the civil law tradition may take a rather different approach to those from, say, common law or Anglo-American backgrounds in determining any legal issues in dispute and in setting the procedure to be followed. For example, it might be strange for an arbitration between two parties from civil law countries to be chaired by a US arbitrator, since the latter's approach to matters such as disclosure and deposition evidence could be completely at odds with the legal tradition with which the parties may be familiar. *Force majeure* provides an example in point. It is a creature of contract and has no formal definition under English law. However, in certain civil law jurisdictions it has a different status under the law, independent of contract. Set-off is another legal concept which is subject to varying interpretations under different legal systems – some of which may be helpful, others not. Further, good faith (which can appear as a factor in price review clauses) is not recognised as a legal concept under English law, but is readily accepted as an enforceable obligation in most civil law jurisdictions.

(c) *The process*

The parties have a choice over the rules which will govern any arbitration proceeding. They can select the rules of an established institution or they can opt for an *ad hoc* arbitration with maximum flexibility. By deciding to involve an arbitral institution, such as the International Court of Arbitration of the International Chamber of Commerce or the London Court of International Arbitration, not only do the parties have well-tested arbitration rules at their disposal, but the arbitration will also be administered by the institution which will coordinate the arbitration and provide support services, such as assisting in identifying suitably qualified and experienced arbitrators. In contrast, an *ad hoc* arbitration is run by the parties and arbitrators alone. The parties may create their own rules for the arbitration. Alternatively, the United Nations Commission on International Trade Law has established a set of arbitration rules specifically designed for parties running an *ad hoc* arbitration, which provide a useful structure. The drawback of *ad hoc* arbitration is the lack of institutional 'clout' if one party refuses to cooperate, although this may be less of an issue in the context of a long-term contractual relationship such as those common in the LNG industry.

Despite the large degree of flexibility in international arbitration, there are a few procedural steps which are common to all arbitrations. Arbitrations will generally be commenced with the submission by one party to the other of a request for arbitration, in which the subject matter of the dispute will be set out. The tribunal will then be appointed in accordance with the arbitration agreement and a provisional procedural timetable will be decided by the tribunal. The exact contents of the timetable will vary, but it will set out the steps to be taken for the parties to present their cases and for evidence to be exchanged. It may also set a timeframe for any hearing to take place. Once the parties have had a chance to make their submissions and all the evidence has been heard, it will then fall to the tribunal to make its decision, which will be communicated to the parties in the final award.

2.2 Expert determination

Expert determination is used in a wide variety of commercial contracts and is particularly suited to disputes of a technical nature which are more effectively determined by an industry or financial expert. LNG contracts commonly refer disputes of this kind to expert determination, preferring arbitration for more generalised and wider contractual disputes. In particular, it is common to see disputes relating to measurement or quality specifications referred to expert determination. Another instance where expert determination tends to be used in LNG disputes is where a particular price index is discontinued and the parties cannot agree how it should be replaced in the price formula.

Expert determination is a creature of contract. Usually, there is no right of appeal and the expert's determination is final and binding on the parties. Expert determination does not have a formal and highly regulated structure like litigation; nor does it have machinery that allows for supervision by the courts, like arbitration. The law does not prescribe a set procedure for the manner in which an expert is appointed or conducts a reference. The expert clause in the parties' contract is therefore of fundamental importance in determining the procedure for the expert determination.

Many expert determination clauses in LNG contracts do not provide detailed guidance as to the conduct of the process. In the event that a dispute is to be referred to expert determination, the parties will need to appoint the expert (by reference to the process in the contract) and then determine the procedural directions in consultation with the appointed expert. In practice, parties usually draw up two separate agreements: one that governs the appointment of the expert (the terms of engagement) and another that sets out the procedure for the expert determination (the terms of reference).

(a) Appointment of the expert

For practical reasons, a contract rarely identifies the expert, since a named person may be unavailable or unable to act as expert when a dispute arises. Rather, the expert determination clause usually provides that the parties will attempt to agree the identity of the expert within a certain timeframe, and that if they fail to do so, the expert will be appointed by a specified body. This body is often referred to as the appointing authority and is usually a professional association, such as the International Centre for Expertise of the International Chamber of Commerce.

The expert determination clause may require the expert to have a prescribed technical background and/or a particular professional or academic qualification. Alternatively, the clause may be silent in relation to the expert's experience or provide that a person shall not be appointed as an expert unless "qualified by appropriate education, experience and training to determine the matter in dispute".

(b) ***The expert's terms of engagement***

The terms of engagement usually deal with the nature of the dispute to be determined by the expert, the expert's remuneration and independence and the extent of the expert's liability. It is important to define the expert's reference accurately (particularly if the scope of the expert's task has been refined compared to the provisions of the expert determination clause in the contract) so that all parties, including the expert, know the extent of the expert's jurisdiction.

Experts, in contrast with arbitrators, may be liable in negligence. Often, therefore, an expert will require a clause to be inserted in the terms of engagement which states that he will be immune from suit.

(c) ***Terms of reference***

There is no set procedure for the manner in which an expert should conduct a reference. Where the expert determination clause does not stipulate a procedure, the expert will be free to set out an appropriate procedure in discussion with the parties. However, the expert is not required to secure the parties' agreement to each step of the procedure.

The terms of reference should typically include the following:

- a precise definition of the issue the expert is to decide, regardless of whether that description simply repeats the expert clause or goes further and clarifies or refines the expert's reference;
- a statement (even if it is already included in the expert clause) that "the expert shall act as an expert and not as an arbitrator". The main reason for including this wording is to provide certainty and to exclude the law relating to arbitration;
- a provision that all information supplied to the expert is to be used solely for the purposes of the expert determination and that, save in certain specified circumstances, the parties agree to preserve the confidentiality of the expert determination and all information and documentation disclosed and prepared in the context of the expert determination;
- the procedure to be followed in relation to communications, submissions, the taking of evidence and hearings. Expert determination is an inquisitorial rather than an adversarial process, and the procedure can often be markedly different from that employed in litigation and/or arbitration. An expert is retained for his own expertise; he need not rely on the parties, but can pursue independent investigations into the issue for determination and should be empowered to instruct technical or legal experts if he deems it necessary. Alternatively, an expert may deem witness or expert evidence unnecessary and decide to determine the issue on the papers without any hearing being held; and
- the procedure for delivery of the expert's decision. The parties and the expert may agree that the expert should provide the parties with a draft decision in advance of the final determination. It is likely, however, that the expert will not want such an obligation to be absolute. The terms of reference may therefore provide that the expert "shall use reasonable endeavours" to

furnish the parties with the draft determination by a certain date. The parties will not ordinarily be able to make further submissions in relation to the draft determination, but having an advance copy of the draft provides the parties with an opportunity to point out any fundamental errors before the determination becomes final. The expert need not give reasons for his determination and so the parties should stipulate that reasons should be given, if required. The terms of reference should also reiterate that the expert's decision will be final and binding. If it is important for the expert to make a determination quickly, it would also be prudent to deal with this in the terms of reference.

(d) Enforcement of, and challenges to, the expert's decision

An expert's decision is binding on the parties (as a result of the wording in the expert determination clause and/or the terms of reference). If one party refuses to comply with the expert's decision, the other party can seek to enforce the decision by suing the offending party for breach of contract.

Although there is no right of appeal against the decision of the expert, a court may set aside a decision in certain limited circumstances – namely, mistake, fraud, collusion, partiality or bad faith, or lack of jurisdiction.

3. Price reviews

Price reviews (sometimes also referred to as 'price reopeners') are a prime area of dispute in relation to LNG contracts.

Price review provisions typically appear in long-term contracts where the price of the LNG is indexed to a basket of competing fuels – usually oil and oil products, although sometimes coal and inflation indices are also included. Price review clauses attempt to legislate for future unforeseeable changes in the value of the LNG being sold – perhaps as a result of a change in the value of gas in a particular market – by allowing the parties to commence a process of reviewing the price if certain criteria have been met (sometimes referred to as the 'trigger' for a price review). However, with buyer and seller having inherently different frames of reference, it is difficult to draft a trigger mechanism which does not favour one party over the other. As a result, price review clauses are often drafted in quite general terms. This leaves significant potential for dispute.

Typically, both regular and special price reviews will be included in a long-term LNG contract. Regular price reviews take place at regular intervals during the life of a sale and purchase agreement, while each party may also have the right to initiate a limited number of special price reviews over the life of a contract. For example, a party may have the right to request a regular price review on the third anniversary of the commencement date of deliveries under the sale and purchase agreement and each three-year anniversary thereafter, while retaining the right to call a special price review on three other occasions at any time during the term of the contract. In terms of how price reviews function, the classification of a price review as 'regular' or 'special' will usually mean little, as both will follow the same process. Both have the aim of ensuring that market fluctuations are taken into account over time by comparing the circumstances existing at the commencement date (or the date of the last price review, as the case may be) against the circumstances existing at the time of the review.

Wherever possible, parties will generally consider it preferable to resolve any disputes regarding pricing through negotiation rather than an adversarial dispute resolution process, in order to avoid the inherent uncertainty involved in any such process and the negative effects which a drawn-out dispute may have on a long-term contractual relationship. For this reason, price review provisions usually stipulate a substantial period of negotiation before either party can refer the issue to a formal dispute resolution process. Typically, price review disputes will then be submitted to arbitration, in accordance with the general dispute resolution provisions in the contract. Exceptionally, price review disputes may be referred to expert determination, although this is increasingly rare given the potentially wide scope of issues to be addressed (which often involve legal questions of interpretation of the underlying contract).

The evidence points to price reviews becoming more contentious and to an increasing number being taken to arbitration. Some of this is due to the fact that even small changes to the price formula in a 20-year sale and purchase agreement can have significant financial implications. Consequently, price review arbitrations can result in shifts in value of hundreds of millions of dollars. It was widely reported that the award made in 2010 against Gas Natural in its price review with Sonatrach was set to cost it up to €2 billion ($2.89 billion) in retroactive price increases. That dispute finally settled in June 2011, when Gas Natural agreed to pay $1.897 billion to Sonatrach (the deal also gave Sonatrach the opportunity to acquire a minority stake of up to 3% in Gas Natural).

3.1 Structure of a price review clause

Traditionally, price review clauses are structured so that the trigger event must be established in order for either party to request a revision to the contract price. Typically, the party seeking the price review will need to identify circumstances beyond its control which have given rise to a 'significant' (or 'material') change in the value of the LNG sold under the agreement, in comparison with its value at the date that the price was last set.

The party seeking the price review must normally serve a notice setting out the trigger event (or events) which it believes have given rise to the right to a price review, in which it must also propose a revised price formula and give details of the grounds for the proposed revision. If it is established that there has been a trigger event, a price review clause will then typically direct that the price should be revised in order to reflect the significant change or changes and the effect that such changes have had on the value of the gas sold under the contract.

An example of a typical price review clause is as follows:

Either party may request a review of the method for determining the Contract Sales Price set forth in clause X if it has a good faith basis for believing that, for reasons outside its control, there has been a significant change in the value of imports into Y and/or regasified LNG in the Y market which is anticipated to have lasting effect.

Important aspects of this clause include control, significant change, value, relevant market and lasting effect.

(a) Control

As in *force majeure* provisions, the party requesting a price review is often required to demonstrate that it has no control over the alleged significant change in the market. This brings into question important issues of sovereign control (particularly in liberalised markets) when a state entity is involved. The ability of parties to lobby for market change may also be a factor to consider.

(b) Significant change

In order to avoid the price review clause being restricted to certain specified scenarios which may or may not arise, the level of change required is often described in relatively undefined terms, such as 'material' or 'significant'. Such language is, of course, open to interpretation and disputes can develop as to what circumstances constitute a 'significant' change.

(c) Value, relevant market

It is usually a requirement of a price review clause that a causal link be shown between the significant change or changes identified and a change in the value of the LNG sold under the contract. Sometimes, price review clauses incorporate an objective element in the assessment of the value that can be obtained for the gas sold under the contract, by referring to concepts such as 'the value obtained by a prudent and efficient gas company'.

A price review clause may point towards value being assessed either at the import stage or as the regasified LNG is sold to end users (or, as in the example price review clause set out above, at both of these stages). Depending on the precise wording of a clause, it may suggest a comparison of import prices or, for example, an analysis of the buyer's margin based on netting back from end user gas prices (or, where the gas is used for power generation, from electricity prices). However, often a price review clause will give little clear assistance as to how value is to be measured or how the price formula should be revised to reflect changes in the value of the gas. Again, this reflects the view that the wording of price review clauses should be left relatively open, since they are designed to deal with the unexpected, and if too prescriptive, they might not actually address the particular situation faced by the parties. This has doubtless contributed to some of the uncertainty which has been perceived in price review disputes.

(d) Lasting effect

The use of words such as 'lasting effect' or 'excluding short-lived changes' is intended to ensure that price reviews do not capture market fluctuations of short duration. If reflected in an amended price formula, such fluctuations would be locked in until the next review, even if the market stabilised shortly afterwards. That said, the difficulty in predicting or, indeed, demonstrating that a change is only a short-lived market fluctuation makes this another area ripe for disputes.

A further issue which is frequently problematic is the correct methodology for revising the price formula. For example, it may be unclear from a price review clause whether the economic effect of each significant change should be quantified in turn and then adjustments made based on those calculations or whether, if it is accepted that a significant change or changes have occurred, it is enough to reset the price having a general regard to the new position. The extent of an arbitral tribunal's jurisdiction is also relevant. The question may arise as to whether a tribunal should be limited to resetting the base price and changing the indexation in a price formula, or whether it is possible for a tribunal to tear up the pricing provisions under a contract as drafted and impose a new price on the parties. This issue is considered in more detail below.

3.2 Issues relating to price review arbitrations

A relatively common complaint in relation to price review arbitrations has been the unpredictability of the result. Arbitral tribunals typically have a wide discretion, which may on occasion be used to reach a result that neither party advocated during the arbitration process. A regularly cited example of this is *Gas Natural Aprovisionamientos, SDG, SA v Atlantic LNG Company of Trinidad and Tobago*, an arbitration initiated in 2005 between Atlantic LNG (the seller) and Gas Natural (the buyer). This dispute entered into the public domain due to the fact that it was appealed in the US federal courts. The facts of the case may be summarised as follows:

- Under the sale and purchase agreement entered into by the parties, Gas Natural was able to transport LNG not only to its receiving facilities in Spain, but also to a facility in New England in the United States.
- Following liberalisation in the Spanish market, end user prices fell and Gas Natural entered into a long-term agreement to sell all of the LNG delivered under the contract with Atlantic at the New England receiving facility. Effectively, after 2002, no further deliveries were made to Spain. Atlantic LNG relied on these changes to trigger a price review and initiated arbitral proceedings seeking an increase in the contract price to reflect the higher price of gas in the US market.
- The arbitral tribunal imposed its own solution to the dispute, which had not been suggested by either of the parties. While the tribunal preserved the price formula contained in the contract (with an adjustment to the base price), it also introduced a 'New England Market Adjuster' which would apply when more than a specified percentage of the LNG was delivered to the New England receiving facility during a particular period. The immediate impact of this was that Atlantic LNG had to repay approximately $70 million to Gas Natural for sums which had already been invoiced.
- Atlantic LNG subsequently sought to overturn the arbitration award. It argued that the arbitrators had no power to rewrite the contract price. However, on appeal, the US federal court in New York confirmed that in circumstances where the price review clause contained no restrictions on the extent of the revisions that could be made to the price formula, the

> arbitrators were empowered to set a "fair and equitable revision" of the price, and that in principle there was no limitation on the scope of what they could do.

While the *Atlantic LNG* arbitration is illustrative of the wide discretion which is typically given to tribunals in price review disputes, the case turned on a very specific set of facts. It is not suggested that it is usual for tribunals to impose a solution which does not reflect the position adopted by either party.

A different, and perhaps more consistent, complaint which has been made in relation to price review arbitrations is that when faced with complex legal and technical arguments as to how a price should be revised, arbitrators will shy away from making any substantial change. Rather, a tribunal will seek to split the difference between the parties, in the process potentially alighting on a 'middle way' which is unsupported by any clear commercial or economic reasoning. One danger of this is that parties may be encouraged to exaggerate the increase or decrease in price that they are seeking, in the hope that an arbitrator will simply look to draw a line between the respective parties' positions. There is also the danger that speculative claims might be encouraged by this approach.

There are various ways in which the parties to an LNG sale and purchase agreement may seek to prevent these unpredictable or arbitrary results arising as a result of a price review arbitration. When new agreements are drafted or existing agreements amended, it is possible to limit the scope of the arbitrator's powers. Parameters can be set which would prevent a tribunal from imposing, for example, two prices (as happened in the *Atlantic LNG* case), or a radically different price structure (eg, substituting a gas trading hub price for a price indexed to oil products).

The range of possible outcomes could also be limited to an extent by more precise drafting. An objective methodology might be set out in order to determine whether a change is 'significant' and, if so, how the contract price should be adjusted to reflect it. For example, it could be agreed that a netback from end user prices should be calculated in order to establish whether a buyer's return has changed over time and that, if it has changed by more than a certain amount, it should be restored to what it was previously. While there may be dangers in being too prescriptive, it is possible to clarify the agreed process for a price review without being tied to overly specific formulae or ratios.

It is also possible to structure a price review arbitration so as to limit (to an extent) the risk of an undesirable result. The parties may agree to split the process, so that an initial hearing is held solely to determine whether a trigger event has occurred and therefore whether there is any basis for revising the price formula. Only if the tribunal finds that a trigger event has occurred will the parties then have the opportunity to submit further evidence and arguments as to how the price should be revised. The advantage of such a split approach is that once there is a initial award setting out the tribunal's findings as to whether significant changes have occurred (and, if so, what those significant changes are), a negotiated settlement as to how the price should be revised to reflect those changes becomes much more likely. Such a split approach might also enable the parties to avoid incurring wasted costs, since if the tribunal were to find following an initial hearing that no trigger event had been established, the parties would not then need to incur the costs of instructing experts and lawyers to deal with an adjustment of the price.

Alternatively, the parties to a price review arbitration might agree that the question of how to adjust the price formula should be dealt with by way of a 'pendulum' arbitration (also sometimes known as a 'final offer' or 'baseball' arbitration). In a pendulum arbitration, the tribunal is permitted only to adopt one of the parties' proposed solutions, rather than imposing its own solution. This can effectively constrain a tribunal from alighting on a result which is undesirable for either party, and may also encourage the parties to adopt more reasonable positions in the first place (which, in turn, increases the chances of a negotiated settlement being reached).

4. *Force majeure* disputes

Force majeure clauses are often activated in lengthy LNG contracts and disputes as to the scope of a *force majeure* clause are not uncommon.

A *force majeure* clause is intended to excuse the parties from performance of their contractual obligations, or to suspend performance, upon the occurrence of an event beyond their control which has prevented or hindered performance of the contract. *Force majeure* is a concept recognised in most legal systems but – under English law at least – it does not have a precise legal definition. Therefore, the circumstances which constitute a *force majeure* event and the parties' ensuing rights are matters for negotiation. This often results in a non-exhaustive list of *force majeure* events being included in the contract by way of example.

Acts of God and extreme weather conditions are frequently included within a *force majeure* clause (eg, in 2005, Hurricane Katrina's impact in the Gulf of Mexico triggered many such clauses). Other common factors include mechanical failure, fire, flooding, war, government action or decree, industrial disturbances and refusal of necessary licences. However, whether topical events, such as avian flu outbreaks and terrorist attacks, allow a party to suspend contractual performance will depend on the drafting of the *force majeure* clause in question.

Government or regulatory decisions are another common area of debate, particularly when LNG production and supply are linked to national interests. Given that the extent of the claiming party's control over *force majeure* events is a key consideration, the status of state-owned or related entities can often lead to dispute. At what point is a state-owned entity said to be independent of the state and its decisions so as to enable it to take advantage of *force majeure* provisions on account of a government decision in its home territory? This can be an important factor in jurisdictions where energy markets are less liberalised.

In contrast, changes in market conditions and the parties' financial circumstances are rarely included as instances of *force majeure*. This omission is, of course, mitigated by the presence of a price review clause in the contract.

4.1 Scope

The scope of *force majeure* clauses varies from contract to contract. Notice mechanisms will typically be included and the burden of proof will lie with the party seeking to rely on the clause. That is, a party must prove both that the event occurred and that there was a causal connection between the event and its inability to perform. However, subtle differences in the drafting of these provisions make them susceptible to varying interpretations, and therefore to disputes.

For example, strict compliance with notice provisions might be considered a condition precedent to a party relying on a *force majeure* clause, so it is important that notice provisions are followed closely. Similarly, the seller may be able to allege *force majeure* on account of the failure of its liquefaction equipment only if it has performed routine maintenance to the standard of a reasonable and prudent operator.

Additionally, parties may be contractually required to take certain steps upon the occurrence of a *force majeure* event before they are entitled to suspend performance of their obligations. For instance, where the event affects the buyer's receiving facilities, the contract might allow for destination flexibility whereby a buyer has to use reasonable endeavours to receive the LNG at an alternative safe port. Should a *force majeure* event occur, the parties' original obligations are, in effect, amended. When there is no longer any disruption caused by the *force majeure* event, the parties' rights and obligations are reactivated. However, the manner in which the parties' obligations may be amended is often ill defined, leading to disputes as to the consequential effects of *force majeure*. For example, in a destination flexibility provision, ineffective drafting may beg the question as to who is responsible for additional shipping costs. Should the seller be liable for those costs where the buyer has chosen a port which, although safe, is not necessarily the closest, and therefore cheapest, port to which to transport the LNG?

Force majeure clauses will also afford varying significance to events affecting parties upstream and downstream from the buyer and seller. For example, the *force majeure* clause may define as an event of *force majeure* an act, event or circumstances which primarily affects a third party, including any owner or operator of the seller's facilities, upstream gas suppliers or the owners or operators of LNG tankers. In situations where a buyer has a limited number of end users, the buyer should seek to provide for *force majeure* events affecting not only its own contractual performance, but also that of parties downstream.

4.2 Consequences

The consequence of *force majeure* in a general commercial contract is that performance is temporarily suspended and breach of any obligations is excused. In the context of LNG, a party may also be entitled to terminate a contract for extended *force majeure*. Typically, the *force majeure* event will need to have continued for a specific period of time, such as one or two years, and performance-related tests may also be required. For example, the *force majeure* clause might allow the parties to terminate the contract only where there has been a percentage reduction of loaded LNG over a period of 24 months. The extent to which the performance criteria are met is, of course, open to dispute. That said, a well-drafted clause will also recognise that the relationship under an LNG sale and purchase agreement may continue despite a *force majeure* event. Most notably, provision is often made for the *pro rata* supply by the seller and acceptance by the buyer of LNG which is still capable of delivery despite the *force majeure* event. If the contract does not so provide, disagreements relating to continuing performance under the contract can arise.

5. Conclusion

While there is significant potential for disputes in the LNG sector, arbitration and expert determination provide well-established routes to dispute resolution. The key is for parties to give due consideration to the relevant dispute resolution provisions when drawing up their agreement. This is of increasing importance since there are signs that the LNG industry is becoming more litigious as it continues to develop. Price review arbitrations in particular are becoming part of the commercial landscape for long-term LNG contracts, as a result of the volatility that has been seen in LNG markets. To a certain extent, this may increase the risks associated with the LNG business, although if the dispute resolution process is well managed, the uncertainties inherent in such disputes can be mitigated – even if they will never be eradicated.

LNG regulation

Garry Pegg
Philip Weems
King & Spalding

1. Introduction

The liquefaction and transportation of natural gas are not new processes. It may be surprising to many that liquefying natural gas was technically feasible over 100 years ago and it was technically possible to transport it by ship by 1914.[1]

Commercial development of the technology began in the United States: the first liquefied natural gas (LNG) plant was built in West Virginia in 1912 and the first commercial plant commissioned in Cleveland, Ohio in 1941. Another milestone was the commissioning in the 1960s of the world's first large-scale liquefaction plant in Algeria. In 1964 the United Kingdom signed a 15-year contract to take around 1 million tonnes from the plant, followed by the French the following year. The British, French and Americans had been working on new ship designs since the 1950s and the transportation of gas by sea was now possible. In 1959 the world's first LNG tanker, the Methane Pioneer (a converted World War II Liberty freighter), carried LNG from Lake Charles in the United States to Canvey Island in the United Kingdom, demonstrating that large quantities of LNG could be transported safely across the seas.

Since the early days, the focus of the industry and the international community has been on ensuring safety along the LNG chain – that is, ensuring that liquefaction plants, LNG vessels and regasification plants are designed, constructed and operated to the highest technical standards so as to minimise injury to persons and damage to property and to the environment. Recognised risks[2] include injury and damage caused by the heat of 'pool fires'.[3] Even if no fire occurs, being close to an LNG release could be dangerous: contact with the cloud could cause severe frostbite, death by freezing and, if a person were caught within a dense cloud of evaporated gas, death by asphyxia. While there have been some incidents over the years, the industry's safety record is good.

---

1 Michael Faraday, British chemist and physicist, experimented with liquefying natural gas. German engineer Karl van Linde built the first practical compressor refrigerator machine in Munich in 1873. Godfrey Cabot submitted a patent for a barge to carry liquid gas in 1914, although there is no evidence it was ever built.

2 See "Consequence Assessment Models for Incidents Involving Releases from Liquefied Natural Gas Carriers", ABS Consulting for the US Federal Energy Regulatory Commission, May 2004.

3 LNG does not explode like natural gas. To burn, it must first boil to produce gas. However, the boil-off gas is so dense it cannot burn, as there is not enough oxygen to support combustion. In windy conditions, however, oxygen could be mixed in at the fringes of the gas cloud and allow the gas to burn. While unlikely to cause an explosion (because there is nothing to contain the gas pressure), this could create an intensely hot flame that rapidly evaporates more gas to create a 'pool fire'.

LNG's inherent nature and advances in cryogenic technology have enabled the LNG industry to develop an enviable safety record. LNG is odourless, colourless, non-toxic and non-corrosive. Because LNG is mostly methane, it is lighter than air and will easily disperse if there is a gas leak. When mixed with air or vaporised, it burns in a narrow range of concentration (5% to 15%). Shell and other major LNG companies have spent millions of dollars carrying out research into the physics and chemistry of the dispersion and combustion of LNG, which has resulted in a better understanding of the true hazards of LNG. The LNG industry was also a major beneficiary of the aerospace industry's heightened understanding of cryogenics and cryogenic storage. Liquid gas systems are designed and constructed in compliance with very high standards to ensure their integrity, reliability and safety.

There is a school of thought that the more prominent role of the LNG spot market could potentially open up the LNG business to a greater risk of safety incidents, as operators become more risk averse in order to meet time pressures. There are also increasing concerns that the rapid expansion in fleet size is causing a shortage of skilled personnel. It remains to be seen whether these risks will materialise. Whatever the future holds, over the past 40 years there have been very few incidents associated with either LNG or LNG tankers, and LNG has an outstanding safety record.

This chapter reviews how safety and security policy has been implemented by considering the international conventions applicable to the transit of LNG by LNG tanker from the liquefaction plant across the high seas to a destination port in the United Kingdom or the United States, and the laws and regulations applicable at that destination.

With indigenous European supplies of gas now on the decline and demand for and imports of LNG to European countries correspondingly increasing, the focus is also now on the speedy development and construction of new LNG import facilities.

In the United States, the story has been the opposite. Due to the rise of non-conventional gas supplies such as shale, LNG imports into the country declined by 11.1% between 2009 and 2010.[4] This notwithstanding, there is still demand for LNG imports in the United States, with two new import terminals going onstream during 2010: the Golden Pass terminal near the Sabine Pass, Texas and the Neptune LNG terminal in Massachusetts.

The planning and approval regimes for new-build facilities in the United Kingdom and the United States are therefore also considered in this chapter.

Encouraging new development is one element of national and international strategy; the other is the facilitation of competitive gas markets, in particular the removal of barriers to entry. With new import facilities being built and financed for dedicated LNG shipments, open access to these facilities is now high on the agenda. This chapter also reviews how competition and antitrust regulations are being used in the United Kingdom and United States to implement this strategy.

---

4 *The LNG Industry 2010*, GIIGNL

2. Regulation of LNG tankers in transit

2.1 Introduction

LNG carrier capacities vary from:

- between 20,000 cubic metres (m^3) and 50,000m^3;
- between 50,000m^3 and 100,000m^3;
- between 100,000m3 and 140,000m^3; and
- for large carriers, between 165,000m^3 and 266,000m^3.

Today, more than 350 LNG ocean tankers safely transport more than 220 million metric tons of LNG annually to ports around the world. In 2010 alone LNG tankers completed 3,951 voyages (up from 3,414 in 2009).

LNG vessels have many safety features. A double-hull structure is designed to prevent leakage or rupture in an accident. LNG is stored in either double membrane containment systems located within the vessel's inner hull (with sensor equipment able to detect the smallest presence of methane in the insulation space between membranes) or special three-quarter-inch thick spherical tanks. Other safety features include:

- emergency shutdown systems;
- fire and gas detection and fire-fighting systems;
- radar and positioning systems;
- global maritime distress systems; and
- velocity meters.

Special operating procedures, training and maintenance further contribute to safety.

Transportation of LNG by sea has a long record of safe operation. There have been few reported accidents and none involving a fatality or major release of LNG. To date, of the 40,000-plus voyages by LNG tankers, there have been no pollution incidents involving an escape of LNG and no significant breach of cargo containment. This safety record is attributable not only to the continuous improvement of tanker technology, tanker safety equipment, comprehensive safety procedures, training and equipment maintenance, but also to effective government regulation and oversight.

2.2 International regulation

States have sought to standardise maritime law principally through international maritime conventions.[5] In the context of LNG transportation, the conventions were promulgated to improve the safety of international shipping (regulating construction and operation) and prevent pollution from ships. All LNG vessels in international service must comply with the major maritime treaties; this section

5 'Ratification', 'acceptance' or 'approval' are all acts by a state expressing its agreement to be bound by a convention. A state may 'accept' or 'approve' where its constitution does not require the convention to be ratified by the head of state. Generally speaking, once a state agrees to be bound by a convention, it is obliged to enact it into national law.

highlights the main conventions which are applicable.

Among the agencies of the United Nations, the International Maritime Organisation (IMO) is the most active in promoting maritime conventions. The IMO Maritime Safety Committee is the technical committee addressing safety-at-sea issues, including such things as:

- the construction and equipment of vessels;
- aids to navigation;
- the prevention of collisions;
- the handling of dangerous cargoes;
- safety in ship manning; and
- safety procedures and requirements.

In addition to promoting the adoption of conventions, the IMO adopts recommendations and codes of practices not generally suitable for regulation by formal treaty instruments. Although not usually binding on governments,[6] many governments adopt codes of practice into national legislation in any event.

Other non-governmental entities playing a dominant role in the safe and responsible operation of gas carriers and terminals include:

- the Society of International Gas Tanker and Terminal Operators (SIGTTO);
- the National Fire Protection Association (NFPA);
- the Oil Companies International Maritime Forum;
- the International Association of Ports and Harbours;
- the World Shipping Council; and
- the International Navigation Association.

SIGTTO, for example, is the industry leader in providing guidelines and recommendations to its members for safely handling liquefied gases, hazard analysis and contingency planning. Despite its name, the NFPA is an international organisation that develops, publishes and disseminates consensus codes and standards to reduce fire risks.

So who polices and enforces the law?

International maritime law is enforced and policed at the national level by the flag state and the coastal state. Classification societies also have a significant role in creating and policing maritime safety regulations, and special mention must be made of these.

Classification societies are independent, non-profit organisations. There are around 50 worldwide, the largest of which (based on the number of ships classed) include Lloyd's Register of Shipping[7] and the American Bureau of Shipping. They create specification rules for vessels and monitor compliance with them, with national laws and with international conventions. If a vessel passes inspection, the classification society issues a classification certificate. Major classification societies

---

6 Codes become mandatory when they are part of conventions.

7 Lloyd's Register is particularly active in the classification of LNG vessels. It classed the first-ever LNG ship in the world to bear the notation 'Liquefied Gas Tanker' in 1958.

are members of the International Association of Classification Societies (IACS), which self-regulates its members, and compliance with its Code of Ethics and Quality System Certification Scheme is mandatory. As discussed further below, surveys and certifications required under state laws (often derived from international conventions) are frequently delegated to classification societies.

Under international law, the high seas are not regarded as part of any state's territory.[8] Thus, on the high seas every flag state possesses authority, although not absolute, over its own registered vessels and takes legal responsibility to ensure that its ships comply with accepted standards of international law and conventions to which it has agreed to be bound.[9]

Coastal state jurisdiction over shipping is limited to the internal waters of the state,[10] its territorial sea[11] and its exclusive economic zone.[12] The United Nations Convention on the Law of the Sea (UNCLOS) permits coastal states to legislate for navigation safety – that is, to provide for port state control to inspect ships entering a port and detain any for necessary repairs that do not meet international safety standards.

(a) *Main conventions and codes of practice*

SOLAS: The International Convention on Safety of Life at Sea (SOLAS) came into force in 1965.[13] It provides, among other things, that ships must be designed, constructed and maintained in compliance with the structural, mechanical and electrical requirements of a recognised classification society.[14]

SOLAS (and other conventions) permit the flag state to delegate the inspection and survey of ships to a 'recognised organisation', which includes many IACS members.

MARPOL: The International Convention for the Prevention of Pollution from Ships 1973 (MARPOL) addresses the prevention of pollution of the marine environment by

---

8 The 1982 United Nations Convention on the Law of the Sea (UNCLOS) defines 'high seas' as "all parts of the sea that are not included in the exclusive economic zone, in the territorial sea or in the internal waters of a state".

9 Article 94 of UNCLOS provides that the flag state must "effectively exercise its jurisdiction and control in administrative, technical and social matters over ships flying its flag", and take "such measures for ships flying its flag as are necessary to ensure safety at sea".

10 Ports, harbours, lakes, rivers, canals and waters on the landward side of the baselines from which the breadth of the territorial sea is measured.

11 The territorial sea is established in Article 3 of UNCLOS as "up to a limit not exceeding 12 nautical miles, measured from baselines determined in accordance with this Convention". The normal baseline is the low water line along the coast (Article 5 of UNCLOS).

12 Established under UNCLOS as "an area beyond and adjacent to the territorial sea", being an area of water extending up to 200 miles from a state's shoreline claimed (or established) at the discretion of each state. The coastal state has sovereign rights over natural resources (to explore, exploit, conserve and manage) within the zone, but not over shipping. Thus, as for the high seas, there is freedom of navigation. However, the coastal state has the right to protect and preserve the marine environment within the exclusive economic zone, enforcing pollution regulations related to the dumping of waste, oil spills and other forms of pollution from ships.

13 The 1974 SOLAS Convention modified the 1960 convention by including the amendment procedure whereby SOLAS can be updated to reflect changes in the shipping environment without having to call a conference – that is, amendments adopted by the MSC enter into force on a predetermined date, unless objected to by a specific number of states

14 Chapter VII Part C deals with the construction and equipping of ships carrying liquefied gases in bulk.

ships. It is a combination of two treaties adopted in 1973 and 1978 respectively, and updated by amendments over the years.

Ships engaged on international voyages must carry on board valid international certificates which may be accepted at foreign ports as *prima facie* evidence that the ship complies with the requirements of the convention. However, if there are clear grounds to believe that the condition of the ship or its equipment does not correspond substantially with the particulars of the certificate, or if the ship does not carry a valid certificate, the authority carrying out the inspection may detain the ship until it is satisfied that the ship can proceed to sea without presenting unreasonable threat of harm to the marine environment.

Any violation of MARPOL within the jurisdiction of any party state is punishable either under the law of that state or under the law of the flag state.

IGC Code: As discussed earlier, shippers have long recognised the particular risks of transporting products under cryogenic or pressure conditions. Accordingly, the International Code for the Construction and Equipment of Ships Carrying Liquefied Gases in Bulk (IGC Code) was developed by the IMO to provide an international standard for the safe transport by sea in bulk of liquefied gases (and certain other substances), by prescribing the design and construction standards of ships involved in such transport and the equipment they should carry. The requirements of the code were designed to minimise the risks to the ship, its crew and the environment as far as is practicable, based on present knowledge and technology. Classification societies have incorporated the code into their rules.

The code applies to gas carriers constructed on or after July 1 1986.[15]

ISPS Code: Since the terrorist attacks in the United States in 2001, the threat to transport systems worldwide, including maritime transport, has changed. The IMO responded by developing new security requirements for ships and port facilities to counter the threat of terrorism, placing obligations on contracting governments and the maritime industry. These requirements are set out in amendments to SOLAS[16] and in a new International Ship and Port Facility Security Code (ISPS Code) implemented on July 1 2004.

Under the ISPS Code, each vessel must carry on board a ship security plan (approved by the relevant organisation) establishing:

- access control measures;
- security measures for cargo handling and delivery of ship's stores;
- surveillance and monitoring;
- security communications;
- security incident procedures; and
- training and drill requirements.

---

15 The Code for the Construction and Equipment of Ships Carrying Liquefied Gases in Bulk or the Code for Existing Ships Carrying Liquefied Gases in Bulk applies to gas carriers constructed before July 1 1986.

16 A new Chapter XI-2 entitled "Special measures to enhance maritime security".

The shipowner must appoint a ship security officer responsible for the implementation of the ship security plan, keeping it under review, enhancing security vigilance and awareness on board, ensuring proper training and so on. Each ship must have a company security officer, who must be a member of the shoreside management. His function is to monitor vessel security, ensure modifications to the ship security plan are implemented as required, and arrange for security audits and inspections. Lastly, each shipowner must retain full security records relating to training, drills and exercises, security threats and incidents.

Each contracting state in which a vessel is registered must appoint officers to verify compliance with the code and be responsible for issuing an international ship security certificate, valid for a period of five years.

Harmonisation of certification: Pre-certification surveys can often result in vessels being out of service for several days, made worse by intervals between surveys and survey dates which do not coincide. Accordingly, the IMO developed a harmonised system of survey and certification to alleviate these problems. The system, which entered into force on February 3 2000, covers survey and certification requirements of certain conventions, including SOLAS and MARPOL,[17] as well as the ICG Code and others.[18]

In short, the system provides for:

- a one-year standard interval between surveys, based on initial, annual, intermediate, periodical and renewal surveys, as appropriate;
- a scheme for providing the necessary flexibility for the execution of each survey, with the provision that the renewal survey may be completed in the three months before the expiry date of the existing certificate with no loss of its period of validity;
- a maximum period of validity of five years for all certificates for cargo ships; and
- a system for the extension of certificates (limited to three months) to enable a ship to complete its voyage (or one month for ships engaged on short voyages).

Liability for marine incidents: For centuries, to encourage trade and exploration, shipowners were afforded by many legal regimes limitation on their liability for damage they caused by maritime accidents. International standardisation was pursued first by the Convention on the Limitation of Liability of Owners of Sea-going Ships 1957. This was acceded to by a number of countries (including, and of interest in the LNG context, India and Algeria). Another convention – the Convention on Limitation of Liability for Maritime Claims (otherwise known as the London Convention) – was promulgated in 1976 and has been acceded to by countries such as the United Kingdom, Spain and Belgium. A protocol to amend the

17 It also covers the International Convention on Load Lines 1966.

18 Namely, the International Code for the Construction and Equipment of Ships Carrying Dangerous Chemicals in Bulk and the Code for the Construction and Equipment of Ships Carrying Dangerous Chemicals in Bulk.

London Convention has also been signed by a number of states (including, in this context, the United Kingdom and Norway) (the 1996 protocol). In complete contrast, the United States, for example, is not a party to any of the international conventions governing limitation of liability for maritime casualties, but has legislated domestically.[19]

Therefore, the particular liability regime applicable to any accident in the territorial waters of a state will depend on the domestic law of that state and which convention that state has acceded to. Generally, under the London Convention and 1996 protocol, the shipowner's liability may be limited up to certain thresholds (based on vessel gross tonnage) depending on the damage, personal injury and loss of life claims having a higher limit than other claims (damage to property and pollution).[20]

With respect to damage caused by a typical large LNG tanker today, the shipowner's liability under the 1996 protocol for personal injury and death would be limited to approximately $100 million, and for property damage around half that again.[21] A shipowner is not permitted to rely on these limitations if "it is proved that the loss resulted from his personal act or omission, committed with the intent to cause such a loss, or recklessly and with knowledge that such loss would probably result". When a claim is instigated the owner must set up a limitation fund in the appropriate court against which the claim can be pursued.

This regime will change radically if the International Convention on Liability and Compensation for Damage in Connection with the Carriage of Hazardous and Noxious Substances by Sea 1996 is ratified and adopted by the required number of states.[22]

This convention is the first to impose strict liability on the LNG industry, among others, for damage caused by maritime incidents. Liability arises to the registered shipowner without the need to establish fault or negligence, but only causation. The owner of the LNG involved in the incident is not liable under the convention.

It covers damage caused by hazardous and noxious substances (HNS)[23] in the territory or territorial sea of a party state. It also covers pollution damage in the exclusive economic zone or equivalent area of a party state and damage (other than

---

19 The Limitation of Vessel Owner's Liability Act, 46 USCA §§ 30503 *et seq* (2006). See below.

20 The shipowner is entitled to limit its liability to an amount calculated on the basis of the units of gross tonnage of the ship. The thresholds are measured in special drawing rights (SDR), an artificial monetary unit established by the International Monetary Fund (IMF) based on a basket of currencies. The daily conversion rates for SDR can be found on the IMF website at http://www.imf.org.

21 This assumes an LNG tanker with a gross tonnage of 120,000 and 1 SDR =S$1.60927 (the rate on July 28 2011).

22 The convention will enter into force 18 months after ratification by at least 12 states subject to the following conditions:
- Four states must each have a registered ship's tonnage of at least 2 million units of gross tonnage; and
- Contributors in the states that have ratified the convention must, between them, have received during the preceding calendar year a minimum of 40 million tonnes of cargo consisting of bulk solids and other hazardous and noxious substances liable for contributions to the general account.
- Canada, Denmark, Finland, Germany, Netherlands, Norway, Sweden and the United Kingdom have signed the convention subject to ratification.

23 The convention defines the concept of HNS largely by reference to lists of individual substances that have been previously identified in a number of international conventions and codes designed to ensure maritime safety and prevention of pollution, including LNG.

pollution damage) caused by HNS carried on board ships registered in, or entitled to fly the flag of, a state party outside the territory or territorial sea of any state. Costs of measures taken to prevent or minimise damage are also covered wherever taken. The convention does not cover damage caused during the transport of HNS to or from a ship. Cover starts from the time when the HNS enters the ship's equipment or passes its rail, on loading, and ends when the HNS ceases to be present in any part of the ship's equipment or passes its rail on discharge.

'Damage' is defined to include:

- loss of life or personal injury on board or outside the ship carrying HNS;
- loss of or damage to property outside the ship;
- loss of income in fishing and tourism; and
- loss or damage by contamination of the environment.

It also includes the costs of preventive measures and further loss or damage caused by those measures. These would therefore include measures such as clean-up or removal of HNS from a wreck if the HNS presented a hazard or pollution risk.

Under the convention, the shipowner is liable for loss or damage up to a certain amount, which is covered by protection and indemnity insurance (first tier). The limit is based on vessel tonnage.[24] The typical larger LNG vessel of today will currently mean a maximum liability of around $160.926 million.[25] This limitation of liability will not apply if it is proved that the damage resulted from the shipowner's personal act or omission committed either with intent to cause damage or recklessly and with knowledge that damage would probably result. The shipowner is required by the convention to provide evidence of insurance cover upon the ship's entry into port of any party state by production of a certificate, regardless of whether the flag state is party to the convention. The certificates will be issued by the flag state or, if that state is not party to the convention, by a party state.

However, the shipowner will not be liable if it can prove that the damage was:

- caused by acts of war, hostilities, civil war, insurrection or certain natural phenomena;
- wholly caused by an act or omission done with intent to cause damage by a third party (eg, a terrorist strike); or
- wholly caused by the negligence or other wrongful act of any governmental authority responsible for the maintenance of lights or other navigational aids.

A compensation fund will provide additional compensation when claimants do not obtain full compensation from the shipowner or its insurer (second tier). Once the first tier limit is reached, compensation will be paid from the fund up to a maximum of 250 million SDR (including compensation paid under the first tier). The

---

24 The shipowner is entitled to limit its liability to an amount calculated on the basis of the units of gross tonnage of the ship with the aggregate amount of the shipowner's liability not exceeding 100 million SDR. Liability for damage caused by a 120,000 gross tonnage LNG tanker will therefore be capped at this limit.

25 100 million SDR @ US$1.60927.

compensation fund will also pay compensation if the shipowner is exonerated from liability or is financially incapable of meeting its obligations.

Claimants must prove only that there is a reasonable probability that the damage resulted from an incident involving one or more ships, so the fund may in such cases be liable to pay compensation even if the particular ship causing the damage cannot be identified.

The compensation fund will not be liable to pay compensation if the damage was caused by an act of war, hostilities and so on, or by HNS discharged from a warship or other ship owned or operated by a state and used, for the time being, only on government non-commercial service. If the fund proves that the damage resulted wholly or partly either from an act or omission done with intent to cause damage by the person who suffered the damage or from the negligence of that person, it may be exonerated, wholly or partially, from its obligation to pay compensation (although not damages incurred by taking preventive measures).

The amount available for compensation from the shipowner and the compensation fund will be distributed among claimants in proportion to their established claims. However, claims for loss of life and personal injury have priority over other claims. Up to two-thirds of the available compensation amount is reserved for these claims.

The compensation fund will be financed by contributions levied on persons who have received, in a calendar year, contributing cargoes after sea transport in a member state in quantities above the thresholds laid down in the convention. However, any contributions shall be made by the person that, immediately prior to its discharge, held title to an LNG cargo discharged in a port or terminal of that member (the titleholder) where:

- the titleholder has entered into an agreement with the receiver that the titleholder shall make such contributions; and
- the receiver has informed the member party that such an agreement exists.

If the aforementioned titleholder does not make the contributions or any part thereof, the receiver shall make the remaining contributions.

For each contributor the levies will be in proportion to the quantities of HNS received by that person each year. Contributions will be made post-incident and may be spread over several years in the case of a major incident.

By 2009, the convention still had not entered into force, due to an insufficient number of ratifications. A second international conference, held in April 2010, adopted a protocol to the HNS Convention (2010 HNS Protocol), which was designed to address practical problems that had prevented many states from ratifying the original convention.

Once the 2010 HNS Protocol enters into force, the 1996 convention, as amended by the 2010 protocol, will be called the International Convention on Liability and Compensation for Damage in Connection with the Carriage of Hazardous and Noxious Substances by Sea, 2010 (2010 HNS Convention).

The 2010 HNS Protocol will enter into force 18 months after the date on which it is ratified by at least 12 states, including four states each with not less than 2 million units of gross tonnage, and having received during the preceding calendar year a total quantity of at least 40 million tonnes of cargo that would be contributing to the general account.

A special consultative meeting took place in Rotterdam, the Netherlands, on June 14 and 15 2011, which brought together nine states that had expressed an interest in ratification of the 2010 HNS Protocol. In light of the changes introduced by the 2010 HNS Protocol, the meeting provided an occasion to revisit the outcomes of the 2003 Ottawa meeting, mentioned in Resolution 4 of the 2010 HNS Protocol, as the best approach for the implementation of the 1996 HNS Convention.

The delegates at the Rotterdam meeting were reminded that the 2010 HNS Protocol is open for signature until October 31 2011.

2.3 UK regulation

The UK Department for Transport is responsible for the transportation of dangerous goods by all modes of transport but delegates to various authorities – in the case of sea transportation, to the Maritime Coastguard Agency.

The Department for Transport is supported by the Health and Safety Executive, whose role in this context originates from the Health and Safety at Work Act 1974. The Health and Safety Executive currently maintains main responsibility for enforcement and compliance.

The Maritime Coastguard Agency is the competent authority with responsibility for the safe transport of dangerous goods by sea. Its powers are derived from the Coastguard Act 1925, the Merchant Shipping Act 1995, the Merchant Shipping and Maritime Security Act 1997 and associated secondary legislation. The Merchant Shipping Acts regulate, among other things, the safe design, construction and operation of vessels, and qualifications of the master and crew, giving effect to the international maritime conventions referred to above.

(a) Vessel certifications

In addition to the many vessel certificates and required documents on board,[26] UK LNG vessels are required to hold an international certificate of fitness for the carriage of liquefied gases in bulk.

The IGC Code was implemented in the United Kingdom by the Merchant Shipping (Gas Carriers) Regulations 1994 (and subsequently amended by the Merchant Shipping (Gas Carriers) (Amendment) Regulations 2004).[27] They apply to vessels constructed on or after July 1 1986 but before October 1 1994 (1986-1994 gas carriers) and to gas carriers constructed on and after that date (new gas carriers).

---

26 The certificates required will depend on the vessel and how it is equipped. Typical certificates required will include an international load line certificate, a safety radio certificate, a safety equipment certificate, a safety construction certificate and a safety management certificate.

27 The 2004 regulations amend the 1994 regulations to give effect to amendments to the IGC Code adopted by the IMO on May 24 1990. Resolution MSC 17(58) of the IMO introduced a harmonised and simplified system of survey and certification for ships constructed to the IGC Code, standardising the period of validity and intervals between surveys for the nine main convention certificates (issued under the various provisions of SOLAS 1974, MARPOL 73/78 and the Loadline Convention 1966). In essence, the 2004 regulations harmonise the survey and certification requirements of that code with the requirements of SOLAS.

The regulations incorporate by reference the IGC Code requirements on the construction, equipping and operation of gas carriers,[28] and introduce a system for surveying UK vessels and issuing an international certificate of fitness for the carriage of liquefied gases in bulk. The first survey comprises a complete examination of the vessel's structure and equipment (to the extent covered by the ICG Code). Renewal surveys to ensure compliance are then conducted every five years. Also, intermediate surveys must be conducted three months before or after the second or third anniversary of the issue of the original certificate. Surveys must also be conducted after any repairs carried out after a compliance investigation, or after any other repairs or renewals. The owner and master must also ensure that, after a survey, no modifications are made to the ship, and they are required to report significant accidents or defects to the specified authorities.

Where a ship does not comply with the regulations (or if the integrity of the ship is substantially compromised), the ship may be detained. The regulations also make it an offence for failure to comply with the IGC Code and to carry LNG without having in place a valid certificate. The owner and the master can be prosecuted for these offences, which are punishable by fine. The power to detain is also extended to non-UK ships where a full and proper report of an accident or defect has not been made, or where the secretary of state is not satisfied that action has been taken to restore the ship.

In 2003, with the objective of minimising duplication and disruption to crew routines, the Maritime Coastguard Agency introduced, on a trial basis, the voluntary Alternative Compliance Scheme to allow owners, operators, designers and builders of UK flagged vessels to achieve statutory certification through a streamlined process. Under the scheme, certain classification societies[29] are empowered to undertake the majority of surveys required on UK registered vessels (including an international certificate of fitness for the carriage of liquefied gases in bulk). A certificate of inspection is issued by the Maritime Coastguard Agency to the ship to signify this. This certificate is valid for five years, subject to a further satisfactory mid-term inspection by the agency.

The agency retains responsibility for audits and surveys under the International Safety Management Code and International Labour Organisation Convention 178, as well as for safe manning certification and approval of equivalencies and exemptions from maritime conventions. However, these audits and surveys are linked with the certificate of inspection.

(b) Health and safety on board

The Merchant Shipping (Dangerous Goods and Marine Pollutants) Regulations 1997 implement Article 4 of Council Directive 93/75/EEC[30] relating to vessels bound for or leaving community ports and carrying dangerous or polluting goods and the 1991 and 1994 amendments to the SOLAS Convention.

The regulations impose a general duty on every operator, every employer of

28 1986-1994 gas carriers are subject to the 1983 edition of the ICG Code and new gas carriers to the 1993 edition.

29 These currently include Lloyd's Register of Shipping and the American Bureau of Shipping.

30 (1993) OJ L247 19.

persons aboard a ship and every master of a ship to ensure that, as far as is reasonably practicable, when dangerous goods are being handled, stowed or carried on the ship, nothing in the manner in which those goods are handled, stowed or carried, as the case may be, is such as might create a significant risk to the health and safety of any person. The duty extends to the provision and maintenance of ship's structure, fittings and equipment for the handling, stowage and carriage of dangerous goods and the provision of information, instruction, training and supervision to all employees in connection with the handling, stowage and carriage of dangerous goods in the ship. The regulations are not prescriptive; rather, they require compliance with specific codes (or recommendations applicable to gas carriers). If an employer, operator or master fails to comply, it will be guilty of an offence and liable on summary conviction to a fine.

(c) Vessel security

Following the amendments to SOLAS and the promulgation of the ISPS Code, at a European level, the Council and European Parliament adopted Regulation (EC) 725/2004 (the Maritime Security Regulation) on enhancing ship and port facility security seeking to provide for consistent implementation of the IMO requirements in EU member states.

The Maritime Security Regulation came into force on May 19 2004, giving direct legal effect to the ISPS Code in the United Kingdom. The requirements went further than the ISPS Code – for example, they require all ships intending to enter ports within the European Union to declare intention to enter at least 24 hours in advance and provide security information, including whether the ship is in possession of a valid international ship security certificate.

Although directly applicable in the United Kingdom, certain of its provisions required UK legislation to make them fully effective, and accordingly the United Kingdom introduced the Ship and Port Facility (Security) Regulations 2004. These regulations:

- designate the secretary of state as the United Kingdom's competent authority
- designated authority and focal point for maritime security for the purposes of the Maritime Security Regulation and IMO regime;
- establish an inspection regime for monitoring compliance with the Maritime Security Regulation and IMO requirements, and impose criminal sanctions for unlawful presence in a restricted area of a ship or port facility, and the obstruction of a duly authorised officer acting in exercise of his powers under the security regime;
- establish an enforcement regime and impose criminal sanctions against companies, ships and port facilities for non-compliance with the Maritime Security Regulation and the IMO regime; and
- introduce requirements for the issue and revocation of detention notices for ships that fail to comply with the security regime.

(d) UK harbour areas

The statutory harbour authorities control marine traffic into and through harbours and the berthing and moving of vessels in accordance with the Marine Safety Code.

The Marine Safety Code (together with the Guide to Good Practice on Port Marine Operations, published by the Department for Transport) provides guidance to harbour authorities to help them understand their duties and how to discharge them. The code outlines the general duties and powers of harbour authorities and the measures with which they must comply. Compliance is monitored by the Maritime Coastguard Agency.

The carriage, loading, unloading and storage of LNG in harbour areas is controlled by the Dangerous Substances in Harbour Areas Regulations 1987. The regulations confer responsibility on the Health and Safety Executive for enforcing the regulations, apart from the regulations dealing with such things as the marking and navigation of vessels and the fitness of vessels, which are enforced by the statutory harbour authorities. Breaches are subject to penalties levied under the Health and Safety at Work Act.

(e) *Liability for marine accidents*

The London Convention has the force of law in the United Kingdom by virtue of the Merchant Shipping Act 1995. The Merchant Shipping and Maritime Security Act 1997 added Sections 185(2A) to (2E) to the Merchant Shipping Act 1995, providing for the coming into force of the 1996 protocol; the 1996 protocol was brought into force in the United Kingdom by the Merchant Shipping (Convention on Limitation of Liability for Maritime Claims) (Amendment) Order 2004.[31]

The convention allows individual states to make their own regulations in respect of vessels of less than 300 tons. In the United Kingdom, for example, the limits for such vessels has been set at 500,000 SDR for property damage and 1 million SDR for personal injury and death claims.

The United Kingdom has signed the International Convention on Liability and Compensation for Damage in Connection with the Carriage of Hazardous and Noxious Substances by Sea and is currently taking steps to ratify it. Enabling legislation exists in the form of the Merchant Shipping Act 1995, but the implementation of this convention is likely to be protracted.

2.4 US regulation

(a) *Vessel certifications*

As in the United Kingdom, all LNG vessels entering the United States must meet international and domestic regulations. US regulations are now substantially codified under Title 46 of the United States Code which, closely paralleling international regulations, specifies requirements for a vessel's design, construction, equipment and operation. US regulations are more stringent, providing, for example, for enhanced grades of steel in certain areas of the hull and higher allowable stress factors for certain types of tank.

Before arrival, LNG vessel operators must submit detailed vessel plans and other

---

31 SI 1998 No 1258 of May 19 1998.

information to the US Coast Guard's Marine Safety Centre to establish that the vessel has been constructed to these higher US standards. Vessels are certified after satisfactory review of the plan and following an on-site verification by marine inspectors. Once issued, the certificate is valid for two years, subject to annual examination by the marine inspectors.

LNG ships undergo more rigorous and frequent examination than tankers transporting crude. Prior to US port entry, marine safety personnel board LNG ships to verify proper operation of key navigation, safety, fire-fighting and cargo control systems.

(b) *Vessel security*

LNG ships are also subject to additional security measures conducted under the authority of existing port safety and security statutes, such as the Magnuson Act[32] and the Ports and Waterways Safety Act.

These include:

- special traffic control measures for the LNG vessel (implemented when the vessel is in transit or approaching a US port);
- security zones around the vessel to prevent other vessels from approaching it;
- escorts by US Coast Guard patrol craft; and
- coordination with other federal, state and local transportation, law enforcement and/or emergency management agencies to reduce the risks to or interference from other port area infrastructure or activities.

Additional security measures were implemented after September 11 2001, including a requirement that all ships calling in the United States provide the US Coast Guard with a 96 hours' advance notice of arrival (previously 24 hours). The notice must now include information on the vessel's last ports of call, crew identities and cargo. The US Coast Guard will conduct at-sea boarding and 'security sweeps' to ensure 'positive control' of the vessel is maintained throughout the port transit process.

Under the Maritime Transportation Security Act 2002, security measures, closely aligned with the ISPS Code, were developed which are applicable to vessels, marine facilities and maritime personnel. The ISPS Code provides more resources for sea marshals and background checks. It also establishes port safety committees and vessel identification tracking systems, and extends the US Coast Guard's jurisdiction to 12 nautical miles from shore.

(c) ***Liability for marine accidents***

As mentioned earlier, the United States is not a party to any of the international conventions governing limitation of liability for maritime casualties. However, a shipowner may limit its liability pursuant to the Limitation of Vessel Owner's Liability Act[33] to the amount of the "owner's proportionate interest in the vessel and

---

32 50 USC 191 *et seq.*

33 46 USCA §§ 30503 *et seq* (2006).

pending freight".[34] This is determined on termination of the voyage in which the loss or damage occurs. Under US law, where a ship sinks following a collision, the sinking is the termination of the voyage for purposes of the act, and the value of the vessel (and accordingly the limitation of the shipowner's liability) is measured by the value of whatever is saved prior to the sinking.[35] Hull and machinery insurance has no bearing on determining value for these purposes.

Limitation of liability is an action in its own right under the act, rather than a defence to an action for damages. Consequently, a shipowner is best advised to initiate a limitation action by petitioning the relevant federal district court within six months of submission of the claimant's written notice to the owner of the claim.[36]

3. UK regulation of LNG import facilities

3.1 Introduction

The United Kingdom currently has four LNG import facilities – Dragon LNG at Milford Haven; Gasport LNG at Teeside; Qatar Petroleum/ExxonMobil, also located at Milford Haven (known as South Hook); and Grain LNG for the Isle of Grain – and five small-scale LNG storage facilities strategically located across the country (designed only to address winter peak demand requirements and not suitable for LNG imports). The United Kingdom has no liquefaction plant and does not export LNG.

The United Kingdom has three other small-scale LNG facilities strategically located across the country (Glenmavis, Avonmouth and Partington) owned by LNG Storage, a trading division of National Grid Gas plc (previously known as Transco), with a combined capacity of 180 million cubic metres. These facilities are designed only to address winter peak demand requirements and are not suitable for large-scale LNG imports. Their future use is currently under review due to insufficient market interest in their long-term capacity,

3.2 Environmental/health and safety regulation

In contrast to the United States, no specific government authorisations are required to construct or operate LNG facilities. Instead, owners and operators must comply with all relevant environmental, planning and health and safety requirements.

(a) *General planning permissions*

Depending on the particular facilities being constructed, planning permissions will be required under the Planning Act 2008.

(b) *Hazardous substances consent*

Onshore facilities[37] wishing to hold stocks of certain hazardous substances (including LNG) above a threshold quantity (for LNG, 15 tonnes) must apply for consent from the relevant hazardous substances authority (usually the local planning authority)

---

34 *Id* § 30505.
35 See especially *The City of Norwich* 118 US 468, 493 (1886).
36 46 USCA § 30511 (2006).
37 Including connections between the LNG vessel and the jetty and the unloading equipment.

under the Planning (Hazardous Substances) Regulations 1992.[38]

The hazardous substances authority is obliged to consult with 11 separate organisations (including the Health and Safety Executive) on the advisability of locating the facilities. In particular, the Health and Safety Executive provides advice to the hazardous substances authority on the related health and safety issues and may advise that consent be granted subject to certain health and safety conditions. The Health and Safety Executive will set a 'consultation zone' around the site, so that whenever a development is proposed within the zone, the Health and Safety Executive must be consulted on the advisability of locating the particular development in the zone.

The various general planning permissions described above are separate from the hazardous substances consent. The Health and Safety Executive has no role in these applications, unless the development falls in the consultation zone of another site.

(c) Control of Major Accident Hazards Regulations consent

The Control of Major Accident Hazards Regulations 1999 came into force on April 1 1999.[39] In an LNG project, the regulations will apply to:

- connections between the ship and the jetty (including the unloading equipment at the jetty);
- the onshore site for LNG facilities (including storage tanks, pipelines and regasification plant); and
- the out-feed pipelines to the national gas transmission system.

LNG on board ship is not covered by these regulations, but is subject to the Dangerous Substances in Harbour Areas Regulations 1987 (see below).

The Control of Major Accident Hazards Regulations are enforced jointly by the Health and Safety Executive and the Environment Agency.

The proposed owner of the facilities is required to submit to the competent authority a pre-construction safety report, the purpose of which is to ensure that safety is considered fully at the design stage of the facilities. The report must cover all parts of the facilities, including integrity of the connections between the ship and the jetty, the storage tanks and LNG pipelines within the facilities.

A site survey will be required to investigate the geological characteristics of the region. The size of the area surveyed will depend on the site, but is generally limited to 329 kilometres from site.

The competent authority will examine the pre-construction safety report to satisfy itself regarding the incorporation of adequate safety and reliability into the design, the application of good practice and the reduction of risks to "as low as is reasonably practicable". Construction cannot commence until the competent authority has concluded its review.

38 Amended by the Planning (Control of Major Accident Hazards) Regulations 1999.

39 Amended by the Control of Major Accident Hazards (Amendment) Regulations 2005 from June 30 2005, implementing Council Directive 96/82/EC (known as the Seveso II Directive, as amended by Directive 2003/105/EC and replacing the Control of Industrial Major Accident Hazards Regulations 1984 (often referred to as CIMAH).

Regular inspection visits will be made during construction to ensure that the integrity of plant and equipment is in accordance with the pre-construction safety report. Construction activities are also observed to ensure compliance with the health and safety of those working at the site.

Prior to commissioning (when LNG is first introduced to the facilities), the owner must submit a pre-operations safety report. This report must demonstrate that the operator has taken all measures necessary to prevent major accidents, and to limit the consequences to people and the environment of any that occur. Operations cannot commence until the report is approved. The competent authority has the power to prohibit the operation of those parts of the facilities which it considers are deficient in any way. Towards the end of its assessment, the competent authority will organise a system of inspections for future monitoring.

In order to contain and control incidents, and to minimise the effects and limit damage to persons, the environment and property before commissioning, operators are also required by the Control of Major Accident Hazards Regulations to produce on-site emergency plans. Operators must also provide information to the local authority to assist them in producing an off-site emergency plan.

(d) ***The Pipelines Safety Regulations 1996***

Under the Pipelines Safety Regulations 1996, pipeline operators must notify the Health and Safety Executive of any new pipeline which is to be connected to allow gas from a new facility to be connected to the national transmission system. The Health and Safety Executive will assess the pipeline's design and inspect its construction and operation.

(e) ***Gas transporter licence***

The Gas Act 1986 (Exemptions Order) 2005, No 16 exempts persons conveying gas from LNG import facilities to a pipeline system operated by a licensed gas transporter, or to premises associated with an LNG import facility, from the requirement to hold a gas transporter licence under the Gas Act 1986. It also exempts from licensing the conveyance of gas from a vessel and from a licensed gas transporter's pipeline system to an LNG facility and the supply of gas to the facility.

3.3 UK competition regulations

EU and UK domestic regulation of the natural gas sector is underpinned by three sets of competition law instruments:

- the antitrust provisions (Articles 101 and 102 of the Treaty on the Functioning of the European Union and their UK equivalents under the Competition Act 1998);
- merger control provisions (the EU Merger Regulation and the relevant sections of the UK Enterprise Act 2002); and
- the EU law provisions governing state aid (Articles 107 and 108 of the Treaty on the Functioning of the European Union).

These instruments interact to promote competition within the United Kingdom and often have particular relevance to newly liberalised markets, such as those making up the energy sector, where restrictive arrangements, monopolistic positions and generous state subsidies or protection have traditionally been common.

(a) *The UK regulators*

The Office of Fair Trading (OFT) and the Gas and Electricity and Markets Authority (GEMA) (acting through the Office of Gas and Electricity Markets (OFGEM), the UK regulator for electricity and gas) have concurrent powers under the Competition Act 1998 to apply and enforce the UK competition law prohibitions on restrictive agreements and abuse of a dominant position (Chapter I and Chapter II of the Competition Act respectively). Chapter I and Chapter II apply where there may be an effect on trade within the United Kingdom, and are based on the EU competition law prohibitions in Articles 101 and 102 of the Treaty on the Functioning of the European Union, which apply where there may be an effect on trade between member states. The UK authorities may apply Articles 101 and 102 where applicable.

Under the Enterprise Act, OFGEM and the OFT have concurrent powers to make a reference to the UK Competition Commission where they have reasonable grounds to suspect that competition is being prevented, restricted or distorted by a feature of a market in the United Kingdom for goods or services in relation to, for example, the exploration, production, transportation or supply of gas.

However, OFGEM has no jurisdiction under the merger provisions of the Enterprise Act, although it is routinely invited to comment on mergers that affect its particular sectors. Where the OFT is reviewing restrictions which may be directly related and necessary to the implementation of a merger, OFGEM will be consulted in relation to such matters. In general, where restrictions are not considered directly related and necessary (and therefore fall to be examined under the Competition Act), and relate to the industry sector of a regulator, then that particular regulator will be involved.

OFGEM also has powers to prevent or prosecute anti-competitive practices under the Gas Act 1986. Overall, the OFT coordinates with the gas specific regulators to ensure competition in the relevant markets and the continuing ability of licensee(s) to fulfil their duties under relevant legislation.

Regulation of the gas market in Northern Ireland is carried out by the Northern Ireland Authority for Energy Regulation.

(b) *Potentially restrictive arrangements*

Article 101 prohibits agreements and concerted practices which have as their object or effect the prevention, restriction or distortion of competition in the common market and which may affect trade between member states, unless certain exemption criteria are met. Chapter I of the Competition Act, the domestic equivalent to Article 101, applies to agreements which may affect trade within the United Kingdom.

Article 101 and Chapter I deal with both vertical relationships (those between undertakings at different levels within the market, such as supplier and retailer), and horizontal relationships (those existing between undertakings at the same level of production or supply). In terms of horizontal relationships, issues of potential concern might include joint selling and price fixing, joint production activities and joint construction of infrastructure. At the vertical level, issues might arise with respect to long-term exclusive supply agreements, arrangements which afford territorial protection and take-or-pay contracts. Examples of provisions in gas contracts that might raise issues under Article 101 or Chapter I include the following:

- Destination clauses – these are common in commodity contracts and may have the direct or indirect effect of restricting wholesalers from reselling outside the countries where they are established. The practice provides the supplier with a form of territorial protection, helping to maintain price differentials between national markets. Examples as seen in gas cases (and which would be equally applicable to LNG contracts) include:
  - restriction on resale in another member state;[40]
  - indirect restrictions on resale such that different prices are set by reference to the type of resale customer;[41]
  - restrictions on use insofar as a particular purpose, such as electricity generation, is specified;[42] and
  - profit splitting mechanisms obliging the purchaser to pass over to the supplier a portion of any profit made on gas resale outside national borders or where the gas is resold for purposes other than that which were agreed.
- Volume adjustment mechanisms – these are provisions which reduce the purchaser's obligation to take and/or pay as a result of the supplier making sales in the purchaser's home territory, and may be viewed as an indirect means of deterring competition between the parties.
- Take-or-pay clauses – these may have foreclosure effects, in that they prevent other suppliers from offering a more competitive supply to the purchaser depending on the duration of the contract.
- Consent before sales – these are clauses which require the prior consent of the purchaser before the supplier can sell into the purchaser's home territory.[43]
- First refusal rights – these give the purchaser a right of first refusal on any further gas to the home territory of the purchaser.[44]
- Joint selling – this is a common practice in the gas sector whereby the number of owners of gas group together to provide a single source of gas for downstream sales.[45]

(c) *Abuse of dominant position*

Article 102 prohibits the abuse of a dominant market position in the European Union or a substantial part thereof. Chapter II of the Competition Act, the domestic equivalent to Article 102, applies if the conduct in question may affect trade within

---

40 See *Gazprom/OMV* (February 2005).
41 For example, DUC/DONG (April 2003).
42 *Gas Natural/Endesa* (March 2003).
43 *Gazprom/ENI* (October 2003).
44 *Gazprom/OMV* (February 2005).
45 *DUC/DONG* (April 2003).

the United Kingdom. The mere holding of a dominant position is not prohibited – only its abuse.

Dominant companies, according to European case law, enjoy positions of economic power that enable them to behave, to a large extent, independently of effective competition pressures. Dominance may be acquired through economic success or may be conferred through government policy (as has often been the case in the European energy sector). There are no market share thresholds for determining whether dominance exists, although as a general rule dominance concerns typically do not arise at a level of market share below 40%. However, market share is not the only indicator of dominance; other factors are considered, such as:

- the relative market shares of the presumed dominant company and its competitors;
- the degree of market concentration; and
- the existence of barriers to entry or expansion into the relevant market.

Under Article 102/Chapter II, a dominant company abuses its position when it conducts its business in a manner that restricts competition which remains in the market with the effect of exploiting its commercial partners and customers, or excluding competitors from markets, and where there is a likelihood of consumer harm. Article 102/Chapter II constrains the behaviour only of entities that are already in a dominant position and does not regulate the way in which such dominance is achieved. Examples of such abuse include:

- denying access to essential infrastructure;
- refusing to supply an existing customer without objective (ie, not anti-competitive) reason;
- charging unfair prices or imposing unfair trading conditions;
- limiting production, markets or technical development; and
- applying different conditions to similar transactions or the same conditions to dissimilar transactions.

(d) State aid

The substantive EU state aid provisions are set out in Articles 107 and 108 of the Treaty on the Functioning of the European Union (formerly Articles 87 and 88 of the EC Treaty), as accompanied by a procedural regulation and in-depth guidelines. Measures constituting state aid (which covers a wide range of financial and economic assistance to commercial parties) are unlawful under EU law if they are not notified to the European Commission and compatible with the common market. Any form of aid – whether provided directly by the state or provided indirectly through 'state resources' – is incompatible with the common market if it distorts or threatens to distort competition within the common market.

State aid issues are of particular significance to the energy sector, as many of the industry participants are former monopolies with residual ties to government. The European Commission has wide-ranging powers to investigate any such state aid (however received), and to order member state governments to recover unlawful state aid.

(e) Consequences of infringement

Parties to anti-competitive agreements or undertakings that abuse their position of dominance on a relevant market may be vulnerable to public and private sanction. Undertakings may be fined up to 10% of their worldwide turnover for an infringement of Article 101 or 102, and their officers may be exposed to other civil penalties and/or criminal sanctions in the United Kingdom under domestic provisions. Where individuals engage in 'hardcore' cartel activity (including price fixing, bid rigging and market sharing), this can also amount to a criminal offence under the criminal cartel provisions of the Enterprise Act. Offenders may be subject to up to five years in prison or an unlimited fine.

In addition, unlawful agreements will be void and unenforceable (in whole or in part), while undertakings found to be abusing a position of dominance may be ordered to terminate the abusive behaviour or to take positive steps to remedy the infringement(s). Finally, undertakings acting in breach of Articles 101 and 102/Chapter I and II may be vulnerable to private suits for injunction and/or damages brought by third parties which consider that they have suffered loss as a result of such infringements.

In monitoring the licensing regime, OFGEM has the power to enforce compliance with the licensing and statutory provisions through the imposition of an enforcement order where there is evidence that a licence holder is contravening relevant conditions or requirements. Where non-compliance continues despite the enforcement order, OFGEM may revoke the licence and impose fines of up to 10% of the licence holder's turnover.

In state aid cases, a finding of unlawful state aid will lead to an order by the European Commission requiring the member state to recover any unlawful aid already paid. Damages may be available to competitors that suffer loss as a result of the grant of unlawful state aid.

(f) Exceptions

Public service obligations may, under certain circumstances, be permitted to override competition concerns; this consideration is particularly relevant to the energy sector, where a seemingly unavoidable tension exists between promoting competition within the internal market and ensuring vital public policy goals (ie, security of supply of essential economic inputs). Article 106(2) of the Treaty on the Functioning of the European Union provides the possibility for removing arrangements from the scope of Articles 101 and 102 (and the state aid rules) where services of general economic interest and revenue-producing monopolies are involved. This would apply if the competition rules can be shown to obstruct the performance of essential tasks assigned to the undertakings in question. This was seen in the *Campus Oil* case, where concerns over security of national supply of oil were viewed by the Court of Justice of the European Union as justifying what would otherwise have been constituted as competition law infringements.

In addition, otherwise restrictive arrangements may fall outside the scope of Article 101 if they:

- are considered *de minimis*;

- come within the scope of a relevant block exemption (the European Commission has adopted certain block exemptions which remove standard categories of agreements from the prohibition in Article 101 where the requirements of the block exemption are met);
- benefit from exemption under Article 101(3); or
- can be construed as a concentration (and therefore fall for review under the EU Merger Regulation or domestic merger provisions rather than Article 101/Chapter 1).

However, these conditions may prove to have limited effect in the context of restrictive arrangements operating in the natural gas sector. Few deals, if any, concerning gas are likely to involve parties small enough to fall within the *de minimis* exemption. Furthermore, the European Commission is sceptical about gas companies placing too much reliance on the Specialisation Block Exemption, while the Vertical Restraints Block Exemption will tend to have limited value to many of the sector-specific arrangements where the restrictions involved are often 'hardcore' in nature (and therefore not eligible for exemption under any circumstances). Finally, assessing restrictive agreements against the Article 101(3) criteria has become a much more precarious exercise since the coming into force of the EU Modernisation Regulation (Regulation 1/2003), with parties having to self-assess the legality of agreements without the benefit of a formal notification system.

3.4 Third-party access

The regulation of access to gas and LNG facilities within the United Kingdom has evolved over a number of years. For most of that time, the practical reality was that LNG was a strategically irrelevant alternative in the United Kingdom's mix of energy sources. While some small facilities did exist, the proximity and productivity of the North Sea gas fields meant that there was little need for LNG.

However, with the decline of North Sea reserves, the importance of LNG – and as such, the regulation attached to it – has achieved increasing importance. Regulation of third-party access today is guided by European law, although the aim of much of that legislation runs parallel with the objectives and approach of the United Kingdom in encouraging the liberalisation of energy markets.

(a) EU legislation

The EU legislation which establishes a third-party access regime for LNG facilities is the 2009 EU Third Gas Directive.[46] The directive developed themes within earlier legislation and aimed generally to increase the level and speed of liberalisation within the gas markets of member states. The purpose of the exemption is to promote investment in new facilities (or in the significant expansion of existing facilities), so as to develop new sources of gas supply and improve security of supply in doing so.

46 Directive 2009/73/EC of the European Parliament and of the Council Concerning Common Rules for the Internal Market in Natural Gas.

Under the directive, each member state is required to implement a system of regulated third-party access in relation to LNG facilities (eg, as well as gas transmission networks). This requires owners of infrastructure to provide access to third parties on a non-discriminatory, transparent basis, applying published charges which have been previously approved by the national regulator.

The directive applies the regulated third-party access requirements to LNG loading and regasification facilities. Accordingly, access to berths is captured, as well as access to the regasification plant itself. However, the directive takes a slightly different approach for any gas storage facilities which may be related to those LNG facilities: owners are given the choice of either regulated third-party access or 'negotiated third-party access', whereby access to the facility is granted on the basis of bilateral negotiations between the facility owner and the potential user (which is generally thought to provide more opportunity for obstruction than regulated third-party access).

The directive does set out a few limited grounds under which access may be refused:

- lack of capacity; and
- financial difficulties due to take-or-pay contracts.

However, the most important feature of the regime in respect of LNG facilities is not these, but the possibility of securing an exemption from the regulated third-party access regime as a whole.

While regulated third-party access addresses a fundamental concern in the liberalisation of energy markets, it is not a panacea. One downside is that it may actually discourage developers from constructing new infrastructure. There are competing supplies through regasification plants to the same market. Regulated third-party access presents a significant investment risk by creating an obstacle to ensuring the recovery of costs and obtaining a return on the significant investment required. In the absence of a monopoly buyer for any throughput, investors in regasification facilities will look for other ways to secure their objectives of recovery and return, such as long-term contracts for capacity.

It is for this reason that the directive provides in Article 36 for an exemption regime for 'new infrastructure' – that is, any infrastructure which is constructed or has its capacity significantly expanded after July 2003.

In order to be granted an exemption, the directive requires sponsors to demonstrate that the project satisfies the following criteria:

- The facility will enhance competition in gas supply and enhance security of supply;
- The investment would not occur without an exemption, due to levels of risk involved;
- The infrastructure will not be owned by the transmission system operator of the gas network to which it is linked;
- Charges for use are levied by the owner; and
- The exemption is not detrimental to competition, the internal gas market or the market of the relevant member state.

The European Commission has published a guidance note on exemptions from the third-party access regime. This guidance emphasises that applications will be dealt with on "a case-by-case basis...on their merits". However, it goes on to provide some detail relating to a range of issues, such as:

- what constitutes new, major, high-risk infrastructure to which exemptions may apply;
- what types of exemption may be available;
- detail regarding the criteria for awarding an exemption; and
- the supporting information to be provided.

National regulators should have regard to this guidance when considering any application for an exemption. In addition, the European Commission has the right under the directive to require such national regulators to amend or withdraw any exemption that they have granted.

Legal theory expounded by the commission is interesting, of course, but how has this been put into practice in the United Kingdom?

(b) Implementation in the United Kingdom

The legislation within the United Kingdom relating to access to LNG facilities was largely compliant with the requirements of the Third Gas Directive when it came into force on March 3 2011. Only a few modifications to the Gas Act 1986 by the Gas (Third Party Access) Regulations 2004 were required to ensure full compliance.[47]

The monitoring and enforcement of the new regime in Great Britain is the responsibility of OFGEM. There is a separate regulator for Northern Ireland, the Office for Regulation of Electricity and Gas. However, due to the absence of a developed gas market at this time, the regime currently has little practical application in respect of LNG facilities in Northern Ireland. For this reason, this section concentrates on the implementation of the regime by OFGEM.

OFGEM has set out the approach it will take in relation to applications for exemptions.[48] As one would expect, this involves a consideration of the impact of the facility on the matters set out in the criteria, such as security of supply. In addition, however, OFGEM will analyse the access conditions to the facility, to take account of the characteristics and anticipated effects of those rules.

Characteristics which OFGEM would regard positively include:

- an initial offer of capacity to market;
- capacity allocation rules that enable effective secondary trading of capacity;
- anti-hoarding measures such as 'use it or lose it' regimes in respect of capacity; and
- measures to ensure the provision of information to the market.

There have been three applications within Great Britain for exemptions from the

---

47 The relevant provisions are set out in Sections 19C and D of the Gas Act 1986.

48 "LNG facilities and interconnectors: EU legislation and regulatory regime – DTI/OFGEM final views", Department of Trade and Industry and OFGEM (June 2003).

regulated third-party access regime for LNG facilities: Dragon, South Hook and Isle of Grain. All of these projects have been granted an exemption by OFGEM which was not overturned by the European Commission. The term of the exemptions ranged from 20 to 25 years from the construction or planned expansion of the relevant facilities. Isle of Grain is the only terminal which is not 'own use' – that is, where the capacity is initially contracted to sponsors of the project.

These projects all share particular characteristics:

- They have all made a recognised contribution to security of supply – most were identified as priority projects under the Trans-European Networks programme;
- There is a significant level of competition for supply in the final market – as such, there may be a potentially significant risk to cost recovery by investors if a regulated third-party access regime is imposed; and
- In terms of market share analysis, the relevant projects involve the entry of new players or result in no significant increase in market power of established operators.

Although exemptions may be fixed term, they may (and will) include reopeners, allowing OFGEM to reconsider the terms or even the existence of the exemption. To decrease regulatory risk, any reopeners need to be clearly set out – the current apparent willingness of OFGEM to grant applications for exemptions means that the drafting of any reopener will be extremely significant for project sponsors, and close communication with OFGEM should be maintained both during the application process and following the grant of any exemption. In addition, project sponsors must remember that even where an exemption applies, other requirements of the Gas Act continue to apply.

By way of example, the exemption in respect of the Isle of Grain terminal was subject to the condition that it have in place effective 'use it or lose it' arrangements so that other companies could import gas through the facility if the contracted through-putters were not using 100% of the capacity. OFGEM had the power to amend or revoke this exemption if the anti-hoarding arrangements were not satisfactory. After issues arose with the measures in place, OFGEM expressed concerns to Isle of Grain about transparency and the effectiveness of the 'use it or lose it' arrangements. Isle of Grain sought to improve the information available on the website to companies that might want to use the facility and introduced modifications to the access arrangements. OFGEM stated that it would continue to monitor the effectiveness of the revised terms and expressed a willingness to "move straight to reviewing the exemption order" if these did not address OFGEM's concerns in practice.

The conditions attached to Dragon and South Hook are similar, but since the construction process is less progressed, they are required to confirm that the material provided by the owner to OFGEM is accurate in all material respects. On this basis, if the timing or characteristics of the plant are different from those originally intended, then the exemption can be revisited.

A review of the successful applications demonstrates that, in addition to seeking to show that the Section 19C criteria have been satisfied, a number of other factors are considered. Examples of the lines of reasoning adopted include the following:

- Regulated third-party access is designed to address market failures that are not a feature of the UK market; and
- Not only will the granting of an exemption fail adversely to affect the UK market, but the failure to grant an exemption will send a message to the sponsors and the market in general that may jeopardise security of supply by pushing projects to those countries with a lighter-touch regulatory regime, such as the United States.

(c) *Implementation in the European Union*

Despite the March 3 2011 deadline for implementation of the Third Gas Directive, many EU countries have yet to transpose the directive in national legislation.

4. US regulation of LNG projects

4.1 Introduction

Until very recently, domestic natural gas production was on the decline and the United States was expected to become heavily dependent on foreign-sourced LNG Based on this expectation, the rules and procedures regulating the development of LNG terminals and access to them were streamlined to facilitate the expansion. However, the discovery of abundant shale gas reserves and the development of the hydraulic fracturing technology necessary to exploit these reserves have changed the LNG outlook of the United States. Where only a few short years ago more than 40 LNG import terminals were being planned, a number of existing LNG import terminals plan to become bi-directional to support the expected export of LNG from the continental United States to foreign markets.

The approach taken by the United States in the regulation of LNG projects differs in a number of important respects from that taken in the European Union and the United Kingdom.

4.2 Environmental/health and safety regulation

US regulation of LNG import and export facilities depends in part on where they are sited. Onshore US facilities, offshore LNG facilities in federal waters and onshore facilities built in another country but connected to US markets by pipeline are regulated in different ways. The regulation of the latter is outside the scope of this chapter.

(a) *US onshore facilities*

Currently, 10 out of the 13 US (including Puerto Rico) LNG terminals are onshore.

Various regulatory bodies are engaged in the approval process for this type of facility and their functions often overlap. The following is a description of the key agencies and the permits required.

FERC approval: The US Federal Energy Regulatory Commission (FERC) is responsible for authorising the siting, construction and operation of onshore LNG terminals,[49] including terminals offshore but within state waters and not within the jurisdiction of the Deepwater Port Act. FERC's policy goal has been to remove economic and regulatory barriers to the development of LNG terminals. As a result, its authorisation process for terminals focuses almost exclusively on safety and environmental issues. FERC also exercises detailed regulation under its jurisdiction over interstate natural gas pipelines, which will generally apply to the transport of regasified LNG away from an import terminal or natural gas to a liquefaction facility.

State agency approvals: Certain federal permitting is delegated to the individual states. The key agencies and permits required are set out below:

- State land agencies are responsible for approval under the Coastal Zone Management Act (Section 307(c)).[50] The applicant must certify that the proposed activity in a designated coastal zone complies with the enforceable policies of the affected state's coastal zone management programme. Appeals against adverse decisions can be made to the US secretary of commerce.
- State and local environmental agencies exercise delegated authority under the Clean Air Act and Clean Water Act to issue permits and approvals. Key permits are a Clean Water Act permit[51] (issued when the state agency certifies compliance with the state's water quality standards for any activity that may result in discharge into navigable waters), and a Clean Air Act permit,[52] which is required for any person to operate a source of air pollution.
- State and local wildlife agencies have responsibilities under state and local threatened and endangered species regulations, and permits may be required.
- Likewise, state and local historical preservation agencies may need to be consulted under the National Historic Preservation Act.
- Permits will also likely be required from other state and local agencies, such as road crossing permits and construction permits.

A state is also designated a 'cooperating agency' with FERC during a review of the project under the National Environmental Policy Act, and can contribute to the complete environmental review of the proposal. In February 2004 FERC, the Coast Guard and the Department of Transportation entered into an interagency agreement for compiling a single environmental impact statement, with FERC acting as the lead agency and the other regulatory bodies acting as cooperating agencies.

US Coast Guard: The US Coast Guard must be consulted on marine issues and safety pursuant to the Deepwater Port Act 1974, which gives the US Coast Guard jurisdiction over LNG terminals and pipelines outside state waters. FERC has jurisdiction where pipelines are above the high water mark and come onshore.

When building an LNG waterfront terminal, the owner must submit a letter of

49 15 USC § 717b (2006).
50 16 USC § 1456 (2006).
51 See 33 USC § 1341 (2006).
52 See 42 US § 7661b (2006).

intent to the Coast Guard detailing the physical location, describing the facility, explaining vessel characteristics and frequency of shipments, and including charts showing waterways.

The Coast Guard is also responsible for issuing a letter of recommendation concerning the suitability of a waterway for LNG marine traffic to a proposed terminal, considering:

- the density of traffic;
- man-made obstructions in waterways;
- water depth;
- tidal range;
- protection from high seas;
- natural hazards;
- underwater pipes and cables;
- distance of berthed vessels from the channel; and
- the width of the channel.

In January 2011 the Coast Guard updated its Navigation and Vessel Inspection Circular, providing guidance on the process utilised to make its waterway suitability determination.

The Coast Guard has the right to make reasonable inspections and examinations of the facility, and may order an operator to suspend operation to prevent damage to bridges or other structures or adjacent land or to protect navigable waters.

Regulations developed under the authority of the Ports and Waterways Safety Act also assign to the Coast Guard responsibility for safety issues within the 'marine transfer area' of LNG terminals (the area that is part of a waterfront facility between the vessel, or where the vessel moors, and the first shut-off valve on the pipeline immediately before the receiving tanks). This responsibility covers, for example:

- electric power systems;
- lighting;
- communications;
- transfer hoses and piping systems;
- alarms and fire-fighting equipment; and
- operational matters.

An LNG plant operator must also submit operations and emergency manuals to the captain of the port of the relevant zone at least 30 days prior to commissioning. The operations manual must contain:

- a description of the transfer system to be used and the procedures;
- the duties of each person involved in the transfer;
- relief valve settings; and
- a description of the security procedures, communication systems and training programme for all employees.

The emergency manual must cover:

- LNG release response procedures;

- emergency shutdown procedures;
- emergency lighting and power;
- personnel shelters;
- first aid procedures;
- emergency procedures for mooring and unmooring vessels; and
- contact numbers.

Other agencies and permits: Under the Clean Water Act and the Rivers and Harbors Act 1899, the US Army Corps of Engineers has authority for the review and issue of permits to discharge dredging or fill materials into US waters, and for activities that occupy, fill or grade land in a floodplain, streambed or stream channel.

A certificate of consistency with any coastal zone management programme (if the state has one) will be required from the US Department of Commerce. The state can object, but the objection can be overruled by the secretary of commerce. Consultation with the National Oceanic and Atmospheric Administration may also be necessary regarding compliance with the Endangered Species Act and the Magnuson-Stevens Fishery Conservation and Management Act.

The US Fish and Wildlife Service, a bureau in the US Department of the Interior, has responsibilities for wildlife endangerment and will need to be consulted pursuant to the Endangered Species Act and the Migratory Bird Treaty Act.

National Environmental Policy Act filing process: Applicants for permits must use the pre-filing process under the National Environmental Policy Act.[53] This procedure allows for FERC, state and local agencies and the public to give their opinions before a formal application is submitted.[54] The procedure may be summarised as follows:

- Before making a pre-filing submission, applicants must consult with the director of FERC's Office of Energy Projects on the project, pre-filing request and how much information the applicant has gathered for the initial submission.
- Following consultation, applicants must file environmental, design, siting and stakeholder information with FERC, and a letter of intent and waterway suitability assessment with the US Coast Guard.
- The applicant and FERC then provide relevant agencies and the public with information on the project and invite comment.
- FERC initiates the preparation of a preliminary environmental assessment or draft environmental impact statement, and other agencies may be consulted regarding the review process. Although an environmental assessment is normally the first step under the National Environmental Policy Act, FERC will normally commence an environmental impact statement when considering applications for the siting, construction and operation of an LNG facility.[55]

---

53 42 USCA §§ 4321 et seq (2011).

54 FERC has issued final regulations to implement the pre-filing process: Pre-Filing Procedures and Review Process for LNG Terminal Facilities and Other Natural Gas Facilities Prior to Filing of Applications; 18 CFR § 157.21 (2011).

55 18 CFR § 380.6 (2011).

- Following this, the applicant files a formal application with FERC.
- Following review and consultation, FERC undertakes the draft environmental impact statement. Among other items, the environmental impact statement addresses:
  - alternatives;
  - shoreline erosion;
  - water resource;
  - impact on residencies;
  - wetlands and vegetation;
  - wildlife and aquatic resources;
  - fish habitat;
  - endangered or threatened species;
  - air quality and noise impacts;
  - land use;
  - soils and sediment;
  - cultural resources; and
  - safety.
- After considering agency comments, FERC issues a second draft environmental impact statement and invites public comment.
- FERC then issues a final environmental impact statement (based on comment from agencies and the public).
- FERC approves or denies the project.
- If the project is approved, FERC imposes conditions which must be fulfilled before construction begins.
- When these conditions are met, FERC authorises construction.
- If the application is denied, the applicant can request a rehearing. If the applicant's request for rehearing is denied, the applicant can seek judicial review in the appropriate US Court of Appeals.

Applications typically take around 18 months to two years from the start of the pre-filing process.

Ongoing monitoring: FERC and the federal agencies monitor construction through progress reports and site visits. Operations may not be commenced until FERC believes the facility meets the conditions of FERC's authorisation. Once operations have commenced, the terminal will still be subject to periodic inspections by FERC, the Coast Guard and the Office of Pipeline Safety.

Acquisition of property rights: Developers must secure all property rights by way of privately negotiated transactions. This is because an LNG terminal, as defined in the Natural Gas Act (as amended by the 2005 Energy Policy Act), is authorised under Section 3, which does not confer eminent domain authority (unlike the authorisation for pipelines and other interstate facilities, which receive certificates of public convenience and necessity under Section 7).

Security: Maritime security regulations require an LNG terminal operator to conduct a facility security assessment and develop a threat-scalable security plan that addresses the risks identified in the assessment. Much like the requirements prescribed for vessels, the facility security plan establishes:

- access control measures;
- security measures for cargo handling and delivery of supplies;
- surveillance and monitoring;
- security communications;
- security incident procedures; and
- training and drill requirements.

(b) US offshore facilities

The offshore LNG import facility off the coast of Louisiana was the first of its kind in the world. LNG is vaporised on board and is piped to shore; however, the facility plans to cease operations. As of April 2011, there are three operational deepwater LNG port facilities in the United States and four other such projects have been approved.

The Maritime Administration, part of the Department of Transportation, and the US Coast Guard have joint authority to process applications for the development of deepwater LNG facilities. The Maritime Administration is responsible for issuing deepwater port licences. The Maritime Administration and the US Coast Guard are obliged to consult with agencies, adjacent coastal states and the public.

The other agencies involved in the issue of environmental and safety permits are very similar to those involved with onshore facilities, with the addition of the Department of the Interior's Bureau of Ocean Energy Management, Regulation and Enforcement. The state's involvement focuses on the pipeline bringing the gas to shore.

From the date of publication of a Federal Register notice of a complete application, the Maritime Administration has 330 days to approve or deny the project, although in practice, time limits are frequently extended due to incomplete applications or the absence of required information. Once approved, a licence for the development is issued typically within two to five months. The licence must be held by a US citizen (meaning a company with a chief executive or president and chairman of the board of directs and majority of the board of directors who are US citizens).

Department of Energy authorisation for importation and exportation: An importer or exporter in the United States must also have an authorisation to import or export gas issued by the Office of Fossil Energy within the Department of Energy. Blanket authorisation is available for spot market sales or contracts of two years or less. Long-term authorisation should be sought if an importer or exporter has a definite contract for a term of longer than two years, but long-term authorisations to export have been granted in the absence of any such contracts.

There is a strong presumption in favour of granting an import or export authorisation, and it will be granted unless importation or exportation "will not be consistent with the public interest".[56] The Office of Fossil Energy has emphasised that it is its policy to "minimize federal control and involvement in energy markets and to promote free and open trade".[57] Further, if the application for authorisation is for imports from or exports to a country with which the United States has a free trade agreement requiring national treatment for trade in natural gas, then the importation or exportation is deemed to be consistent with the public interest and the application receives expedited review. Recent applications for export authorisation subject to such expedited review have been approved in less than three months.

4.3 US competition regulation

The primary sources of US law dealing with anti-competitive practices come from federal antitrust statutes: Sections 1 and 2 of the Sherman Act, and Section 7 of the Clayton Act. In addition, most states have statutes governing business conduct that provide jurisdiction for the state to take enforcement action against 'unfair competition'. Many state statutes are similar to Sections 1 and 2 of the Sherman Act and have been interpreted consistently with the federal laws.

Although regulation by FERC addresses many competition-related concerns similar to those addressed by the antitrust laws, compliance with FERC regulations does not confer immunity from the antitrust laws. Courts are likely to be deferential to a practice that has been explicitly approved by FERC. Business conduct by an LNG firm that was not specifically directed by FERC, even though related to FERC regulation, would likely be reviewed under the general antitrust standards summarised below.

(a) Section 1 of the Sherman Act

Section 1 of the Sherman Act applies to conduct involving more than one actor. It prohibits contracts, combinations or conspiracies in restraint of trade. In order for Section 1 to apply, two conditions must be met:

- There must be proof of an agreement to engage in the allegedly anti-competitive conduct; and
- The conduct must 'unreasonably' restrain trade.

A small set of activities – price fixing, bid rigging, group boycotts, customer allocations, horizontal territorial divisions – are considered as *per se* violations of Section 1. These activities are deemed inherently harmful to consumer welfare and as such are condemned without opportunity for defence. The remainder – indeed the majority – of the conduct falling under Section 1 is analysed under the 'rule of reason', which enables a defendant to proffer affirmative business reasons for the conduct. The types of business arrangement that might involve LNG facilities that could give rise to a Section 1 challenge include:

---

56 15 USC § 717b(a) (2006).

57 Opinion and Order Denying Request for Review Under Section 3(c) of the Natural Gas Act, FE Docket No 10-111-LNG, at 5.

- most-favoured nation clauses or other arrangements that commit the seller to lower its price or improve the terms of its arrangement with one buyer in the event that the seller agrees to provide better prices or terms to another buyer;
- tying arrangements that force a purchaser to buy one product it does not want in order to purchase a second product it does want;
- restrictions on the business activities of a joint venture participant that affect activities outside the scope of the venture;
- exclusive dealing provisions and requirements contracts;
- non-compete agreements;
- agreements relating to the timing, size or scope of an entering facility;
- agreements restricting the take-away capacity of pipelines connecting to the facility;
- vertical restrictions on the resale;[58] and
- agreements and/or provisions of agreements that limit access to a facility used by competing firms.[59]

Under a rule of reason analysis, such business arrangements may be deemed lawful even if they impose anti-competitive limitations on business conduct, provided that they are ancillary to a legitimate, efficiency-enhancing purpose. All provisions of this sort should be examined carefully by antitrust counsel before they are adopted, as their legality will be highly dependent on specific facts of the market served by an LNG facility.

(b) Section 2 of the Sherman Act

Section 2 applies to the conduct of individual firms. It prohibits monopolisation, attempts to monopolise and conspiracies to monopolise. Section 2[60] does not condemn or prevent the existence of monopolies. Rather, it prohibits monopolisation (or attempted monopolisation) by unlawful means.

There is no established threshold for determining that a firm enjoys a monopoly. Monopoly power is generally understood to confer the power profitably to raise price, reduce output or exclude competitors. Market share is an important factor in assessing whether a firm has monopoly power, but courts also consider other factors such as barriers to entry and the nature of the anti-competitive conduct. In addition, modern courts have not imposed Section 2 liability for business conduct by a firm that did not control at least a 50% share of a properly defined 'relevant market', and many courts have required a much higher market share (in excess of 70%) before making a determination that a monopoly exists.

58 Although the Supreme Court in 2007 in *Leegin Creative Leather Products, Inc v PSKS, Inc*, 551 US 877, 880 (2007) repealed the *per se* rule against minimum resale price maintenance (ie, when a seller requires a buyer to charge at least a certain price for goods it sells), many states (eg, California) continue to consider such agreements to be *per se* violations of the antitrust laws.

59 Take-or-pay provisions have been challenged under the US antitrust laws from time to time, but have generally been upheld as legitimate commercial arrangements that enable the parties to an agreement to manage price and supply risks.

60 *Verizon Communications Inc v Law Offices of Curtis Trinko*, 540 US 398 (2004).

Depending on the level of concentration (as well as the presence of the factors listed above) in the market served by an LNG facility, certain aspects of its business conduct may be subject to challenge under Section 2. The most likely aspects of an LNG's business practices to be scrutinised by Section 2 are:

- denials of access;
- predatory pricing, which is defined as pricing below an appropriate measure of cost;
- terms and conditions of access that have the effect of excluding competitors;
- exclusive dealing arrangements that foreclose a substantial portion of the market to competitors;
- operating terms, conditions and procedures that increase costs or effectively deny access to competitors;
- limitations on capacity that are unrelated to legitimate business concerns; and
- operating terms, conditions and procedures that restrict trade in downstream commodity markets.

The Supreme Court has clarified that Section 2 does not impose an independent duty on a competitor to provide access to a facility. Moreover, the Supreme Court decision also cast doubt on the existence of an 'essential facilities' doctrine under US antitrust law. However, the case noted that Section 2 condemns certain actions by monopolists that have no legitimate business purpose other than to exclude new entry or remove competition from the relevant market. There remains considerable debate as to the definition of unlawful exclusionary conduct under Section 2. The courts and the enforcement agencies have applied several different tests to analyse allegedly exclusionary conduct:

- The 'no economic sense' test evaluates whether the challenged conduct makes no economic sense for the monopolist but for the anticipated benefits from elimination of competition;
- The 'profit sacrifice' test considers that a decision by a monopolist to sacrifice short-term profits in order to exclude or harm a rival is unlawful, provided that the monopolist reasonably could have anticipated recoupment of the sacrificed profits once the competitor had exited the market;
- The 'equally efficient rival' test examines whether the challenged conduct has the effect of eliminating competition by a rival of equal or greater efficiency; and
- The 'consumer welfare' test balances the efficiency benefits claimed by the monopolist for the challenged conduct with the harm to consumers resulting from the conduct.

Since the legal standard for judging conduct that falls under Section 2 is unsettled, it is important for firms that control facilities with potential monopoly attributes to seek advice from antitrust counsel in connection with commercial terms and conditions, as well as operating practices and procedures, to assess the antitrust risk associated with them.

(c) *Enforcement bodies*

Competition laws in the United States are enforced at the federal level by the Federal Trade Commission (FTC) and the US Department of Justice Antitrust Division. Both agencies have the authority to enforce the Sherman Act and Clayton Act. In addition, the FTC has broader powers under Section 5 of the FTC Act to exercise enforcement powers for conduct that is deemed an unfair method of competition. States also have their own state competition laws enforced by the state attorneys general. These laws vary from one jurisdiction to another, but most jurisdictions grant the state attorney general broad powers to prosecute unfair competition.

(d) *Third-party access*

Like their European counterparts, new US LNG facilities[61] do not need to offer their capacity to the market. However, in contrast to Europe, this exemption is available automatically without the need to fulfil any preconditions. As such, a developer does not need to concern itself with third-party access issues in respect of a terminal (although some issues may still arise in respect of the pipeline linking the facility to the gas network).

The current regime in respect of third-party access to onshore LNG facilities dates back to an order issued by FERC in 2002[62] in respect of the Hackberry LNG Terminal (which later became Cameron LNG). Prior to this, facilities were subject to an open access regime under which capacity had to be made available by open season, with the resulting terms and tariffs regulated by FERC. Under Hackberry, FERC eliminated the requirement for open access and, developing this further in its final approval of the facility in September 2003, also removed any obligation on the terminal operator to publish the terms of any capacity agreements entered into.[63] FERC has clarified that it has not surrendered its jurisdiction and has not fettered its discretion to issue regulatory orders in the future if its detects any inappropriate or anti-competitive behaviour.

The law was subsequently partially codified in the Energy Policy Act 2005.

The third-party access regime for offshore terminals in the United States is the same, although the route to the conclusion differs. Under the Marine Transportation Security Act 2002, a developer of an offshore LNG terminal in federal waters would not be subject to open access requirements or regulation of rates or terms. This position continues today.

The US approach of minimal regulatory interference has encouraged new investment. Without doubt, the United States currently has the upper hand with its regulatory certainty and flexibility to allow suppliers to adopt the best approach for them.

5. The future

The future for regulation of LNG in the United Kingdom and the United States may

---

61 This includes expansions of existing facilities.
62 Hackberry LNG Terminal, LLC, 101 FERC ¶ 61,294 (2002).
63 Hackberry LNG Terminal, LLC, 104 FERC ¶ 61,269 (2003).

be viewed from two different perspectives. The regulation of environmental and health and safety risks will continue to develop in order to ensure that where LNG is imported and regasified (or, in the case of the United States, liquefied and exported), this is done as safely as possible. However, regulators will also seek to ensure that the curse of NIMBY (Not in My Back Yard) or, even worse, BANANA (Build Absolutely Nothing Anywhere Near Anyone), is not allowed to disguise their protectionist policies as apparent safety concerns and create unjustifiable barriers for legitimate projects.

On the issue of competition and third-party access regulation, the picture is more fluid. The evolution of 'economic' regulation holds up a mirror to the characteristics of the market to which it is applied. As such, as the market changes, so will the regulation.

Currently both the European Commission and FERC have been willing to accept a 'light-touch' regulatory regime for LNG to ensure that security of supply is not compromised. However, neither regulator has abandoned jurisdiction in respect of the LNG sector, and as the commission's recent sector enquiry demonstrates, as the situation changes so can we expect the approach of the regulators to modify to ensure that their policy objectives continue to be achieved.

The authors would like to thank Hywel Jones for his assistance in the preparation of this chapter.

About the authors

Donna J Bailey
General counsel, Chevron Gas and Midstream, Chevron USA Inc
djbailey@chevron.com

Donna Bailey is general counsel of Chevron Gas and Midstream, a division of Chevron USA Inc (CGM). Through its shipping, pipeline and natural gas and LNG business units, CGM supports the commercialisation of Chevron's significant natural gas resource base, including significant holdings in Africa, Australia, Southeast Asia, the Caspian region, Latin America and North America. Ms Bailey has 30 years of experience in the energy industry, with a primary focus on natural gas and LNG commercial transactions and US natural gas and LNG infrastructure issues.

James Baily
Partner, Herbert Smith LLP
James.Baily@herbertsmith.com

James Baily is a solicitor advocate and partner in Herbert Smith's litigation and arbitration division, advising in relation to a wide range of commercial disputes within the energy sector. His experience includes working with oil and gas companies in relation to upstream projects in West Africa, the Middle East, Central Asia and the North Sea. He has advised a number of major energy companies on price review disputes under long-term LNG sale and purchase agreements, pipeline gas sales agreements and disputes concerning gas trading and marketing in the United Kingdom. In 2001 he was seconded to work for a major international oil company in London for nine months.

Vivek Bakshi
Partner, Fraser Milner Casgrain LLP
Vivek.bakshi@fmc-law.com

Vivek Bakshi is a partner in the energy group at Fraser Milner Casgrain LLP, one of Canada's leading national law firms, and is based in its Toronto office. Mr Bakshi focuses on mergers and acquisitions and project development in the energy, natural resources and infrastructure sector. His expertise in the LNG sector comes from a number of years spent living in Japan, working with LNG traders and sponsors in connection with the trade of LNG and development of LNG liquefaction plants and regasification terminals worldwide. Mr Bakshi is qualified to practise Ontario law and is also a member of the Law Society of England and Wales.

Steven Paul Barra
Senior counsel, Eni SpA
Steven.paul.barra@eni.com

Steven Paul Barra is an Australian-qualified lawyer based in Milan, Italy. He joined the Eni group in 2002 and has a longstanding interest in LNG issues. In his position of senior ounsel at Eni SpA, he has been extensively involved in the provision of legal advice in relation to LNG and gas transportation projects, the purchase and sale of LNG on a short and long-term basis, energy sector acquisitions, shipping issues and contractual arrangements at regasification terminals.

James Douglass
Partner, Linklaters LLP
james.douglass@linklaters.com

James Douglass is a partner in the energy, resources and infrastructure group of Linklaters in Beijing. He holds degrees in law and arts, and has specialised in energy, resources and infrastructure projects for over 18 years while based in London, Hong Kong and Beijing.

Mr Douglass has particular experience in the oil and gas (upstream, LNG and pipelines, refining and petrochemicals) and power sectors, advising majors, national oil companies, banks, development finance institutions and export credit agencies.

He has particular experience in emerging markets, particularly Russia, Turkey, Central and Eastern Europe, Africa and Asia.

Notable transactions including the Sakhalin 2 LNG Phase II project financing and the Bujagali power plant in Uganda.

Susan H Farmer
Partner, Holman Fenwick Willan LLP
Susan.Farmer@hfw.com

Susan Farmer joined Holman Fenwick Willan in October 2011 as a partner in the corporate, projects and finance group. She has over 25 years' experience as an energy lawyer, working in London for English and US law firms, and prior to that as in-house counsel for Texaco Inc, Amoco Corp and British Gas. Her practice has recently focused on advising clients on the negotiation of sales and purchase, pipeline access, terminal use and ship charter agreements for LNG and pipeline gas projects. Ms Farmer is mentioned in the International *Who's Who of Business Lawyers (Oil and Gas)*, the *Guide to the World's Leading Energy and Natural Resources Lawyers* and in *Chambers Guide to UK Lawyers 2010* and *2011* in the projects and energy: oil and gas sector. She obtained her JD from UCLA Law School and is admitted to the Bar in the states of California and Texas.

David Gardner
Partner, Energy Transact LLP
davidgardner@energytransact.co.uk

David Gardner has 18 years' experience in the oil and gas industry and over 13 years' experience of LNG transportation and production projects. Mr Gardner has advised on projects involving the transportation of LNG produced in Australia, Angola, Egypt, Trinidad, Qatar and Yemen into destinations in Europe and the United States. His experience extends from the construction of newbuilding LNG carriers to be let on long-term charter or contract of affreightment to joint venture, sale and purchase and short-term time and voyage charters. In recent years he has advised extensively on trading arrangements including the sale and purchase of spot cargoes of LNG and the use of LNG carriers as floating storage and regasification or production units.

Daniel Gosewisch
Senior corporate lawyer, Queensland Corporate Treasury
Daniel.Gosewisch@gmail.com

Daniel Gosewisch holds degrees in law and environmental science, and has worked in in-house legal roles in the Australian resources industry for over 10 years. Mr Gosewisch was involved in the Queensland coal seam gas industry from 2006 until late 2011. As vice president legal and company secretary of Arrow Energy, he advised on the development of coal seam gas projects from exploration through to development and production. Arrow was acquired by an incorporated joint venture of Petrochina and Shell in 2010, which expanded the business to include the development of a two-train coal seam gas to LNG plant at Gladstone, Queensland. He has recently moved to the Queensland Treasury Corporation as senior corporate lawyer, providing risk management services to Queensland's public infrastructure projects.

Paul Griffin
Partner, Allen & Overy LLP
paul.griffin@allenovery.com

Paul Griffin is recognised as one of the world's leading energy lawyers, with over 25 years' experience in the international oil, gas and LNG sectors. He focuses on M&A transactions and large-scale commercial agreements for energy clients. He has also been involved in disputes and matters of public law in relation to the oil and gas sector.

Mr Griffin was named the Global Oil and Gas Lawyer of the Year 2010 (second consecutive year) by the Who's Who Legal Awards and the World's leading Energy Lawyer 2010 by *Expert Guides: Best of the Best*.

Matthew Griffiths
Managing counsel, global LNG, upstream international, Royal Dutch Shell plc
matthew.griffiths@shell.com

Matthew Griffiths is the managing counsel for global LNG, upstream international at Royal Dutch Shell plc. His role includes being the lead lawyer for Shell's floating LNG commercial team.

Toby Hewitt
General counsel, Dart Energy Limited
thewitt@dartcbm.com

Toby Hewitt is general counsel of Dart Energy Limited, based in Singapore. He has over 15 years' experience in the energy and resources sector in both international private legal practice and in-house roles, specialising in oil and gas, LNG and mining law. Prior to joining Dart, he was a consultant in Herbert Smith's global energy practice. A geologist by first degree, he has lived and worked in the United Kingdom, Australia, Indonesia and Singapore.

Dart Energy is ASX listed and was formed following the demerger of Arrow Energy Limited's international business from its Queensland business in 2010. Dart has coal bed methane assets in Australia, the United Kingdom, Belgium, Poland, China, India, Indonesia and Vietnam.

Paula Hodges
Partner, Herbert Smith LLP
Paula.Hodges@herbertsmith.com

Paula Hodges co-heads the firm's global arbitration practice. She has extensive experience of advising on international disputes, particularly in the energy, telecommunications and technology sectors, and specialises in international arbitration. She has represented clients in many jurisdictions (including London, Paris, Geneva, Zurich, Stockholm, the United States, Canada, the United Arab Emirates, Africa, Asia, Russia and the Commonwealth of Independent States), and has conducted proceedings under the auspices of the major arbitral institutions. She also sits as an arbitrator.

Joanna Kay
Legal adviser, Tullow Oil plc
joanna.kay@tullowoil.com

Joanna Kay is qualified to practise as a solicitor in England and Wales. She is a legal adviser at Tullow Oil plc and until recently was an associate in the energy, transport and infrastructure department of Ashurst LLP in London.

She has practised in the upstream and downstream oil and gas sectors as well as the LNG sector, advising a wide range of clients – including state oil and gas companies, international oil and gas companies, contractors, project companies and financial institutions – on a wide range of upstream and downstream developments, LNG regasification projects and shipping projects.

Anthony Patten
Partner, Allens Arthur Robinson
anthony.patten@aar.com.au

Anthony Patten is a partner in the energy and resources group of the Perth office of international law firm Allens Arthur Robinson. Mr Patten is a projects and corporate lawyer specialising in the international oil and gas business, with a particular emphasis on LNG. He has advised on LNG projects throughout the Asia-Pacific region, the Middle East and in Europe. His work has seen him advise on projects relating to all aspects of the LNG value chain (upstream development, liquefaction, shipping, marketing and trading and regasification). Prior to joining Allens Arthur Robinson, Mr Patten worked for a major international oil company in both London and Dubai.

Garry Pegg
Co-managing partner, King & Spalding
GPegg@KSLAW.com

Garry Pegg is the co-managing partner of King & Spalding's London office. His practice, which extends over 25 years, focuses on corporate and commercial transactions, joint ventures, project finance, mergers, share, and asset acquisitions and disposals. He is recognised for the depth of his experience in a number of market sectors, most notably oil, gas, LNG, power and electricity. In the energy sector, Mr Pegg advises governments and public and private companies on the full range of international corporate and commercial matters.

Peter Roberts
Partner, Ashurst LLP
Peter.roberts@ashurst.com

Peter Roberts is a partner in Ashurst's global energy team, based in London. He is an experienced midstream and upstream energy projects lawyer, specialising principally in project structuring, project development, M&A and construction, procurement and operations contracts. Mr Roberts was formerly the general counsel of Centrica Energy and has also been in-house counsel with Exxon. He has previously also worked in Hong Kong and Singapore. He is the author of two leading energy sector textbooks (on gas sales and joint operating agreements), the editor of the Association of International Petroleum Negotiators' *Journal of World Energy Law & Business* and an honorary lecturer at Dundee University. He has worked on LNG sales, export and import projects in more than 20 countries.

Jeff Scobie
Partner, Fraser Milner Casgrain LLP
jeff.scobie@fmc-law.com

Jeff Scobie is a partner in the energy group of Fraser Milner Casgrain LLP. He has over 20 years of experience in domestic and international energy law – mostly in the midstream area, but also in the area of upstream acquisitions and dispositions (including those involving large shale gas fields). He is a former general counsel of Qatar Petroleum, the state-owned oil and gas concern of the state of Qatar. His LNG experience on behalf of numerous clients involves both import and export projects in Canada, Qatar, Russia, Mexico, the United States, the United Kingdom, Italy, China, India and Belgium, as well as shipping-related matters. Mr Scobie was, for several years, an instructor in the Faculty of Law at the University of Calgary. He is a member of the Law Society of Alberta and a non-practising member of the Law Society of England and Wales.

Ron Stuber
Partner, Fraser Milner Casgrain LLP
ron.stuber@fmc-law.com

Ron Stuber is a partner of Fraser Milner Casgrain LLP. Mr Stuber is the national co-lead of the firm's renewable energy team. His practice focuses on the development and financing of projects and major commercial transactions, particularly in the energy and infrastructure sectors. He has extensive international experience advising proponents, lenders and others involved in major projects and transactions (including renewable and other electricity generation, transmission, distribution and supply; upstream, mid-stream and downstream oil and gas; petrochemicals and LNG; hospitals, road, rail, ports and other infrastructure). Mr Stuber has spent many years based in London, United Kingdom with major UK law firms, where he advised extensively on energy and infrastructure projects and transactions throughout the United Kingdom, Europe, the Middle East and Africa. Mr Stuber is a member of the Alberta Bar, the British Columbia Bar and the Law Society of England and Wales.

Harry W Sullivan, Jr
Senior counsel, international E&P legal group, Conocophillips
harry.w.sullivan@conocophillips.com

Harry W Sullivan, Jr is senior counsel in the international E&P legal group of ConocoPhillips. He has a JD from Louisiana State University School of Law and an LLM from Southern Methodist University's School of Law. He is licensed to practise law in the states of Louisiana and Texas and before the US Supreme Court; and is board certified in oil, gas and mineral law in Texas. His experience includes negotiating and working in more than 25 different countries, including Algeria, Australia, Azerbaijan, China, Egypt, Georgia, Indonesia, Kazakhstan, Kuwait, Malaysia, Mozambique, Myanmar, Nigeria, Oman, Pakistan, Qatar, Russia, Saudi Arabia, Turkey, the United Arab Emirates and the United States. Additionally, he has taught oil and gas law at Texas Wesleyan School of Law, SMU's Cox School of Business and Dallas's El Centro College; business law at the University of Texas at Arlington; and various subjects at numerous seminars and short courses.

Philip Thomson
Partner, Ashurst
philip.thomson@ashurst.com

Philip Thomson is a partner with international law firm Ashurst, based in its Singapore office. He specialises in all aspects of the development and financing of LNG projects and other large-scale projects in the oil and gas, petrochemical and power sectors. He has advised on a large number of LNG projects at all segments of the LNG value chain. He has also spent several months on secondment at a major oil and gas company working exclusively on a natural gas liquefaction project.

Philip R Weems
Managing partner, King & Spalding
PWeems@KSLAW.com

Philip R Weems is co-head of King & Spalding's global energy practice and managing partner of the firm's Singapore office. Mr Weems formerly served as managing partner of the firm's three offices in the Middle East, based in Dubai from 2007 to 2010. He has specialised in LNG matters since 1990, having formerly served as an in-house LNG lawyer in Indonesia from 1990 to 1999. He is a former president of the Association of International Petroleum Negotiators (2003-2004).